WITHDRAWN

by Rene Fester Kratz, PhD,
and Donna Rae Siegfried

John Wiley & Sons, Inc.

**Biology For Dummies®, 2nd Edition**

Published by
**John Wiley & Sons, Inc.**
111 River St.
Hoboken, NJ 07030-5774
www.wiley.com



Published by John Wiley & Sons, Inc., Hoboken, New Jersey

Published simultaneously in Canada



For general information on our other products and services, please contact our Customer Care Department within the U.S. at 877-762-2974, outside the U.S. at 317-572-3993, or fax 317-572-4002.

For technical support, please visit www.wiley.com/techsupport.

Wiley publishes in a variety of print and electronic formats and by print-on-demand. Some material included with standard print versions of this book may not be included in e-books or in print-on-demand. If this book refers to media such as a CD or DVD that is not included in the version you purchased, you may download this material at http://booksupport.wiley.com. For more information about Wiley products, visit www.wiley.com.

Library of Congress Control Number: 2010926846

ISBN 978-1-119-17378-6

# About the Authors

**Rene Fester Kratz, PhD,** teaches cellular biology and microbiology. She is a member of the North Cascades and Olympic Science Partnership, where she helped create inquiry-based science courses for future teachers. Kratz is also the author of *Molecular and Cell Biology For Dummies* and *Microbiology The Easy Way*.

**Donna Rae Siegfried** has written about pharmaceutical and medical topics for 15 years in publications including *Prevention, Runner's World, Men's Health,* and *Organic Gardening*. She has taught anatomy and physiology at the college level. She is also the author of *Anatomy & Physiology For Dummies.*

# Dedication

To the memory of Cindy Fuller Kratz Berdan, RN. Thanks for all of your encouragement over the years. —Rene Kratz


# Author's Acknowledgments

Thanks to Matt Wagner, of Fresh Books, Inc., for helping me (Rene) find the opportunity to work on the second edition of this book. And thanks to all the great people at Wiley who made it happen: editors Tim Gallan and Jennifer Tebbe, acquisitions editor Erin Calligan Mooney, art coordinator Alicia South, and technical reviewers Michael Pratt and Medhane Cumbay. Thanks also to Sheree Montgomery, the project coordinator in Composition, and Kathryn Born, who worked on the art. On the home front, thanks to my husband, Dan, and my sons, Hueston and Dashiel, for all of their love and support. —Rene Kratz

## Publisher's Acknowledgments

We're proud of this book; please send us your comments through our Dummies online registration form located at http://dummies.custhelp.com. For other comments, please contact our Customer Care Department within the U.S. at 877-762-2974, outside the U.S. at 317-572-3993, or fax 317-572-4002.

Some of the people who helped bring this book to market include the following:

***Acquisitions, Editorial, and Media Development***

**Senior Project Editor:** Tim Gallan

**Acquisitions Editor:** Lindsay Lefevere

**Copy Editor:** Jennifer Tebbe

**Senior Editorial Assistant:** David Lutton

**Technical Editors:** Medhane G. Cumbay, Michael W. Pratt

**Editorial Manager:** Michelle Hacker

**Editorial Assistants:** Jennette ElNaggar, Rachelle S. Amick

**Art Coordinator:** Alicia B. South

**Cover Photos:** © Digital Art/Corbis

**Cartoons:** Rich Tennant (www.the5thwave.com)

***Composition Services***

**Project Coordinator:** Sheree Montgomery

**Illustrator:** Kathryn Born

**Layout and Graphics:** Ashley Chamberlain

**Proofreaders:** Laura Bowman, Lindsay Littrell

**Indexer:** Potomac Indexing, LLC

---

**Publishing and Editorial for Consumer Dummies**

**Kathleen Nebenhaus,** Vice President and Executive Publisher

**Kristin Ferguson-Wagstaffe,** Product Development Director

**Ensley Eikenburg,** Associate Publisher, Travel

**Kelly Regan,** Editorial Director, Travel

**Publishing for Technology Dummies**

**Andy Cummings,** Vice President and Publisher

**Composition Services**

**Debbie Stailey,** Director of Composition Services

# Contents at a Glance

# Table of Contents

# Introduction

Life is all around you, from invisible microbes and green plants to the other animals with whom you share the Earth. What's more, these other living things aren't just around you — they're intimately interconnected with your life. Plants make your food and provide you with oxygen, microbes break down dead matter and recycle materials that all living things need, and insects pollinate the plants you rely on for food. Ultimately, all living beings rely on other living beings for their survival.

What makes biology so great is that it allows you to explore the interconnectedness of the world's organisms and really understand that living beings are works of art and machines rolled into one. Organisms can be as delicate as a mountain wildflower or as awe-inspiring as a majestic lion. And regardless of whether they're plants, animals, or microbes, all living things have numerous working parts that contribute to the function of the whole being. They move, obtain energy, use raw materials, and make waste, whether they're as simple as a single-celled organism or as complex as a human being.

Biology is the key you need to unlock the mysteries of life. Through it, you discover that even single-celled organisms have their complexities, from their unique structures to their diverse metabolisms. Biology also helps you realize what a truly miraculous machine your body is, with its many different systems that work together to move materials, support your structure, send signals, defend you from invaders, and obtain the matter and energy you need for growth.

## About This Book

*Biology For Dummies,* 2nd Edition, takes a look at the characteristics all living things share. It also provides an overview of the concepts and processes that are fundamental to living things. We put an emphasis on looking at how human beings meet their needs, but we also take a look at the diversity of life on planet Earth.

# Conventions Used in This Book

To help you find your way through the subjects in this book, we use the following style conventions:

- *Italics* highlight new words or terms that are defined in the text. They also point out words we want to emphasize.
- **Boldface** indicates key words in bulleted lists or the action parts of numbered steps.
- `Monofont` points out Web addresses so you can recognize them easily.
- Sidebars are gray-shaded boxes that contain text that's interesting to know but not necessarily critical to your understanding of the chapter or section topic.

Also, whenever we introduce scientific terms, we try to break the words down for you so that the terms become tied to their meanings, making them easier to remember.

# What You're Not to Read

Throughout this book you'll find paragraphs marked with a Technical Stuff icon and sidebars (gray-shaded boxes). The Technical Stuff paragraphs provide more in-depth explanation of a topic or concept, and the sidebars include stories or information related to the main topic. They're fun to read, but they're by no means necessary for a thorough understanding of biology. So skip over them if you want to or read them to your heart's content — the choice is yours!

# Foolish Assumptions

As we wrote this book, we tried to imagine who you are and what you need in order to understand biology. Here's what we came up with:

- You're a high school student taking biology, possibly in preparation for an advanced placement test or college entrance examination. If you're having trouble in biology class and your textbook isn't making much sense, try reading the relevant section of this book first to give yourself a foundation and then go back to your textbook or notes.
- You're a college student who isn't a science major but is taking a biology class to help fulfill your degree requirements. If you want help following along in class, try reading the relevant sections in this book before you

go to a lecture on a particular topic. If you need to fix a concept in your brain, read the related section after class.

- You're someone who just wants to know a little bit more about the living world around you. Good news . . . this book is your oyster! Read it at your leisure, starting with whatever topic fascinates you most. We include several examples of how biology impacts everyday life to help keep your interest piqued.

# How This Book Is Organized

*Biology For Dummies,* 2nd Edition, is organized so that it mirrors the order of topics covered in a typical biology class. Like all *For Dummies* books, each chapter is self-contained, so you can pick up this book whenever you need it and jump straight into the topic you're working on.

***Note:*** After we explain a subject, we use that information in later topics. If you don't read the book in order, you may occasionally have to refer back to an earlier section for some background information. When that's the case, we refer you to the appropriate chapter.

## Part I: Biology Basics

If biology is the study of life and life is so complex, then you may be wondering where to even begin in your study of biology. Never fear. This part breaks down the all-encompassing field of biology into smaller, more palatable chunks.

First, we take a look at the living world and then explain exactly how biology is studied (hint: the scientific method is a huge part of it). Next, we give you a review of the types of molecules that are important to a cell's functioning (yes, this means delving into some basic chemistry; sorry!). Then we spotlight the most basic unit of life — the cell. Every organism, whether it's a human, a dog, a flower, a strep throat bacterium, or an amoeba, has at least one cell (most actually have millions). Finally, because cells need energy to function, we explain just where that energy comes from.

## Part II: Cell Reproduction and Genetics: Let's Talk about Sex, Baby

How do you get a multicellular human from a one-celled embryo? Cellular reproduction, of course! Cells can make exact copies of themselves in order to repair, grow, or produce offspring that are genetically identical to the parent cell. You find out all about that in this part, as well as how some organisms mix things up by engaging in sexual reproduction, creating offspring that have combinations of genes that are different from those of their parents.

Regardless of whether organisms reproduce asexually or sexually, the traits of the parents are visible in the offspring because parents pass DNA on to their offspring. As you discover in this part, DNA contains the blueprints for proteins that do the work in cells and thus determine the characteristics of the offspring.

## Part III: It's a Small, Interconnected World

All the amazingly diverse forms of life on Earth interact with each other (if they didn't, life on this planet would be in big trouble). This part allows you to explore all the ways life on Earth is connected, as well as how biologists classify organisms. You also get to discover how yesterday's living beings are connected to today's living beings through biological evolution.

## Part IV: Systems Galore! Animal Structure and Function

Organisms respond to changes in their environment, trying to maintain their internal conditions within a range that supports life. Animals have many different systems designed to support this struggle for balance. In this part of the book, we present most of the systems that support the structure and function of the human body, as well as those of other animals. These systems coordinate many functions in animals, such as digestion, movement, circulation, gas exchange, and defense.

## Part V: It's Not Easy Being Green: Plant Structure and Function

Plants, your green neighbors, often get overlooked in the hustle and bustle of animal life. However, the importance of plants to life on Earth simply can't be overstated. After all, without them, you wouldn't have any food. When you take the time to study plants, you find that they're actually pretty interesting. Just like animals, they're made of cells and have systems to transport materials around their bodies and exchange matter and energy with their environment, all of which you find out in this part.

### Part VI: The Part of Tens

No *For Dummies* book would be complete without The Part of Tens and its chapters containing fun and interesting facts. When you venture to this part, prepare to find out about ten great biology discoveries and ten ways biology affects your life.

## Icons Used in This Book

We use some of the familiar *For Dummies* icons to help guide you and give you new insights as you read the material. Here's the scoop on what each one means.

The information highlighted with this icon is stuff we think you should permanently store in your mental biology file. If you want a quick review of biology, scan through the book reading only the paragraphs marked with Remember icons.

Next to these icons lie paragraphs that give you extra information but aren't necessary to understanding the material in the chapter. If you want to take your understanding of biology to a higher level, or if you just want to build your knowledge base of interesting facts, incorporate these paragraphs into your reading. If you just want the basics and don't want to bother with nonessential information, skip them.

This bull's-eye symbol offers pointers that help you remember the facts presented in a particular section so you can better commit them to memory.

## Where to Go from Here

Where you start reading is up to you. However, we do have a few suggestions:

- If you're currently in a biology class and having trouble with a particular topic, jump right to the chapter or section featuring the subject that's confusing you.
- If you're using this book as a companion to a biology class that's just beginning, you can follow along with the topics being discussed in class with one small exception. Many biology classes work from the smallest to the largest, beginning with molecules and then moving on to cells. We prefer to start with cells to give you an idea of where everything is happening before moving on to the molecules.

Whatever your situation, the table of contents and index can help you find the information you need.

# Part I
# Biology Basics

## In this part . . .

Biology is the study of living things — how they reproduce, how they change and respond to the environment, and how they obtain the energy and matter they need to grow. One goal of this part is to immerse you in the world of biology so you can understand how biologists go about studying living things and know what chemical components make up all forms of life.

Living things with many cells, like you, are made up of organ systems, organs, tissues, and cells. Cells are the smallest entities that show all the properties of life, so that's where we begin zeroing in on things. The other goal of this part is to acquaint you with the structure of cells and how they obtain the energy they need to function.

# Chapter 1

# Exploring the Living World

### In This Chapter

- Seeing how cells are part of all living things
- Finding out the fundamentals of where babies come from and why you have the traits you do
- Recognizing that all of Earth's ecosystems are interconnected
- Surveying animal anatomy and physiology
- Exploring the similarities and differences between plants and people

*Biology* is the study of life, as in the life that covers the surface of the Earth like a living blanket, filling every nook and cranny from dark caves and dry deserts to blue oceans and lush rain forests. Living things interact with all of these environments and each other, forming complex, interconnected webs of life. For many people, a hike in the forest or a trip to the beach is a chance to reconnect with the natural world and enjoy the beauty of life.

In this chapter, we give you an overview of the big concepts of biology. Our goal is to show you how biology connects to your life and to give you a preview of the topics we explore in greater detail later in this book.

## It All Starts with a Cell

Quick. What's the smallest unit of life you can think of? (Here's a hint: Try to recall the basic properties of life; if you can't, head to Chapter 2 to discover what they are.) Your mind may automatically call up images of ants, amoebas, or bacteria, but that's not quite the answer. The absolute smallest unit of life is a single cell.

Everything an organism's body does happens because its cells make those actions happen, whether that organism is a single-celled *E. coli* bacteria or a human being made up of approximately 10 trillion cells.

Of course, the number of cells you have isn't the only difference between you and *E. coli.* The structure of your cells is a little bit different — your cells have more specialized internal compartments, such as the nucleus that houses your DNA (we cover cell structure in Chapter 4). Yet you have some distinct similarities as well. Both you and *E. coli* are made up of the same raw materials (flip to Chapter 3 to find out what those are) and have DNA as your genetic material (more on DNA in Chapter 8). You also use food the same way (see Chapter 5), and you build your proteins in the same manner (see Chapter 8).

# Life Begets Life: Reproduction and Genetics

You began life as a single cell, when a sperm cell from your dad met an egg cell from your mom. Your parents made these reproductive cells through a special type of cell division called meiosis (we explain meiosis in detail in Chapter 6). When their reproductive cells combined, your dad and mom each donated half of your genetic information — 23 chromosomes from mom and 23 from dad — for a total of 46 chromosomes in each of your cells. The genes on those 46 chromosomes determined your characteristics, from your physical appearance to much of your behavior. The science of genetics tracks the inheritance of genes and studies how they determine traits (see Chapter 7). Through genetics, you can understand why your skin is a certain color or why some traits seem to run in your family.

Your genes are found in your DNA, which is in turn found in your chromosomes. Each chromosome consists of hundreds of different blueprints that contain the instructions for your cells' worker molecules (which are mostly proteins). Each type of cell in your body uses the blueprints found in your genes to build the proteins it needs to do its particular job. So what exactly does all that mean? Here it is, plain and simple: DNA determines your traits because it contains the instructions for the worker molecules (proteins) that make your traits happen.

Scientists are discovering more and more about DNA; they're also developing tools to read and alter the DNA in cells (see Chapter 9). Chances are you're already experiencing the impacts of scientists' work with DNA, even if you don't know it. Why? Because scientists use *recombinant DNA technology* to alter organisms used in food and medicines. This technology allows them to take genes from one organism and place them into the cells of another, changing the characteristics of the receiving organism. For example, scientists alter the cells of bacteria with human genes, turning them into tiny living factories that produce human proteins needed to treat diseases.

# Making the Connection between Ecosystems and Evolution

As you discover in Chapter 10, the amazing diversity of life on Earth helps ensure that life continues in the face of environmental change. Each type of organism plays a role in the environment, and each one is connected to the other. Green organisms such as plants combine energy and matter to make the food on which all life depends, predators hunt prey, and decomposers such as bacteria and fungi recycle dead matter so it becomes available again to other living things. (For more on the interconnectedness of all living things on Earth, head to Chapter 11.)

Humans are part of the natural world, and like all living things, we use resources from the environment and produce wastes. However, the human species is unusual in its ability to use technology to extend its reach, drawing heavily on the natural resources of the Earth and changing environments to suit its needs. The human population has expanded to cover most of the Earth, and the numbers just keep on growing.

Yet as humans draw more heavily upon the Earth's resources, we're putting stress on many other species and possibly driving them to extinction. The great lesson of biological evolution (a topic we cover in Chapter 12) is that not only do populations change over time but they're also capable of going extinct. The challenge that humans face today is discovering ways to get what we need but still live in balance with the Earth's various ecosystems.

# Getting Up Close and Personal with the Anatomy and Physiology of Animals

All animals work hard to maintain *homeostasis*, or internal balance, as change occurs in the environment around them (see Chapter 13 for more on homeostasis). In a complex, multicellular animal like you, all of your organ systems must work together to maintain homeostasis.

Following is a rundown of all of your organ systems, including what they do and what they consist of:

- **Skeletal system:** Provides support, helps with movement, and forms blood cells. Made up of your bones (see Chapter 14).
- **Muscular system:** Enables movement. Consists of your skeletal and smooth muscles (see Chapter 14).

- **Respiratory system:** Brings in oxygen and expels carbon dioxide. Made up of your lungs and airways (see Chapter 15).
- **Circulatory system:** Transports materials throughout the body. Consists of your heart, blood, and blood vessels (see Chapter 15).
- **Digestive system:** Takes up nutrients and water and eliminates wastes. Made up of your stomach, intestines, liver, and pancreas (see Chapter 16).
- **Excretory system:** Maintains the balance of water and electrolytes in your body and removes wastes. Consists of your kidneys and bladder (see Chapter 16).
- **Integumentary system:** Serves as your first line of defense against infection. Made up of your skin (see Chapter 17).
- **Immune system:** Defends against foreign invaders. Consists of your thymus, spleen, and lymph nodes (see Chapter 17).
- **Nervous system:** Controls your body functions via electrical signals. Made up of your brain, spinal cord, and nerves (see Chapter 18).
- **Endocrine system:** Produces hormones that control your body functions. Consists of your glands (see Chapter 18).
- **Reproductive system:** Is responsible for sexual reproduction. Made up of ovaries, fallopian tubes, a uterus, a cervix, a vagina, and a vulva if you're female, and testes, a scrotum, vas deferens, a prostate gland, seminal vesicles, and a penis if you're male (see Chapter 19).

# Comparing Plants to People

At first glance, plants seem pretty different from people, but actually humans and plants occupy nearby branches on the tree of life. Both humans and plants engage in *sexual reproduction,* meaning they produce new offspring from the fusion of sperm and eggs that contain half the genetic material of the parents (see Chapter 20 for more information on how plants reproduce). Also like you, plants have systems for moving materials throughout their bodies (flip to Chapter 21 for the scoop on this), and they even control their functions with hormones.

Of course, plants also have major differences from humans. Most importantly, they make their own food using carbon dioxide, water, and energy from the Sun, whereas humans have to eat other organisms to survive. As a byproduct of their food production, plants give off oxygen as waste. Humans gladly breathe oxygen in and return the favor by breathing out carbon dioxide that the plants can use to make food (see Chapter 5 for more on photosynthesis and respiration and how they lead to this gas exchange between humans and plants).

# Chapter 2

# How Life Is Studied

### In This Chapter

- Studying life
- Using observations to solve the world's mysteries
- Recognizing science as an always-changing thing
- Discovering where to find scientists' research and conclusions

Biology wouldn't have gotten very far as a science if biologists hadn't used structured processes to conduct their research and hadn't communicated the results of that research with others. This chapter explores the characteristics that distinguish living things from the nonliving materials in the natural world. It also introduces you to the methods scientists (whether they're biologists, physicists, or chemists) use to investigate the world around them and the tools they use to communicate what they've discovered.

## Living Things: Why Biologists Study Them and What Defines Them

Biologists seek to understand everything they can about living things, including

- The structure and function of all the diverse living things on planet Earth
- The relationships between living things
- How living things grow, develop, and reproduce, including how these processes are regulated by DNA, hormones, and nerve signals
- The connections between living things, as well as the connections between living things and their environment
- How living things change over time
- How DNA changes, how it's passed from one living thing to another, and how it controls the structure and function of living things

An individual living thing is called an *organism*. Organisms are part of the natural world — they're made of the same chemicals studied in chemistry and geology, and they follow the same laws of the universe as those studied in physics. What makes living things different from the nonliving things in the natural world is that they're alive. Granted, life is a little hard to define, but biologists have found a way.

All organisms share eight specific characteristics that define the properties of life:

- **Living things are made of cells that contain DNA.** A *cell* is the smallest part of a living thing that retains all the properties of life. In other words, it's the smallest unit that's alive. *DNA,* short for *deoxyribonucleic acid,* is the genetic material, or instructions, for the structure and function of cells. (We fill you in on cells, including the differences between plant and animal cells, in Chapter 4, and we tell you all about the structure of DNA in Chapter 3.)
- **Living things maintain order inside their cells and bodies.** One law of the universe is that everything tends to become random over time. According to this law, if you build a sand castle, it'll crumble back into sand over time. You never see a castle of any kind suddenly spring up and build itself or repair itself, organizing all the particles into a complicated castle structure. Living things, as long as they remain alive, don't crumble into little bits. They constantly use energy to rebuild and repair themselves so that they stay intact. (To find out how living things obtain the energy they need to maintain themselves, turn to Chapter 5.)
- **Living things regulate their systems.** Living things maintain their internal conditions in a way that supports life. Even when the environment around them changes, organisms attempt to maintain their internal conditions. Think about what happens when you go outside on a cool day without wearing a coat. Your body temperature starts to drop, and your body responds by pulling blood away from your extremities to your core in order to slow the transfer of heat to the air. It may also trigger shivering, which gets you moving and generates more body heat. These responses keep your internal body temperature in the right range for your survival even though the outside temperature is low. (When living things maintain their internal balance, that's called *homeostasis;* you can find out more about homeostasis in Chapter 13.)
- **Living things respond to signals in the environment.** If you pop up suddenly and say "Boo!" to a rock, it doesn't do anything. Pop up and say "Boo!" to a friend or a frog, and you'll likely see him or it jump. That's because living things have systems to sense and respond to signals. Many animals sense their environment through their five senses just like you do, but even less familiar organisms, such as plants and bacteria, can sense and respond. (Have you ever seen a houseplant bend and grow toward sunlight? Then you've seen one of the responses triggered by a plant cell detecting the presence of light.) Want to know more about the systems that help plants and animals respond to signals? Flip

to Chapter 18 to read all about the human nervous system and Chapter 21 to discover the details about plant hormones.

- **Living things transfer energy among themselves and between themselves and their environment.** Living things need a constant supply of energy to grow and maintain order. Organisms such as plants capture light energy from the Sun and use it to build food molecules that contain chemical energy. Then the plants, and other organisms that eat the plants, transfer the chemical energy from the food into cellular processes. As cellular processes occur, they transfer energy back to the environment as heat. (For more on how energy is transferred from one living thing to another, check out Chapter 11.)
- **Living things grow and develop.** You started life as a single cell. That cell divided to form new cells, which divided again. Now your body is made of approximately 100 trillion cells. As your body grew, your cells received signals that told them to change and become special types of cells: skin cells, heart cells, liver cells, brain cells, and so on. Your body developed along a plan, with a head at one end and a "tail" at the other. The DNA in your cells controlled all of these changes as your body developed. (For the scoop on the changes that occur in animal cells as they grow and develop, see Chapter 19.)
- **Living things reproduce.** People make babies, hens make chicks, and plasmodial slime molds make plasmodial slime molds. When organisms reproduce, they pass copies of their DNA onto their offspring, ensuring that the offspring have some of the traits of the parents. (Flip to Chapter 6 for full details on how cells reproduce and Chapter 19 for insight into how animals, particularly humans, make more animals.)
- **Living things have traits that evolved over time.** Birds can fly, but most of their closest relatives — the dinosaurs — couldn't. The oldest feathers seen in the fossil record are found on a feathered dinosaur called *Archaeopteryx.* No birds or feathers have been found in any fossils that are older than those of *Archaeopteryx.* From observations like these, scientists can infer that having feathers is a trait that wasn't always present on Earth; rather, it's a trait that developed at a certain point in time. So, today's birds have characteristics that developed through the evolution of their ancestors. (Ready to dig into the nitty-gritty details of evolution? See Chapter 12.)

# Making Sense of the World through Observations

The true heart of science isn't a bunch of facts — it's the method that scientists use to gather those facts. Science is about exploring the natural world, making observations using the five senses, and attempting to make sense of those observations. Scientists, including biologists, use two main approaches when trying to make sense of the natural world:

- **Discovery science:** When scientists seek out and observe living things, they're engaging in *discovery science,* studying the natural world and looking for patterns that lead to new, tentative explanations of how things work (these explanations are called *hypotheses*). If a biologist doesn't want to disturb an organism's habitat, he or she may use observation to find out how a certain animal lives in its natural environment. Making useful scientific observations involves writing detailed notes about the routine of the animal for a long period of time (usually years) to be sure that the observations are accurate.

  Many of the animals and plants you're familiar with were first identified during a huge wave of discovery science that took place in the 1800s. Scientists called *naturalists* traveled the world drawing and describing every new living thing they could find. Discovery science continues today as biologists attempt to identify all the tiniest residents of planet Earth (bacteria and viruses) and explore the oceans to see the strange and fabulous creatures that lurk in its depths.

- **Hypothesis-based science:** When scientists test their understanding of the world through experimentation, they're engaging in *hypothesis-based science,* which usually calls for following some variation of a process called the scientific method (see the next section for more on this). Modern biologists are using hypothesis-based science to try and understand many things, including the causes and potential cures of human diseases and how DNA controls the structure and function of living things.

Hypothesis-based science can be a bit more complex than discovery science, which is why we spend the next two sections introducing you to two important elements of hypothesis-based science: scientific method and experiment design.

## Introducing the scientific method

The *scientific method* is basically a plan that scientists follow while performing scientific experiments and writing up the results. It allows experiments to be duplicated and results to be communicated uniformly. Here's the general process of the scientific method:

1. **First, make observations and come up with questions.**

   The scientific method starts when scientists notice something and ask questions like "What's that?" or "How does it work?" just like a child might when he sees something new.

2. **Then form a hypothesis.**

   Much like Sherlock Holmes, scientists piece together clues to try and come up with the most likely hypothesis (explanation) for a set of observations. This hypothesis represents scientists' thinking about possible answers to their questions. Say, for example, a marine biologist is exploring some rocks along a beach and finds a new worm-shaped creature he has never seen before. His hypothesis is therefore that the creature is some kind of worm.

   One important point about a scientific hypothesis is that it must be testable, or *falsifiable*. In other words, it has to be an idea that you can support or reject by exploring the situation further using your five senses.

3. **Next, make predictions and design experiments to test the idea(s).**

   Predictions set up the framework for an experiment to test a hypothesis, and they're typically written as "if . . . then" statements. In the preceding worm example, the marine biologist predicts that if the creature is a worm, then its internal structures should look like those in other worms he has studied.

4. **Test the idea(s) through experimentation.**

   Scientists must design their experiments carefully in order to test just one idea at a time (we explain how to set up a good experiment in the later "Designing experiments" section). As they conduct their experiments, scientists make observations using their five senses and record these observations as their results or data. Continuing with the worm example, the marine biologist tests his hypothesis by dissecting the wormlike creature, examining its internal parts carefully with the assistance of a microscope, and making detailed drawings of its internal structures.

## Discovery science of the 20th century

Although discovery science about the types of plants and animals on Earth had its heyday in the 1800s, discovery science about life on a level that's too small to see with the naked eye is ongoing. One incredibly important project that employed modern discovery science is the Human Genome Project, which set out to map where each trait is found on the 46 human chromosomes.

Instead of traveling across the oceans to explore the world and catalog living things like the discovery scientists of 200 years ago, scientists from all over the world set out to explore the very tiny, but very complex, landscape of the 46 human chromosomes that contain the collection of all the genes found in humans. Each of the 25,000 genes they located provides information about inherited traits. The traits range from little things, such as whether you can curl your tongue or not, to truly important things, such as whether you have a genetic risk for developing breast cancer or cystic fibrosis. By finding out where genes are located, scientists can turn their attention to using this newfound information to develop hypotheses about cures and gene therapies.

5. **Then make conclusions about the findings.**

   Scientists interpret the results of their experiments through *deductive reasoning,* using their specific observations to test their general hypothesis. When making deductive conclusions, scientists consider their original hypothesis and ask whether it could still be true in light of the new information gathered during the experiment. If so, the hypothesis can remain as a possible explanation for how things work. If not, scientists reject the hypothesis and try to come up with an alternate explanation (a new hypothesis) that could explain what they've seen. In the earlier worm example, the marine biologist discovers that the internal structures of the wormlike creature look very similar to another type of worm he's familiar with. He can therefore conclude that the new animal is likely a relative of that other type of worm.

6. **Finally, communicate the conclusions with other scientists.**

   Communication is a huge part of science. Without it, discoveries can't be passed on, and old conclusions can't be tested with new experiments. When scientists complete some work, they write a paper that explains exactly what they did, what they saw, and what they concluded. Then they submit that paper to a scientific journal in their field. Scientists also present their work to other scientists at meetings, including those sponsored by scientific societies. In addition to sponsoring meetings, these societies support their respective disciplines by printing scientific journals and providing assistance to teachers and students in the field.

## Designing experiments

Any scientific experiment must have the ability to be duplicated because the "answer" the scientist comes up with (whether it supports or refutes the original hypothesis) can't become part of the scientific knowledge base unless other scientists can perform the exact same experiment and achieve the same results.

When a scientist designs an experiment, he tries to develop a plan that clearly shows the effect or importance of each factor tested by his experiment. Any factor that can be changed in an experiment is called a *variable.*

Three kinds of variables are especially important to consider when designing experiments:

- **Experimental variables:** The factor you want to test is an *experimental variable* (also called an *independent variable*).
- **Responding variables:** The factor you measure is the *responding variable* (also called a *dependent variable*).
- **Controlled variables:** Any factors that you want to remain the same between the treatments in your experiment are *controlled variables.*

Scientific experiments help people answer questions about the natural world. To design an experiment:

1. **Make observations about something you're interested in and use inductive reasoning to come up with a hypothesis that seems like a good explanation or answer to your question.**

   *Inductive reasoning* uses specific observations to generate general principles.

2. **Think about how to test your hypothesis, creating a prediction about it using an "if . . . then" statement.**

3. **Decide on your experimental treatment, what you'll measure, and how often you'll make measurements.**

   The condition you alter in your experiment is your experimental variable. The changes you measure are your responding variables.

4. **Create two groups for your experiment: an experimental group and a control group.**

   The experimental group receives the experimental treatment; in other words, you vary one condition that might affect this group. The control group should be as similar as possible to your experimental group, but it shouldn't receive the experimental treatment.

5. **Set up your experiment, being careful to control all the variables except the experimental variable.**

6. **Make your planned measurements and record the quantitative and qualitative data in a notebook.**

   *Quantitative data* is numerical data, such as height, weight, and number of individuals who showed a change. It can be analyzed with statistics and presented in graphs. *Qualitative data* is descriptive data, such as color, health, and happiness. It's usually presented in paragraphs or tables.

   Be sure to date all of your observations.

7. **Analyze your data by comparing the differences between your experimental and control groups.**

   You can calculate the averages for numerical data and create graphs that illustrate the differences, if any, between your two groups.

8. **Use deductive reasoning to decide whether your experiment supports or rejects your hypothesis and to compare your results with those of other scientists.**

9. **Report your results, being sure to explain your original ideas and how you conducted your experiment, and describe your conclusions.**

As an example of how you design an experiment, imagine you're a marathon runner who trains with a group of friends. You wonder whether you and your friends will be able to run marathons faster when you eat pasta the night before the race. To answer your question, follow the scientific method and design an experiment.

1. **Form your hypothesis.**

   Your hunch is that loading up on pasta will give you the energy you need to run faster the next day. Translate that hunch into a proper hypothesis, which looks something like this: The time it takes to run a marathon is improved by consuming large quantities of carbohydrates prerace.

2. **Treat one group with your experimental variable.**

   In order to test your hypothesis, convince half of your friends to eat lots of pasta the night before the race. Because the factor you want to test is the effect of eating pasta, pasta consumption is your experimental variable.

3. **Create a control group that doesn't receive the experimental variable.**

   You need a comparison group for your experiment, so you convince half of your friends to eat a normal, nonpasta meal the night before the race. For the best results in your experiment, this control group should be as similar as possible to your experimental group so you can be pretty sure that any effect you see is due to the pasta and not some other factor. So, ideally, both groups of your friends are about the same age, same gender, and same fitness level. They're also eating about the same thing before the race — with the sole exception being the pasta. All the factors that could be different between your two groups (age, gender, fitness, and diet) but that you try to control to keep them the same are your controlled variables.

4. **Measure your responding variable.**

   Race time is your responding variable because you determine the effect of eating pasta by timing how long it takes each person in your group to run the race. Because scientists carefully record exact measurements from their experiments and present that data in graphs, tables, or charts, you average the race times for your friends in each of the two groups and present the information in a small table.

5. **Compare results from your two groups and make your conclusions.**

   If your pasta-eating friends ran the marathon an average of two minutes faster than your friends who didn't eat pasta, you may conclude that your hypothesis is supported and that eating pasta does in fact help marathon runners run faster races.

## One man's error is another man's starting point

In the early 1900s, a Russian researcher named A.I. Ignatowski fed rabbits a diet of milk and eggs. He found that the rabbits' aortas developed the same kind of plaques that form in people with atherosclerosis. Ignatowski wasn't ignorant, but he assumed that the atherosclerosis was caused by the proteins in the milk and eggs. He was wrong.

A younger researcher who was working in the same pathology department at the time, a Russian named Nikolai Anichkov, knew of Ignatowski's work. Anichkov and some of his colleagues repeated Ignatowski's study with one small change: They split the rabbits into three different groups. The first group was fed a supplement of muscle fluid, the second group was fed only egg whites, and the third group was fed only egg yolks. Only the yolk-eating rabbits developed plaques in their aortas. The young researchers ran the experiment again; this time they analyzed the atherosclerotic plaques to look for any concentrated chemical substances. In 1913, Anichkov and his colleagues discovered that cholesterol in the egg yolk was responsible for creating plaques in the aorta. Their discovery may not have been possible if Ignatowski had never conducted his experiment (or if he'd beaten them to the punch!).

Before you can consider your research complete, you need to look at a few more factors:

- **Sample size:** The number of individuals who receive each treatment in an experiment is your *sample size.* To make any kind of scientific research valid, the sample size has to be rather large. If you had only four friends participate in your experiment, you'd have to conduct your experiment again on much larger groups of runners before you could proudly proclaim that consuming large quantities of carbohydrates pre-race helps marathon runners improve their speed.
- **Replicates:** The number of times you repeat the entire experiment, or the number of groups you have in each treatment category, are your *replicates.* Suppose you have 60 marathon-running friends and you break them into six groups of 10 runners each. Three groups eat pasta, and three groups don't, so you have three replicates of each treatment. (Your total sample size is therefore 30 for each treatment.)
- **Statistical significance:** The mathematical measure of the validity of an experiment is referred to as *statistical significance.* Scientists analyze their data with statistics in order to determine whether the differences between groups are significant. If an experiment is performed repeatedly and the results are within a narrow margin, the results are said to be

significant. In your experiment, if the race times for your friends were very similar within each group, so that pretty much all of your pasta-eating friends ran faster than your non-pasta-eating friends, then that two-minute difference actually meant something. But what if some pasta-eating friends ran slower than non-pasta-eating friends and one or two really fast friends in the pasta group lowered that group's overall average? Then you might question whether the two minutes was really significant, or whether your two fastest friends just got put in the pasta group randomly.

- **Error:** Science is done by people, and people make mistakes, which is why scientists always include a statement of possible sources of error when they report the results of their experiments. Consider the possible errors in your experiment. What if you didn't specify anything about the content of the normal meals to your non-pasta-eating friends? After the race, you might find out that some of your friends ate large amounts of other sources of carbohydrates, such as rice or bread. Because your hypothesis was about the effect of carbohydrate consumption on marathon running, a few non-pasta-eating friends eating rice or bread would represent a source of error in your experiment.

Whether the scientist is right or wrong isn't as important as whether he or she sets up an experiment that can be repeated by other scientists who expect to get the same result.

# Seeing Science as the Constant Sharing of New Ideas

The knowledge gathered by scientists continues to grow and change slightly all the time. Scientists are continually poking and prodding at ideas, always trying to get closer to "the truth." They try to keep their minds open to new ideas and remain willing to retest old ideas with new technology. Scientists also encourage argument and debate over ideas because the discussion pushes them to test their ideas and ultimately adds to the strength of scientific knowledge. Following are some of the facts about scientific ideas that illustrate how science is ever-evolving:

- **Today's scientists are connected to scientists of the past because new scientific ideas are built upon the foundations of earlier work.** For instance, a scientist working in a particular area of biology reads all the scientific publications he can that relate to his work to be sure he has the best understanding possible of what has already been done and what's already known.

That way, he can plan research that will advance the understanding in his field and add new knowledge to the scientific knowledge base.

- **Some scientific ideas are very old but still applicable today.** Occasionally, new technology enables scientists to test old hypotheses in new ways, leading to new perspectives and changes in ideas. Case in point: Up until the 1970s, scientists looking through microscopes thought only two main types of cells made up living things. When scientists of the '70s used new technology to compare the genetic code of cells, they realized that living things are actually made up of three main types of cells — two of the types just happen to look the same under a microscope. Of course, old ideas aren't always proved completely wrong — for example, scientists still recognize the two structural types of cells — but big ideas can shift slightly in the face of new information.

REMEMBER

- **When many lines of research support a particular hypothesis, the hypothesis becomes a scientific theory.** A *scientific theory* is an idea that's supported by a great deal of evidence and hasn't been proven false despite repeated tests. Scientific theories don't change as often as scientific hypotheses due to the significant evidence backing them up, but even scientific theories can shift in light of new evidence. Ideally, scientists always keep an open mind and look at new evidence objectively.

## Conflicting reports mean science is working

Sure, it's aggravating when the media reports conflicting findings — such as margarine is better for your cholesterol level but it also produces harmful fatty acids that contribute to heart disease — but conflicting news reports are a sign that science is alive and well. For example, when scientists figured out that high cholesterol levels contributed to heart disease, they correctly determined that a product created from vegetable oil rather than animal fat — in other words, margarine rather than butter — was a healthier choice if you were trying to lower your cholesterol level.

But scientists don't just leave things alone. They keep wondering, questioning, and pondering. They're curious guys and gals, which is why they kept researching margarine. Recently, they discovered that when margarine breaks down, it releases trans fatty acids, which were found to be harmful to the heart and blood vessels. So, margarine has bad aspects that may outweigh the good. Yes, this can make decisions at the grocery store more confusing, but it can also lead to better health for everyone. Case in point: After the information about trans fatty acids became known, food companies started developing new ways to make margarine and other foods so that they don't contain trans fatty acids.

# Tracking Down Scientific Information

Scientists publish their work in part because scientists in different areas of the world may be trying to answer the same questions and could benefit from seeing how someone else approached the problem. The other part is that if scientists didn't put their work out there, flaws and all, no one would ever know the work was being done. The sections that follow provide an overview of the different sources scientists use to communicate with each other (and the rest of the world).

## Journals: Not just for recording dreams

Hundreds of scientific journals cover every topic and niche imaginable in the fields of biology, chemistry, physics, engineering, and so on. They're published by numerous organizations, including professional groups, universities or medical centers, and medical and scientific publishing companies. Regardless of their subject matter or where they come from, all scientific journals have one common characteristic: They're all considered a *primary source* of scientific information, meaning they contain a full description of the original research written by the original researchers.

Anyone researching a topic, whether he's a student or a scientist, consults the journals first. They contain the original research papers, which means you can always find the latest information in a specific field in a journal. The research papers are written following the scientific style of an *abstract* (summary). First, there's a statement of the hypothesis. Next come a description of the materials used; a description of how the experiment was designed and performed; and the results of the experiment, including raw data, graphs, and tables. Finally, the paper notes the author's conclusions and errors that occurred during the experiment(s).

Scientific journals undergo a peer-review process to help ensure the reliability of published scientific information. Here's how the process works: The editor of a scientific journal sends a research paper out to other scientists who work in the same field as the scientist(s) who wrote the paper to examine and comment on the work. They're tasked with making sure the science is thorough and that the research adds to the scientific knowledge base. The editor then decides whether or not to publish the paper based on the other scientists' comments. If the reviewers' and editor's stringent criteria aren't meant, the research paper can't be published in the journal.

Although many scientific journals are available online, you may not be able to access the articles without paying a fee. However, if you're a student at a college or university, your school library may subscribe to various scientific journals. Ask a member of the library staff whether the school subscribes to any scientific journals and, if so, how you can access them via the library's computers.

## Textbooks: A student's go-to source

Textbooks. Whether you love 'em or hate 'em (or could care less about 'em), textbooks are considered *secondary sources* of scientific information, meaning they compile or discuss information taken from primary sources. Secondary sources aren't usually written by the original researchers. They present the knowledge base of a specific topic or field at a certain point in time, which makes them a good source to turn to for the history of a topic, basic facts about a certain subject, and summaries of important research that has furthered the field.

## The popular press: Not always accurate

The popular press — regular ol' newspapers, magazines, and television and radio programs, are considered *tertiary sources* (meaning they're twice removed from the original source of information). The popular press provides information, of course, but the validity of that information isn't guaranteed. There's always a chance that the journalist doing the reporting may have misunderstood the scientist's research or something he said. It's like that old childhood game where the information given to the first person is usually changed by the time it gets relayed to the last person.

You're always better off citing an article in a journal or textbook before one from a major media outlet.

## The Internet: A wealth of information, not all of it good

Lots of scientific information is available on the Internet, and much of it is available for free. The trick is distinguishing the good stuff from the bad. To find good-quality scientific information:

- **Visit government Web sites.** These Web sites end in `.gov`. Some primary literature is available on government Web sites, but even the secondary literature is usually of high quality. If you use the advanced search feature on your web browser, you should be able to restrict your searches to the types of domains you're interested in.
- **Surf university Web sites.** These Web sites end in `.edu`. Some university scientists post copies of their papers, which are examples of primary literature, and others post good-quality lecture notes and articles. (Better yet, if you're a student at a college or university, access the primary literature — scientific journals — through your school library's subscription service.)

- **Be careful when visiting organization Web sites.** These Web sites end in `.org`. Large organizations with good reputations, such as the American Heart Association, usually have good-quality secondary information on their sites; they may even post links to primary sources. However, smaller organizations that don't have an established reputation aren't good sources for scientific information.

Avoid commercial Web sites (those with `.com` endings) when you're looking for scientific information. People and organizations operating commercial sites are trying to sell you something. They have an agenda of their own, which means you can't trust that they're completely unbiased. The information they present may be one-sided or not accepted as reliable by the scientific community.

# Chapter 3

# The Chemistry of Life

### *In This Chapter*

- Seeing why matter is so important
- Distinguishing among atoms, elements, isotopes, molecules, and compounds
- Getting to know acids and bases
- Understanding the structure and function of important molecules for life

Everything that has mass and takes up space, including you and the rest of life on Earth, is made of matter. Atoms make up molecules, which make up the substance of living things. Carbohydrates, proteins, nucleic acids, and lipids are four kinds of molecules that are especially important to the structure and function of organisms. In this chapter, we present a bit of the basic chemistry that's essential for understanding biology.

## Exploring Why Matter Matters

Matter is the stuff of life — literally. Every living thing is made of matter. In order to grow, living things must get more matter to build new structures. When living things die, be they plants or animals, microbes such as bacteria and fungi digest the dead matter and recycle it so that other living things can use it again. In fact, pretty much all the matter on Earth has been here since the planet formed 4.5 billion years ago; it has just been recycled since then. So, the stuff that makes up your body may once have been part of *Tyrannosaurus rex,* a butterfly, or even a bacterium.

### Invisible matter

What looks like nothing but is really something? Air! Earth's atmosphere may seem like nothing, but it's made of gases such as nitrogen, carbon dioxide, and oxygen. These gases interact with living things in many ways. Plants, for example, take in carbon dioxide to make food and then use that food to build their structures. It's hard to believe, but the tallest tree in the redwood forest grows and grows from the result of invisible carbon dioxide gas being taken in and incorporated into the body of the tree. Obviously the redwood tree takes up space and has mass, but those invisible carbon dioxide molecules are matter too.

Following are a few facts you should know about matter:

- **Matter takes up space.** Space is measured in *volume,* and volume is measured in *liters* (L).
- **Matter has mass.** *Mass* is the term for describing the amount of matter that a substance has. It's measured in *grams* (g). Earth's gravity pulls on your mass, so the more mass you have, the more you weigh.
- **Matter can take several forms.** The most familiar forms of matter are solids, liquids, and gases. *Solids* have a definite shape and size, such as a person or a brick. *Liquids* have a definite volume. They can fill a container, but they take the shape of the container that they fill. *Gases* are easy to compress and expand to fill a container.

To understand the difference between mass and weight, compare your weight on Earth versus your weight on the Moon. No matter where you are, your body is made of the same amount of stuff, or matter. But the Moon is so much smaller than Earth that it has a lot less gravity to pull on your mass. So, your weight on the Moon would be just one-sixth of your weight on Earth, but your mass would remain the same!

## Recognizing the Differences between Atoms, Elements, and Isotopes

All matter is composed of elements. When you break down matter into its smallest components, you're left with individual elements that themselves break down into atoms consisting of even smaller pieces called subatomic particles. And sometimes the number of those subatomic particles within a particular atom differs, creating isotopes. Whew. That's a lot to take in, which is why we break the concepts of atoms, elements, and isotopes down for you in the sections that follow.

## "Bohr"ing you with atoms

An *atom* is the smallest whole, stable piece of an element that still has all the properties of that element. It's the smallest "piece" of matter that can be measured. Every atom actually contains even smaller pieces known collectively as *subatomic particles*. These include protons, neutrons, and electrons (and even quarks, mesons, leptons, and neutrinos). Subatomic particles can't be removed from an atom without destroying the atom.

Here's the basic breakdown of an atom's structure (see Figure 3-1):

- **The core of an atom, called the nucleus, contains two kinds of subatomic particles: protons and neutrons.** Both have mass, but only one carries any kind of charge. *Protons* carry a positive charge, but *neutrons* have no charge (they're neutral). Because the protons are positive and the neutrons have no charge, the net charge of an atom's nucleus is positive.
- **Clouds of electrons surround the nucleus.** *Electrons* carry a negative charge but have no mass.

Atoms become ions when they gain or lose electrons. In other words, *ions* are essentially charged atoms. *Positive (+) ions* have more protons than electrons; *negative (–) ions* have more electrons than protons. Positive and negative charges attract one another, allowing atoms to form bonds, as explained in the later "Molecules, Compounds, and Bonds" section.

## Elements of elements

An *element* is a substance made of atoms that have the same number of protons. Think of them as "pure" substances all made of the same thing. All the known elements are organized into the periodic table of elements (shown in Figure 3-2), which has the following properties:

- **Each row of the table is called a period.** Moving across the table horizontally, you go from metals to nonmetals, with heavy metals in the middle.
- **Each column is called a family or group.** Elements within the same family/group have similar properties. The size of the atom increases from top to bottom within each column.

**A.** Bohr's model of an atom: carbon used as an example.

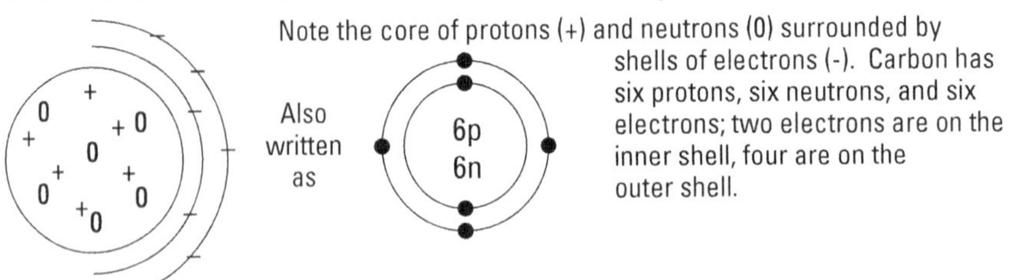


**B.** Sodium and chloride ions joining to form table salt. The sodium ion has a positive charge because there's one more proton than electrons, so the overall charge is positive. The chloride ion is negative because after it accepts the electron from sodium, it then has one more electron than protons (18 versus 17), so the overall charge is negative. Together, though, NaCl is neutral because the "plus 1" charge is balanced by the "minus 1" charge.

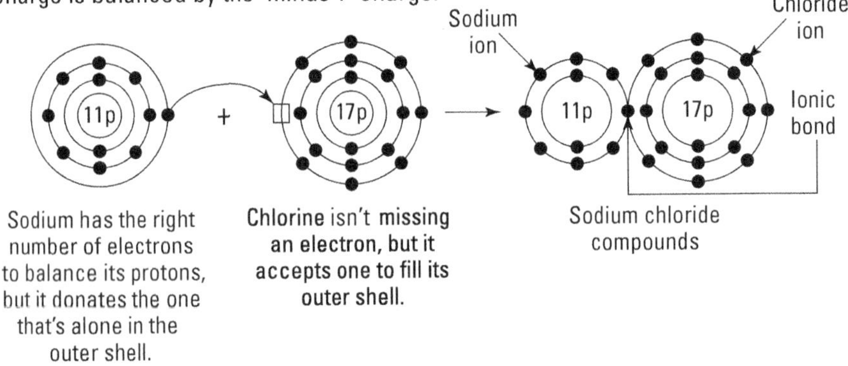


**C.** Two atoms of oxygen joining to form oxygen gas.

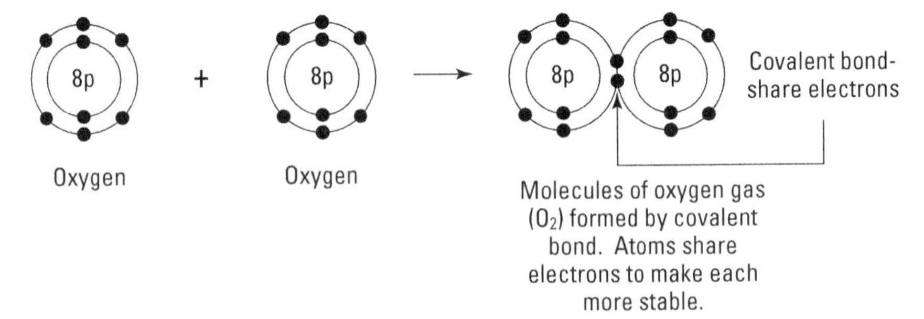


**Figure 3-1:** The Bohr model of an atom's structure.

| 1<br>H<br>Hydrogen<br>1.01 | | | | | | | | | | | | | | | | | 2<br>He<br>Helium<br>4.00 |
|---|---|---|---|---|---|---|---|---|---|---|---|---|---|---|---|---|---|
| 3<br>Li<br>Lithium<br>6.94 | 4<br>Be<br>Beryllium<br>9.01 | | | | | | | | | | | 5<br>B<br>Boron<br>10.81 | 6<br>C<br>Carbon<br>12.01 | 7<br>N<br>Nitrogen<br>14.01 | 8<br>O<br>Oxygen<br>16.00 | 9<br>F<br>Fluorine<br>19.00 | 10<br>Ne<br>Neon<br>20.18 |
| 11<br>Na<br>Sodium<br>22.99 | 12<br>Mg<br>Magnesium<br>24.31 | | | | | | | | | | | 13<br>Al<br>Aluminum<br>26.98 | 14<br>Si<br>Silicon<br>28.09 | 15<br>P<br>Phosphorus<br>30.97 | 16<br>S<br>Sulfur<br>32.06 | 17<br>Cl<br>Chlorine<br>35.45 | 18<br>Ar<br>Argon<br>39.95 |
| 19<br>K<br>Potassium<br>39.10 | 20<br>Ca<br>Calcium<br>40.08 | 21<br>Sc<br>Scandium<br>44.96 | 22<br>Ti<br>Titanium<br>47.90 | 23<br>V<br>Vanadium<br>50.94 | 24<br>Cr<br>Chromium<br>52.00 | 25<br>Mn<br>Manganese<br>54.94 | 26<br>Fe<br>Iron<br>55.85 | 27<br>Co<br>Cobalt<br>58.93 | 28<br>Ni<br>Nickel<br>58.71 | 29<br>Cu<br>Copper<br>63.55 | 30<br>Zn<br>Zinc<br>65.38 | 31<br>Ga<br>Gallium<br>69.72 | 32<br>Ge<br>Germanium<br>72.59 | 33<br>As<br>Arsenic<br>74.92 | 34<br>Se<br>Selenium<br>78.96 | 35<br>Br<br>Bromine<br>79.90 | 36<br>Kr<br>Krypton<br>83.80 |
| 37<br>Rb<br>Rubidium<br>85.47 | 38<br>Sr<br>Strontium<br>87.62 | 39<br>Y<br>Yttrium<br>88.91 | 40<br>Zr<br>Zirconium<br>91.22 | 41<br>Nb<br>Niobium<br>92.91 | 42<br>Mo<br>Molybdenum<br>95.94 | 43<br>Tc<br>Technetium<br>(99) | 44<br>Ru<br>Ruthenium<br>101.07 | 45<br>Rh<br>Rhodium<br>102.91 | 46<br>Pd<br>Palladium<br>106.42 | 47<br>Ag<br>Silver<br>107.87 | 48<br>Cd<br>Cadmium<br>112.41 | 49<br>In<br>Indium<br>114.82 | 50<br>Sn<br>Tin<br>118.69 | 51<br>Sb<br>Antimony<br>121.75 | 52<br>Te<br>Tellurium<br>127.60 | 53<br>I<br>Iodine<br>126.90 | 54<br>Xe<br>Xenon<br>130.30 |
| 55<br>Cs<br>Cesium<br>132.91 | 56<br>Ba<br>Barium<br>137.34 | 57<br>La<br>Lanthanum<br>138.91 | 72<br>Hf<br>Hafnium<br>178.49 | 73<br>Ta<br>Tantalum<br>180.95 | 74<br>W<br>Tungsten<br>183.85 | 75<br>Re<br>Rhenium<br>186.21 | 76<br>Os<br>Osmium<br>190.2 | 77<br>Ir<br>Iridium<br>192.22 | 78<br>Pt<br>Platinum<br>195.09 | 79<br>Au<br>Gold<br>196.97 | 80<br>Hg<br>Mercury<br>200.59 | 81<br>Tl<br>Thallium<br>204.37 | 82<br>Pb<br>Lead<br>207.19 | 83<br>Bi<br>Bismuth<br>208.98 | 84<br>Po<br>Polonium<br>(210) | 85<br>At<br>Astatine<br>(210) | 86<br>Rn<br>Radon<br>(222) |
| 87<br>Fr<br>Francium<br>(223) | 88<br>Ra<br>Radium<br>(226) | 89<br>Ac<br>Actinium<br>(227) | 104<br>Rf<br>Rutherfordium<br>(257) | 105<br>Db<br>Dubnium<br>(260) | 106<br>Sg<br>Seaborgium<br>(263) | 107<br>Bh<br>Bohrium<br>(262) | 108<br>Hs<br>Hassium<br>(265) | 109<br>Mt<br>Meitnerium<br>(266) | | | | | | | | | |

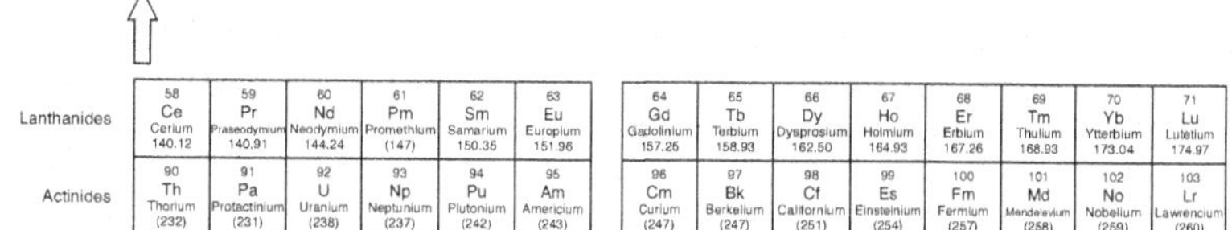

| | | | | | | | | | | | | | | |
|---|---|---|---|---|---|---|---|---|---|---|---|---|---|---|
| Lanthanides | 58<br>Ce<br>Cerium<br>140.12 | 59<br>Pr<br>Praseodymium<br>140.91 | 60<br>Nd<br>Neodymium<br>144.24 | 61<br>Pm<br>Promethium<br>(147) | 62<br>Sm<br>Samarium<br>150.35 | 63<br>Eu<br>Europium<br>151.96 | 64<br>Gd<br>Gadolinium<br>157.25 | 65<br>Tb<br>Terbium<br>158.93 | 66<br>Dy<br>Dysprosium<br>162.50 | 67<br>Ho<br>Holmium<br>164.93 | 68<br>Er<br>Erbium<br>167.26 | 69<br>Tm<br>Thulium<br>168.93 | 70<br>Yb<br>Ytterbium<br>173.04 | 71<br>Lu<br>Lutetium<br>174.97 |
| Actinides | 90<br>Th<br>Thorium<br>(232) | 91<br>Pa<br>Protactinium<br>(231) | 92<br>U<br>Uranium<br>(238) | 93<br>Np<br>Neptunium<br>(237) | 94<br>Pu<br>Plutonium<br>(242) | 95<br>Am<br>Americium<br>(243) | 96<br>Cm<br>Curium<br>(247) | 97<br>Bk<br>Berkelium<br>(247) | 98<br>Cf<br>Californium<br>(251) | 99<br>Es<br>Einsteinium<br>(254) | 100<br>Fm<br>Fermium<br>(257) | 101<br>Md<br>Mendelevium<br>(258) | 102<br>No<br>Nobelium<br>(259) | 103<br>Lr<br>Lawrencium<br>(260) |

**Figure 3-2:** The periodic table of elements.

Notice in Figure 3-2 how each element has a number associated with it? That number is the *atomic number* — the number of protons in the nucleus of an atom of a particular element. For example, carbon (the letter C in the periodic table) has six protons in the nucleus of one atom, so its atomic number is 6. Periodic law states that the properties of elements are a periodic function of their atomic numbers. In other words, when elements are arranged by their atomic number, they form groups with similar properties. The number of electrons in one atom of an element is also equal to the atomic number because atoms are neutral (the positively charged particles are offset by the negatively charged particles one for one).

Of all the elements in the periodic table, living things use only a handful. The four most common elements found in living things are hydrogen, carbon, nitrogen, and oxygen, all of which are found in air, plants, and water. (Several other elements exist in smaller amounts in organisms, including sodium, magnesium, phosphorus, sulfur, chlorine, potassium, and calcium.)

Most often, the elements sodium, magnesium, chlorine, potassium, and calcium circulate in the body as *electrolytes,* substances that release ions (described in the preceding section) when they break apart in water. For instance, when in the "water" of the body, sodium chloride (NaCl) breaks apart into the ions $Na^+$ and $Cl^-$, which are then used either in organs such as the heart or in cellular processes.

### I so dig isotopes

All atoms of an element have the same number of protons, but the number of neutrons can change. If the number of neutrons is different between two atoms of the same element, the atoms are called *isotopes* of the element.

For example, carbon-12 and carbon-14 are two isotopes of the element carbon. Atoms of carbon-12 have 6 protons and 6 neutrons. These carbon atoms have a mass number of 12 because their mass is equal to 12. Atoms of carbon-14 still have 6 protons (because all carbon atoms have 6 protons), but they have 8 neutrons, giving them a mass number of 14.

The *atomic mass* of an element is the average mass of all the isotopes of that element, taking into account their relative abundance. If you look back at the periodic table in Figure 3-2, you can see that the atomic mass of carbon (written underneath the letter C) is 12.01. This number tells you that if you took the average of the mass of all the carbon atoms on Earth, they'd average out to 12.01. The most stable isotope of carbon is carbon-12, so it's more abundant than carbon-14. (When you average the mass of lots of atoms of carbon-12 with some of carbon-14, you get a number slightly larger than 12.)

## Molecules, Compounds, and Bonds

When you start putting elements together, you get more complex forms of matter, such as molecules and compounds. *Molecules* are made of two or more atoms, and *compounds* are molecules that contain at least two different elements.

One way to sort out the differences between elements, molecules, and compounds is to think about making chocolate chip cookies. First, you need to mix the wet ingredients: butter, sugar, eggs, and vanilla. Consider each of those ingredients a separate element. You need two sticks of the element butter. When you combine butter plus butter, you get a molecule of butter. Before you add the element of eggs, you need to beat them. So, when you add egg plus egg in a little dish, you get a molecule of eggs. When all the wet ingredients are mixed together, the molecule of butter is combined with the molecule of eggs, and you get a compound called "wet." Next, you need to mix together the dry ingredients: flour, salt, and baking soda. Think of each of those ingredients as a separate element. When all the dry ingredients are mixed together, you get a compound called "dry." Only when the wet compound is mixed with the dry compound is the reaction sufficiently ready for the most important element: the chocolate chips.

So what holds the elements of molecules and compounds together? Bonds, of course. Two important types of bonds exist in living things:

- **Ionic bonds** hold ions joined together by their opposite electrical charges. Ionic reactions occur when atoms combine and lose or gain electrons. When sodium (Na) and chlorine (Cl) combine, for example, sodium loses an electron to chlorine. Sodium becomes the positively charged sodium ion ($Na^+$), and chlorine becomes the negatively charged chloride ion ($Cl^-$). These two oppositely charged ions are attracted to each other, forming an ionic bond.
- **Covalent bonds** are formed when atoms share electrons in a covalent reaction. When two oxygen atoms join together to form oxygen gas, they share two pairs of electrons with each other. Each shared pair of electrons is one covalent bond, so the two pairs of shared electrons in a molecule of oxygen gas have a double bond. Covalent bonds are extremely important in biology because they hold together the backbones of all biological molecules.

# Acids and Bases (Not a Heavy Metal Band)

Some substances, such as lemon juice and vinegar, have a real edge when you taste them. Others, such as battery acid and ammonia, are so caustic you don't even want to get them on your skin. These substances are acids and bases, both of which have the potential to damage cells.

- **Acids are molecules that can split apart in water and release hydrogen ions ($H^+$).** A common example is hydrochloric acid (HCl). When HCl is added to water, it splits apart into $H^+$ and $Cl^-$, increasing the number of hydrogen ions in the water/HCl solution.
- **Bases are molecules that can split apart in water and release hydroxide ions ($OH^-$).** The most common example is sodium hydroxide (NaOH). When NaOH is added to water, it splits apart into $Na^+$ and $OH^-$.

Charged particles, like hydrogen and hydroxide ions, can interfere with the chemical bonds that hold molecules together. Because living things are made of molecules, strong acids and bases can release enough of these ions to cause damage.

Even water can split apart to create hydrogen and hydroxide ions. Each water molecule ($H_2O$) can separate into one hydrogen ion ($H^+$) and one hydroxide ion ($OH^-$). These ions can easily recombine to reform water molecules and will keep shifting back and forth between their molecular and ionized forms. In pure water, the number of hydrogen ions and hydroxide ions are balanced, so they don't do any damage to living things.

The relative concentration of hydrogen to hydroxide ions is represented by the pH scale. The following sections explain the pH scale and how organisms regulate themselves when their pH gets out of balance.

## "Ph"iguring out the pH scale

In the early 1900s, scientists came up with the *pH scale,* a system of classifying how acidic or basic a solution is. The term *pH* symbolizes the hydrogen ion concentration in a solution (for example, what proportion of a solution contains hydrogen ions). The pH scale goes from 1 to 14. A pH of 7 is neutral, meaning the amount of hydrogen ions and hydroxide ions in a solution with a pH of 7 is equal, just like in pure water.

A solution that contains more hydrogen ions than hydroxide ions is *acidic,* and the pH of the solution is less than 7. If a molecule releases hydrogen ions in water, it's an acid. The more hydrogen ions it releases, the stronger the acid, and the lower the pH value.

A solution that contains more hydroxide ions than hydrogen ions is *basic,* and its pH is higher than 7. Bases *dissociate* (break apart) into hydroxide ions ($OH^-$) and a positive ion. The hydroxide ions can combine with $H^+$ to create water. Because the hydrogen ions are used, the number of hydrogen ions in the solution decreases, making the solution less acidic and therefore more basic. So, the more hydroxide ions a molecule releases (or the more hydrogen ions it takes in), the more basic it is.

Table 3-1 shows you the pH of some common substances. Use it to help you visually figure out the pH scale.

**Table 3-1 The pH of Some Common Substances**

| *Increasing pH* | *Substances* |
|---|---|
| 0 (most acidic) | Hydrochloric acid (HCl) |
| 1 | Battery acid |
| 2 | Lemon juice, vinegar, stomach acid |
| 3 | Cola, apples |
| 4 | Beer |
| 4.5 | Tomatoes |
| 5 | Black coffee, bananas |

| Increasing pH | Substances |
|---|---|
| 5.5 | Normal rainwater |
| 6 | Urine |
| 6.5 | Saliva, milk |
| 7 (neutral) | Water, tears |
| 7.5 | Human blood |
| 8 | Seawater, eggs |
| 9 | Baking soda, antacids |
| 10 | Great Salt Lake |
| 11 | Ammonia |
| 12 | Bicarbonate of soda, soapy water |
| 13 | Oven cleaner, bleach |
| 14 (most basic) | Sodium hydroxide (NaOH), liquid drain cleaner |

## Buffing up on buffers

In organisms, blood or cytoplasm are the "solutions" in which the required ions (for example, electrolytes) are floating. That's why most substances in the body hover around the neutral pH of 7. However, nothing's perfect, so the human body has a backup system in case things go awry. A system of buffers exists to help neutralize the blood if excess hydrogen or hydroxide ions are produced.

*Buffers* keep solutions at a steady pH by combining with excess hydrogen ($H^+$) or hydroxide ($OH^-$) ions. Think of them as sponges for hydrogen and hydroxide ions. If a substance releases these ions into a buffered solution, the buffers will "soak up" the extra ions.

The most common buffers in the human body are bicarbonate ion ($HCO_3^-$) and carbonic acid ($H_2CO_3$). Bicarbonate ion carries carbon dioxide through the bloodstream to the lungs to be exhaled (see Chapter 15 for more on the respiratory system), but it also acts as a buffer. Bicarbonate ion takes up extra hydrogen ions, forming carbonic acid and preventing the pH of the blood from going too low. If the opposite situation occurs and the pH of the blood gets too high, carbonic acid breaks apart to release some hydrogen ions, which brings the pH back into balance.

If something goes wrong with the buffer system and the pH drops too low, an organism can develop *acidosis* (meaning the blood becomes too acidic). If the reverse happens and the pH gets too high, an organism can develop *alkalosis* (meaning the blood becomes too basic).

# Carbon-Based Molecules: The Basis for All Life

All living things rely pretty heavily on one particular type of molecule: carbon. The little ol' carbon atom, with its six protons and an outer shell of four electrons, is the central focus of *organic chemistry,* which is the chemistry of living things. When carbon bonds to hydrogen (which happens frequently in organic molecules), the carbon and hydrogen atoms share a pair of electrons in a covalent bond. Molecules with lots of carbon-hydrogen bonds are called *hydrocarbons.* Nitrogen, sulfur, and oxygen are also often joined to carbon in organisms.

So where do the carbon-containing molecules come from? The answer's simple: food. Some living things, like people, need to eat other living things to get their food, but some organisms, like plants, can make their own food. Regardless of the food source, all living things use food as a supply of carbon-containing molecules.

Carbon atoms are central to all organisms because they're found in carbohydrates, proteins, nucleic acids, and lipids — otherwise known as the structural materials of all living things. The sections that follow describe the roles of these materials.

## Providing energy: Carbohydrates

Carbohydrates, as the name implies, consist of carbon, hydrogen, and oxygen. The basic formula for carbohydrates is $CH_2O$, meaning the core structure of a carbohydrate is one carbon atom, two hydrogen atoms, and one oxygen atom. This formula can be multiplied; for example, glucose has the formula $C_6H_{12}O_6$, which is six times the ratio, but still the same basic formula.

But what *is* a carbohydrate? Well, *carbohydrates* are energy-packed compounds. Living creatures can break carbohydrates down quickly, making them a source of near-immediate energy. However, the energy supplied by carbohydrates doesn't last long. Therefore, reserves of carbohydrates in the body must be replenished frequently, which is why you find yourself hungry every four hours or so. Although carbohydrates are a source of energy, they also serve as structural elements (such as cell walls in plants).

Carbohydrates come in the following forms:

- **Monosaccharides:** Simple sugars consisting of three to seven carbon atoms are *monosaccharides* (see Figure 3-3a). In living things, monosaccharides form ring-shaped structures and can join together to form longer sugars. The most common monosaccharide is glucose.

- **Disaccharides:** Two monosaccharide molecules joined together form a *disaccharide* (see Figure 3-3b). Common disaccharides include sucrose (table sugar) and lactose (the sugar found in milk).
- **Oligosaccharides:** More than two but just a few monosaccharides joined together are an *oligosaccharide* (see Figure 3-3c). Oligosaccharides are important markers on the outsides of your cells (head to Chapter 4 for more on cells), such as the oligosaccharides that determine whether your blood type is A or B (people with type O blood don't have any of this particular oligosaccharide).
- **Polysaccharides:** Long chains of monosaccharide molecules linked together form a *polysaccharide* (see Figure 3-3d). Some of these babies are huge, and when we say huge, we mean some of them can have thousands of monosaccharide molecules joined together. Starch and glycogen, which serve as a means of storing carbohydrates in plants and animals, respectively, are examples of polysaccharides.

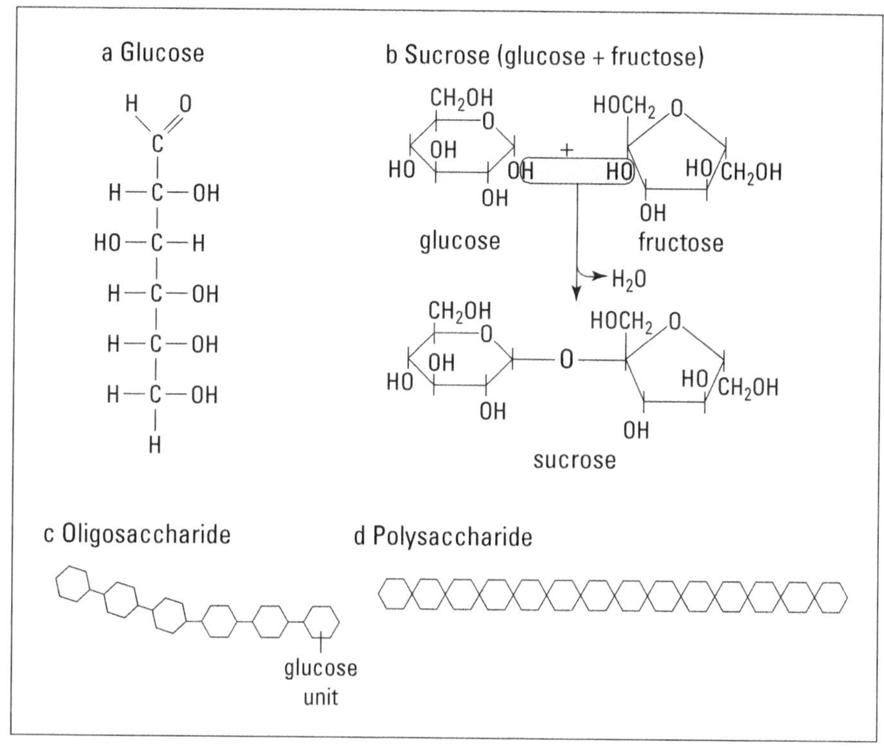


**Figure 3-3:** A variety of carbohydrate molecules.

Note that most of the names of carbohydrates end in *-ose*. Glucose, fructose, ribose, sucrose, maltose — these are all sugars. A *sugar* is a carbohydrate that dissolves in water, tastes sweet, and can form crystals. Just like, well, the sugar in your sugar bowl.

## What's the big deal about "natural" sugars?

Sugars are an important source of energy and carbon for living things. Yet many people these days have a love-hate relationship with sugar. Overconsumption of sugars such as sucrose can lead to obesity, diabetes, and tooth decay, so sometimes people perceive sugar as harmful to the body. This perception has spawned a huge branch of the food industry dedicated to finding alternatives to sucrose. Some products advertise themselves as healthier because they contain "natural" sugars such as fructose (the sugar in fruit). However, fructose really isn't that different from sucrose — in fact, it's one of the two monosaccharides that are part of sucrose. Fructose is a six-carbon sugar molecule that digests a little more slowly than glucose and is sweeter to the taste. So, potentially less fructose can be used to achieve the same effect as sucrose or glucose. However, the sweetening power of fructose depends on the food it's used in. Often there isn't much nutritional difference between foods sweetened with sucrose and those sweetened with fructose. Plus, much of the commercially available fructose is actually derived from sucrose, not extracted from fruit. So although "natural" fruit sugar may sound better, the differences between sugars may be so slight that they don't make much difference to the body at all.

The next sections explain how sugars interact with one another and how the human body stores a particular carbohydrate known as glucose.

### *Making and breaking sugars*

Monosaccharides join together in a process known as *dehydration synthesis,* which involves two molecules bonding and losing a water molecule. Figure 3-3b shows the dehydration synthesis of glucose and fructose to form sucrose.

The term *dehydration synthesis* may sound technical, but it's not at all if you really think about what the words mean. *Dehydration,* as we're sure you know, is what happens when you don't drink enough water. You dry out because water is removed (but not completely) from some cells, such as those in your tongue, to make sure more important cells, like those in your heart or brain, continue to function. *Synthesis* means making something. If you really think about it, in dehydration synthesis, something must be made when water is removed. That's exactly what happens. When glucose and fructose get together, a water molecule is removed from the monosaccharides and given off as a byproduct of the reaction.

The opposite of dehydration synthesis is hydrolysis. A *hydrolysis* reaction breaks down a larger sugar molecule into its original monosaccharides. When something undergoes hydrolysis, a water molecule splits a compound (*hydro* means "water"; *lysis* means "break apart"). When sucrose is added to water, it splits apart into glucose and fructose.

### Converting glucose for storage purposes

Carbohydrates are found in nearly every food, not just bread and pasta. Fruits, vegetables, and even meats also contain carbohydrates, although meats don't contain very many. Basically, any food that contains sugar has carbohydrates, and most foods are converted to sugars when they're digested.

When you digest your food, the carbohydrates from it break down into small sugars such as glucose. Those glucose molecules are then absorbed from your intestinal cells into your bloodstream, which carries the glucose molecules throughout your entire body. The glucose enters each of your body's cells and is used as a source of carbon and energy.

Because glucose provides a rapid source of energy, organisms often keep some on hand. They store it in various polysaccharides that can be quickly broken down when glucose is needed. Consider the following list your primer on the things glucose can be stored as:

- **Glycogen:** Animals, including people, store a polysaccharide of glucose called *glycogen*. It has a compact structure, so lots of it can be stored in cells for later use. Your liver, in particular, keeps a large glycogen reserve on hand for when you exercise.
- **Starch:** Plants store glucose as the polysaccharide *starch*. The leaves of a plant produce sugar during the process of photosynthesis and then store some of that sugar as starch. When the simple sugars need to be retrieved for use, the starch is broken down into its smaller components.

Plants also make a polysaccharide of glucose called *cellulose*. Cellulose plays a structural role for plants rather than a storage role by giving rigidity to the plant cells. Most animals, including people, can't digest cellulose because of the type of bonds between the glucose molecules. Because cellulose passes through your digestive tract virtually untouched, it helps maintain the health of your intestines.

## Making life possible: Proteins

Without proteins, living things wouldn't exist. Many proteins provide structure to cells; others bind to and carry important molecules throughout the body. Some proteins are involved in reactions in the body when they serve as enzymes (see Chapter 4 for more on enzymes). Still others are involved in muscle contraction or immune responses. Proteins are so diverse that we can't possibly tell you about all of them. What we can tell you about, however, are the basics of their structure and their most important functions.

### The building blocks of proteins

*Amino acids,* of which there are 20, are the foundation of all proteins. Think of them as train cars that make up an entire train called a protein. Figure 3-4 shows what one amino acid looks like.

The genetic information in cells calls for amino acids to link together in a certain order, forming chains called *polypeptide chains.* Amino acids link together by dehydration synthesis, just like sugars do (as explained in the earlier "Making and breaking sugars" section), and each polypeptide chain is made up of a unique number and order of amino acids.

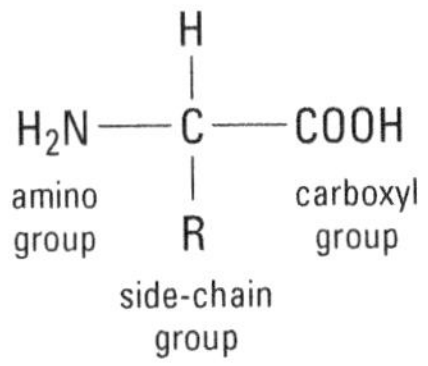

H
|
$H_2N$ — C — COOH
amino group | carboxyl group
R
side-chain group

The central carbon atom is flanked by an amino group and a carboxyl group. The name of the amino acid depends on which one of the 20 side-chain groups is at R. For example, if $CH_2$—C(=O)—$O^-$ was at R, the amino acid would be aspartic acid. Proteins are amino acids joined together by peptide bonds. Specific proteins are created based on the order of amino acids connected together. The order of amino acids is determined by the genetic code.

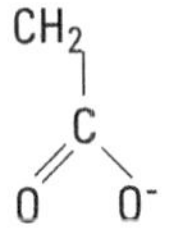

**Figure 3-4:** Amino acid structure.

### The main functions of proteins

One or more polypeptide chains come together to form functional *proteins.* Once formed, each protein does a specific job or makes up a specific tissue in the body.

- **Enzymes are proteins that speed up the rate of chemical reactions.** Metabolic processes don't happen automatically; they require enzymes. For the full scoop on enzymes, head to Chapter 4.
- **Structural proteins reinforce cells and tissues.** *Collagen,* a structural protein found in *connective tissue* (the tissue that joins muscles to bones to allow movement), is the most abundant protein in animals with a backbone. Connective tissue includes ligaments, tendons, cartilage, bone

tissue, and even the cornea of the eye. It provides support in the body, and it has a great capability to be flexible and resistant to stretching.

- **Transport proteins move materials around cells and around the body.** *Hemoglobin* is a transport protein found in red blood cells that carries oxygen around the body. A hemoglobin molecule is shaped kind of like a three-dimensional four-leaf clover without a stem. Each leaf of the clover is a separate polypeptide chain. In the center of the clover, but touching each polypeptide chain, is a heme group with an atom of iron at its center. When gas exchange occurs between the lungs and a blood cell (for more on respiration and circulation, see Chapter 15), the iron atom attaches to the oxygen. Then, the iron-oxygen complex releases from the hemoglobin molecule in the red blood cell so the oxygen can cross cell membranes and get inside any cell of the body.

## Drawing the cellular road map: Nucleic acids

Until as recently as the 1940s, scientists thought that genetic information was carried in the proteins of the body. They thought nucleic acids, a new discovery at the time, were too small to be significant. That all changed in 1953 when James Watson and Francis Crick figured out the structure of a nucleic acid, proving things were the other way around: Nucleic acids created the proteins!

*Nucleic acids* are large molecules that carry tons of small details, specifically all the genetic information for an organism. Nucleic acids are found in every living thing — plants, animals, bacteria, and fungi. Just think about that fact for a moment. People may look different than fungi, and plants may behave differently than bacteria, but deep down all living things contain the same chemical "ingredients" making up very similar genetic material.

Nucleic acids are made up of strands of *nucleotides.* Each nucleotide has three components of its own:

- A nitrogen-containing base called a *nitrogenous base*
- A sugar that contains five-carbon molecules
- A phosphate group

That's it. Your entire genetic composition, personality, and maybe even your intelligence hinge on molecules containing a nitrogen compound, some sugar, and a phosphate. The following sections introduce you to the two types of nucleic acids.

### Deoxyribonucleic acid (DNA)

You may have heard DNA (short for *deoxyribonucleic acid*) referred to as "the double helix." That's because DNA contains two strands of nucleotides arranged in a way that makes it look like a twisted ladder. See for yourself in Figure 3-5.

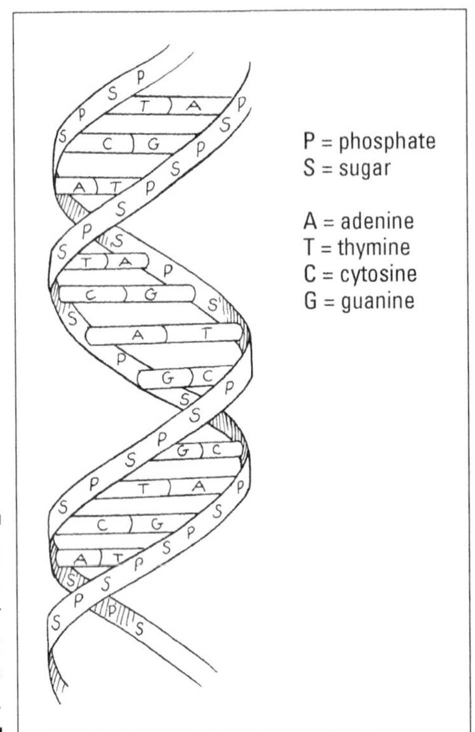


**Figure 3-5:** The twisted-ladder model of a DNA double helix.

The sides of the ladder are made up of sugar and phosphate molecules, hence the nickname "sugar-phosphate backbone." (The name of the sugar in DNA is deoxyribose.) The "rungs" on the ladder of DNA are made from pairs of nitrogenous bases from the two strands.

The nitrogenous bases that DNA builds its double helix upon are adenine (A), guanine (G), cytosine (C), and thymine (T). The order of these chemical letters spells out your genetic code. Oddly enough, the bases always pair in a certain way: Adenine always goes with thymine (A-T), and guanine always links up with cytosine (G-C). These particular *base pairs* line up just right chemically so that hydrogen bonds can form between them.

Certain sections of nitrogenous bases along a strand of DNA form a gene. A *gene* is a unit that contains the genetic information or codes for a particular protein and transmits hereditary information to the next generation. Whenever a new cell is made in an organism, the genetic material is reproduced and put into the new cell. (You can find details about this in Chapter 6.) The new cell can then create proteins and also pass on the genetic information to the next new cell.

But genes aren't found only in reproductive cells. Every cell in an organism contains DNA (and therefore genes) because every cell needs to make proteins. Proteins control function and provide structure. Therefore, the blueprints of life are stored in each and every cell.

The order of the nitrogenous bases on a strand of DNA (or in a section of the DNA that makes up a gene) determines the order in which amino acids are strung together to make a protein. Which protein is produced determines which structural element is produced within your body (such as muscle tissue, skin, or hair) or what function can be performed (such as the transportation of oxygen to all the cells).

Every cellular process and every aspect of metabolism is based on genetic information stored in DNA and thus the production of the proper proteins. If the wrong protein is produced (as in the case of cancer), then disease occurs.

### Ribonucleic acid (RNA)

RNA, short for *ribonucleic acid,* is a chain of nucleotides that serves as an important information molecule. It plays an important role in the creation of new proteins (which we cover in Chapter 8). The structure of RNA is slightly different from that of DNA.

- RNA molecules have only one strand of nucleotides.
- The nitrogenous bases used are adenine, guanine, cytosine, and uracil (rather than thymine).
- The sugar in RNA is ribose (not deoxyribose).

## Supplying structure, energy, and more: Lipids

In addition to carbohydrates, proteins, and nucleic acids, your body needs one more type of large molecule to survive. Yet, if you're like most people, you try to avoid too much of it in your diet. We're talking about fats, which can be both a blessing and a curse because of their incredible *energy density* (the ability to store lots of calories in a small space). The energy density of fats makes them a highly efficient way for living things to store energy — very useful when food isn't always available. But that same energy density makes it really easy to pack in the calories when you eat fatty foods!

Fats are an example of a type of molecule called lipids. *Lipids* are *hydrophobic* molecules, meaning they don't mix well with water. You've probably heard the saying that "oil and water don't mix." Well, oil is a liquid lipid, so the old saying is true; it really doesn't mix with water. Butter and lard are examples of solid lipids, as are waxes, which are valued for their water-repellent properties on snowboards, skis, and automobiles.

Three major types of lipid molecules exist:

- **Phospholipids:** These lipids, made up of two fatty acids and a phosphate group, have an important structural function for living things because they're part of the membranes of cells (see Chapter 4 for more on cell membranes). Phospholipids aren't the type of lipid floating around the bloodstream clogging arteries.
- **Steroids:** These lipid compounds, consisting of four connecting carbon rings and a functional group that determines the steroid, generally create hormones. *Cholesterol* is a steroid molecule used to make testosterone and estrogen; it's also found in the membranes of cells. The downside to cholesterol is that it's transported around the body by other lipids. If you have too much cholesterol floating in your bloodstream, then you have an excess of fats carrying it through your bloodstream. This situation is troubling because the fats and cholesterol molecules can get stuck in your blood vessels, leading to blockages that cause heart attacks or strokes.
- **Triglycerides:** These fats and oils, which are made up of three fatty acid molecules and a glycerol molecule, are important for energy storage and insulation. In people, fats form from an excess of glucose. After the liver stores all the glucose it can as glycogen, whatever remains is turned into triglycerides. (Both sugars and fats are made of carbon, hydrogen, and oxygen, so your cells just rearrange the atoms to convert from one to another.) The triglycerides float through your bloodstream on their way to be deposited into *adipose tissue* — the soft, squishy fat you can see on your body. Adipose tissue is made up of many, many molecules of fat. The more fat molecules that are added to the adipose tissue, the bigger the adipose tissue (and the place on your body that contains it) gets.

  Whether a triglyceride is a fat or an oil depends on the bonds between the carbon and hydrogen atoms.

  - Fats contain lots of single bonds between their carbon atoms. These *saturated bonds* pack tightly (see Figure 3-6), so fats are solid at room temperature.
  - Oils contain lots of double bonds between their carbon atoms. These *unsaturated bonds* don't pack tightly (see Figure 3-6), so oils are liquid at room temperature.

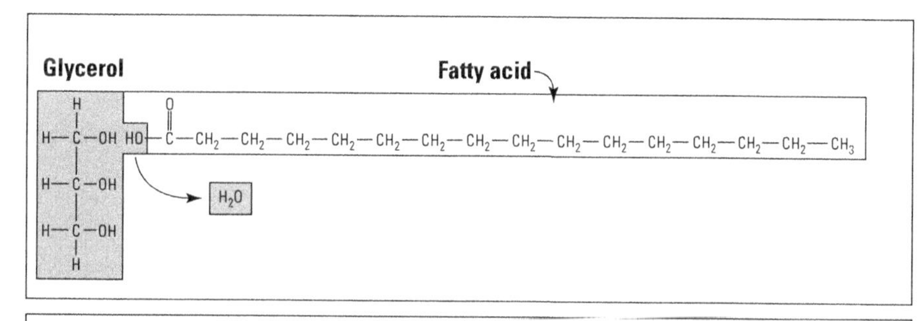


**Figure 3-6:** Saturated and unsaturated bonds in a typical triglyceride.

Fat provides an energy reserve to your body. When you use up all of your stored glucose (which doesn't take long because sugars "burn" quickly in aerobic conditions), your body starts breaking down glycogen, which is stored primarily in the liver and muscle. Liver glycogen stores can typically last 12 or more hours. After that, your body starts breaking down adipose tissue to retrieve some stored energy. That's why aerobic exercise, so long as it's enough to use up more calories than you took in that day, is the best way to lose fat. (Notice we didn't say "pounds" here. Pounds measure everything in your body's composition: fat tissue, muscle tissue, and bone, along with water, your organs, skin, and some incidental stuff.)

# Chapter 4

# The Living Cell

### In This Chapter

- Finding out what makes cells the basic units of life
- Taking a look at the structure of prokaryotic and eukaryotic cells
- Discovering how enzymes accelerate reactions

Every living thing has cells. The smallest creatures have only one, yet they're as alive as you are. What exactly is a cell? In plain and simple terms, it's the smallest living piece of an organism — including you. Without cells, you'd be a disorganized blob of chemicals that'd just ooze out into the environment. And that is why the cell is the fundamental unit of life.

You get to explore the purpose and structure of cells in this chapter. And because cells rely on chemical reactions to make things happen, you also find out all about *enzymes,* which are proteins that help speed up the pace of chemical reactions.

## An Overview of Cells

*Cells* are sacs of fluid that are reinforced by proteins and surrounded by membranes. Inside the fluid float chemicals and *organelles,* structures inside cells that are used during metabolic processes. (Yes, an organism contains parts that are smaller than a cell, but these structures can't perform all the functions of life on their own, so they aren't considered to be alive.)

A cell is the smallest part of an organism that retains characteristics of the entire organism. For example, a cell can take in fuel, convert it to energy, and eliminate wastes, just like the organism as a whole can. Because cells can perform all the functions of life (as shown in Figure 4-1), the cell is the smallest unit of life.

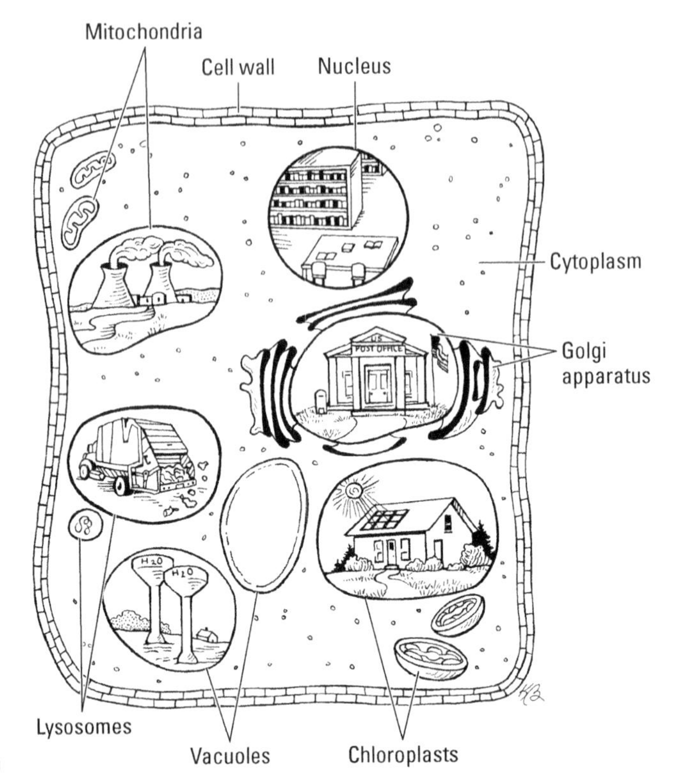


**Figure 4-1:** Cells perform all the functions of life.

Cells can be categorized in different ways, according to structure or function, or in terms of their evolutionary relationships. In terms of structure, scientists categorize cells based on their internal organization:

- **Prokaryotes** don't have a "true" nucleus in their cells. Nor do they have organelles. Bacteria and archaea are all prokaryotes.
- **Eukaryotes** have a nucleus in their cells that houses their genetic material. They also have organelles. Plants, animals, algae, and fungi are all eukaryotes.

## Viruses versus bacteria

Viruses (like those that cause the flu, a cold, or AIDS) aren't made of cells. Although viruses make you sick, just like bacteria do, viruses and bacteria are actually quite different. Viruses are simpler than bacteria and are made of just a few molecules. In fact, viruses have so few parts that they can't even grow and divide by themselves (so they're not considered to be alive). Viruses can only reproduce if they take over a host cell and steal its energy and materials (head to Chapter 17 for the details of how viruses attack cells).

# Peeking at Prokaryotes

Prokaryotes include cells you've probably heard of, such as the bacteria *E. coli* and *Streptococcus* (which causes strep throat), the blue-green algae that occasionally cause lake closures, and the live cultures of bacteria in yogurt, as well as some cells you may never have heard of, called archaeans (see Chapter 10 for more on archaeans).

Whether you've heard of a specific prokaryote or not, you're likely well aware that bacteria have a pretty bad rap. They seem to make the papers only when they're causing problems, such as disease. Behind the scenes, though, bacteria are quietly performing many beneficial tasks for people and the rest of life on planet Earth. Why, if bacteria could get some good headlines, those headlines might read a little something like this:

- **Bacteria are used in human food production!** Yogurt and cheese are quite tasty, humans say.
- **Bacteria can clean up our messes!** Oil-eating bacteria help save beaches, and other bacteria help clean up our sewage.
- **Normal body bacteria help prevent disease!** Bacteria living on the body can prevent disease-causing bacteria from moving in.
- **Bacteria are nature's recyclers!** Bacteria release nutrients from dead matter during decomposition.
- **Bacteria help plants grow!** Nitrogen-fixing bacteria can pull nitrogen out of the air and convert it to a form that plants can use.

The cells of prokaryotes are fairly simple in terms of structure because they don't have internal membranes or organelles like eukaryotic cells do (we cover all the structures present in eukaryotic cells later in this chapter). The majority of prokaryotic cells (like the one in Figure 4-2) share these characteristics:

- A plasma membrane forms a barrier around the cell, and a rigid cell wall outside the plasma membrane provides additional support to the cell.
- DNA, the genetic material of prokaryotes, is located in the cytoplasm, in an area called the nucleoid.
- Ribosomes make proteins in the cytoplasm.
- Prokaryotes break down food using cellular respiration (which requires oxygen, as explained in Chapter 5) and another type of metabolism called fermentation (which doesn't require oxygen).

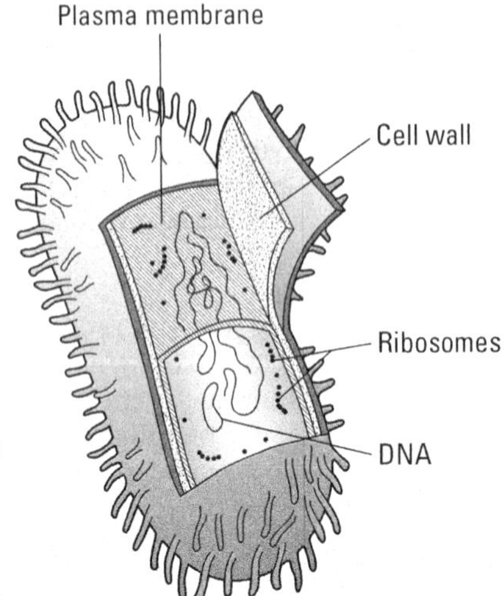


**Figure 4-2:** A prokaryotic cell.

# Examining the Structure of Eukaryotic Cells

The living things you're probably most familiar with — humans, animals, plants, mushrooms, and molds — are all eukaryotes, but they're not the only members of the eukaryote family. Eukaryotes also include many inhabitants of the microbial world, such as algae, amoebas, and plankton.

Eukaryotes have the following characteristics (see Figures 4-3 and 4-4 for diagrams of eukaryotic cells):

- A nucleus that stores their genetic information.
- A plasma membrane that encloses the cell and separates it from its environment.
- Internal membranes, such as the endoplasmic reticulum and the Golgi apparatus, that create specialized compartments inside the cells.
- A cytoskeleton made of proteins that reinforces the cells and controls cellular movements.

- Organelles called mitochondria that combine oxygen and food to transfer the energy from food to a form that cells can use.
- Organelles called chloroplasts, which use energy from sunlight plus water and carbon dioxide to make food. (Chloroplasts are found only in the cells of plants and algae.)
- A rigid cell wall outside of their plasma membrane. (This is found only in the cells of plants, algae, and fungi; animal cells just have a plasma membrane, which is soft.)

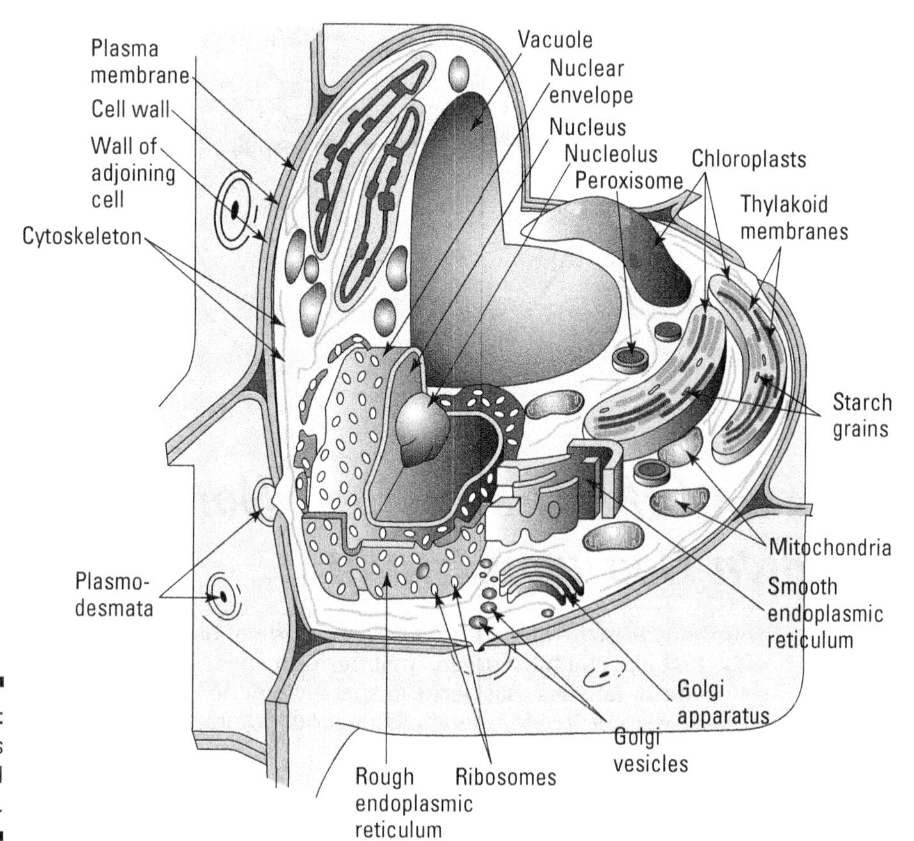


**Figure 4-3:** Structures in a typical plant cell.

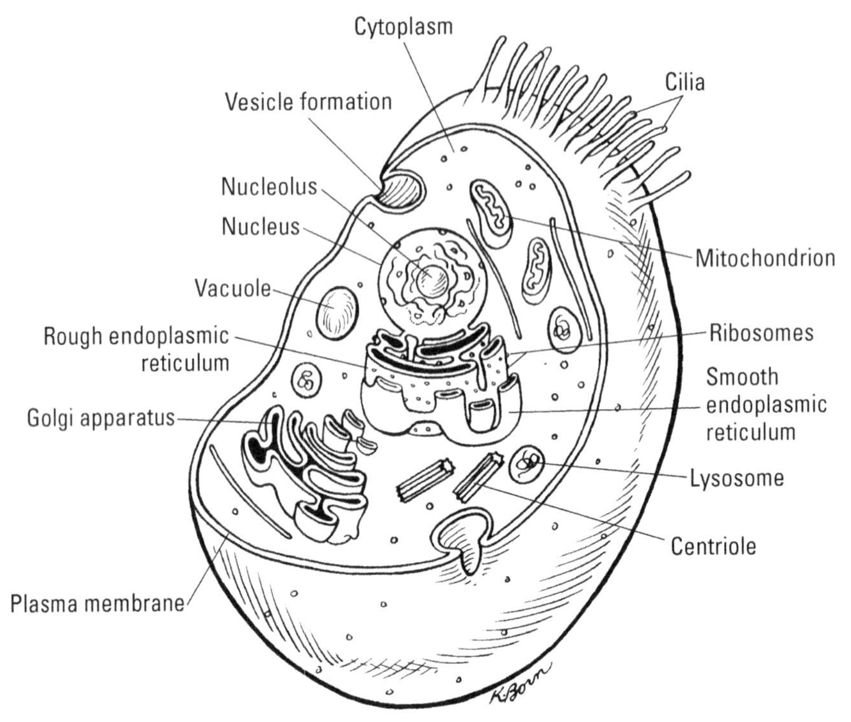


**Figure 4-4:** Structures in a typical animal cell.

# Cells and the Organelles: Not a Motown Doo-wop Group

Your body is made of organs, which are made of tissues, which are made of cells. Just like you have organs that perform specific functions for your body, cells have organelles that perform specific functions for the cell. Some organelles are responsible for metabolizing food; others are responsible for making the structures the cell needs to function.

The sections that follow highlight the organelles found in eukaryotic cells and get you acquainted with their specific functions.

Plant and animal cells are very similar, but they have a few significant differences in their organelles. Plant cells have chloroplasts, large central vacuoles, and cell walls; animal cells don't. What animal cells *do* have that plant cells don't are *centrioles,* small structures that are part of the cytoskeleton and appear during animal cell division.

## Holding it all together: The plasma membrane

The membrane that encloses and defines all cells as separate from their environment is called the *plasma membrane,* or the *cell membrane.* The job of the plasma membrane is to separate the chemical reactions occurring inside the cell from the chemicals outside the cell.

Thinking of the plasma membrane as an international border controlling what enters and leaves a particular country is a good way of remembering the plasma membrane's function.

The fluid inside a cell, called the *cytoplasm,* contains all the organelles and is very different from the fluid found outside the cell. (*Cyto* means "cell," and *plasm* means "shape." So, *cytoplasm* literally means "cell shape," which is fitting because the plasma membrane is what defines cell shape.)

Animal cells are supported by a fluid protein-and-carbohydrate matrix called the *extracellular matrix.* (*Extra* means "outside," so *extracellular* literally means "outside the cell.") Plant cells are supported by a more solid structure, called a *cell wall,* that's made of the carbohydrate cellulose.

The next sections explain the structure of the plasma membrane in detail and describe how materials move through it in order to keep the cell healthy and allow it to do its job.

### Deciphering the fluid-mosaic model

Plasma membranes are made of several different components, much like a mosaic work of art. Because membranes are a mosaic, and because they're flexible and fluid, scientists call the description of membrane structure the *fluid-mosaic model.* We've drawn the model for you in Figure 4-5 to help you visualize all the parts that make up a plasma membrane.

Moving from the left of Figure 4-5 to the right, notice the phospholipid bilayer. This serves as the foundation of the plasma membrane. *Phospholipids* are a special kind of lipid (see Chapter 3 for more on lipids); they have water-attracting *and* water-repelling parts. At body temperature, phospholipids have the consistency of thick vegetable oil, which allows plasma membranes to be flexible and fluid. Each phospholipid molecule has a hydrophilic head that's attracted to water and a hydrophobic tail that repels water. (*Hydro* means "water," *phile* means "love," and *phobia* means "fear," so *hydrophilic* literally means "water-loving" and *hydrophobic* literally means "water-fearing.")

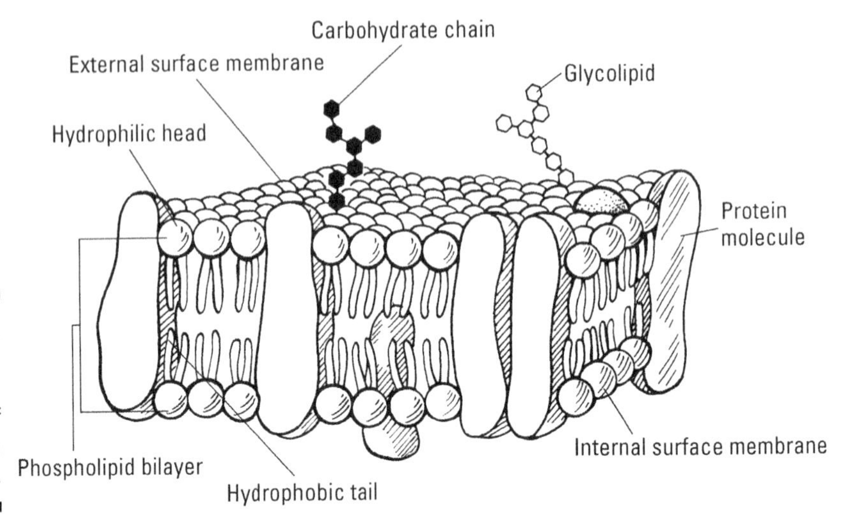


**Figure 4-5:** The fluid-mosaic model of plasma membranes.

In each cell, the hydrophilic heads point toward the watery environments outside and inside the cell, sandwiching the hydrophobic tails between them to form the phospholipid bilayer (see Figure 4-5). Because cells reside in a watery solution (the extracellular matrix), and because they contain a watery solution inside of them (cytoplasm), the plasma membrane forms a sphere around each cell so that the water-attracting heads are in contact with the fluid, and the water-repelling tails are protected on the inside.

In addition to phospholipids, proteins are a major component of plasma membranes. The proteins are embedded in the phospholipid bilayer, but they can drift in the membrane like ships sailing through an oily ocean.

Cholesterol and carbohydrates are minor components of plasma membranes, but they play fairly significant roles.

- Cholesterol makes the membrane more stable and prevents it from solidifying when your body temperature is low. (It keeps you from literally freezing when you're "freezing.")
- Carbohydrate chains attach to the outer surface of the plasma membrane on each cell. When carbohydrates attach to the phospholipids, they form glycolipids (and when they attach to the proteins, they form glycoproteins). Your DNA determines which specific carbohydrates attach to your cells, affecting characteristics such as your blood type.

### Transporting materials through the plasma membrane

Cells are busy places. They manufacture materials that need to be shipped out, and they need to take up materials such as food and signals. These important exchanges take place at the plasma membrane.

Whether or not a molecule can cross a plasma membrane depends upon its structure and the cell. Small, hydrophobic molecules such as oxygen and carbon dioxide are compatible with the hydrophobic tails of the phosopholipid bilayer, so they can easily scoot across membranes. Hydrophilic molecules such as ions can't get through the tails by themselves, so they need help to cross. Larger molecules (think food and hormones) also need help, which comes in the form of transport proteins.

Some of these proteins help form openings called *channels* in the membrane. Small molecules such as hormones and ions may be allowed to pass through these *channel proteins.* Other proteins, called *carrier proteins,* pick up molecules on one side of a membrane and then drop them off on the other side. Other proteins in the membrane act as *receptors* that detect the presence of molecules the cell needs. When these molecules bind to receptors, the cell allows the molecules to cross the plasma membrane with the help of transport proteins.

Because the plasma membrane is choosy about what substances can pass through it, it's said to be selectively permeable. (Permeability describes the ease with which substances can pass through a border, such as a cell membrane. *Permeable* means that most substances can easily pass through. *Impermeable* means substances can't pass through. *Selectively permeable* and *semipermeable* mean that only certain substances are able to pass through.)

Materials can pass through the plasma membrane either passively or actively, as you discover in the following sections.

#### Passively moving along

*Passive transport* requires no energy on the part of the cell. Molecules can move passively across membranes in one of two ways. In both cases, the molecules are moving from where they're more concentrated to where they're less concentrated. (In other words, they spread themselves out randomly until they're evenly distributed.) Here are the two methods of passive transport:

- **Diffusion:** The movement of molecules other than water from an area where they're highly concentrated to an area where they're less concentrated is *diffusion.* To go from a high concentration to a low concentration, the molecules need only diffuse (spread themselves) across the membrane, separating the areas of concentration.
- **Osmosis:** The diffusion of water across a membrane is *osmosis.* It works the same way as diffusion, but it can be a little confusing because the movement of water is affected by the concentration of substances called

*solutes* that are dissolved in the water. Basically, water moves from areas where it's more concentrated (more pure) to areas where it's less concentrated (where it has more solutes).

Try thinking about osmosis in terms of the solutes: Water moves toward the area with the greatest concentration of solutes. For example, the blood in your body contains a certain amount of salt. If the concentration of salt in your blood suddenly rises, water moves out of the blood cells, and your blood cells shrivel up. On the other hand, if too much fluid is in the bloodstream, the blood cells have too many molecules of salt in comparison, so they take in water. If they need to take in too much water to bring everything back into balance, they can swell and burst.

The relative concentration of solutes on either side of a membrane is compared in terms of the tonicity of the solutions. If a solution is *isotonic,* the concentrations of the substances *(solutes)* and water *(solvent)* on both sides of the membrane are equal. If one solution is *hypotonic,* it has a lower concentration of substances (and more water) in it when compared to another solution. If a solution is *hypertonic,* it has a higher concentration of substances in it (and less water) when compared to another solution.

#### Actively helping molecules across

*Active transport* requires some energy from the cell to move molecules that can't cross the phospholipid bilayer on their own from where they're less concentrated to where they're more concentrated. Carrier proteins, called *active transport proteins* or *pumps,* use energy stored in the cell to concentrate molecules inside or outside of the cell.

Active transport is a little like having to pay to take the Staten Island Ferry. The ferry is the carrier protein, and you're the big molecule that needs help getting from the bloodstream (New York Bay) to the inside of the cell (New York City). The fee that you pay is equivalent to the energy molecules expended by the cell.

## Diffusion at work in your lungs

In the human body, one place that diffusion occurs is in the lungs. You breathe in air, and oxygen gets into the tiniest air sacs of the lungs, called the *alveoli.* Surrounding the alveoli are the tiniest blood vessels — the *pulmonary capillaries.* The pulmonary capillaries contain the lowest concentration of oxygen in the body because by the time blood gets to them, most of the oxygen has been used up by other organs and tissues. This means the alveoli have a higher concentration of oxygen than the pulmonary capillaries. Oxygen from the alveoli of the lungs diffuses across the membrane between the air sac and the capillary, getting into the bloodstream so it can travel around the body.

## Supporting the cell: The cytoskeleton

Much like your skeleton reinforces the structure of your body, the *cytoskeleton* of a cell reinforces that cell's structure. However, it provides that reinforcement in the form of protein cables rather than bones. The proteins of the cytoskeleton reinforce the plasma membrane and the nuclear envelope (covered in the next section). They also run through the cell like railroad tracks, helping vesicles and organelles circulate around the cell.

Think of the cytoskeleton as a cell's scaffolding and railroad tracks because it reinforces the cell and allows things to move around within it.

Some cells have whiplike projections that help them swim or move fluids. If the projections are short, like those shown in Figure 4-4, the structures are called *cilia.* If the projections are long, they're called *flagella.* Both cilia and flagella contain cytoskeletal proteins. The proteins flex back and forth, making the cilia and flagella beat like little whips. Cells with cilia exist in your respiratory tract, where they wiggle their cilia to move mucus so you can cough it out; they're also found in your digestive tract, where they help move food along. Flagella are present on human sperm cells; they're what allow sperm to swim rapidly toward an egg during sexual reproduction.

## Controlling the show: The nucleus

Every cell of every living thing contains genetic material called DNA (which is short for *deoxyribonucleic acid,* as explained in Chapter 3). In eukaryotic cells, DNA is contained within a chamber called a *nucleus* that's separated from the cytoplasm by a membrane called the *nuclear envelope* (also known as the *nuclear membrane*). In the nucleus of cells that aren't multiplying, the DNA is wound around proteins and loosely spread out in the nucleus. When DNA is in this form, it's called *chromatin.* However, right before a cell divides, the chromatin coils up tightly into chromosomes. Human cells have 46 *chromosomes,* each one of which is a separate piece of DNA.

DNA contains the instructions for building molecules, mostly proteins, that do the work of the cell. Cell function depends upon the action of these proteins, and organism function depends on cell function. So, ultimately, organism function depends upon the instructions in the DNA.

Consider the nucleus the library of the cell because it holds lots of information. The chromosomes are the library's books, full of instructions for building cells.

Proteins in the nucleus copy the instructions from the DNA into molecules that get shipped out to the cytoplasm, where they direct the behavior of the cell. For example, each nucleus has a round mass inside it called a *nucleolus.*

The nucleolus produces ribosomes, which move out to the cytoplasm to help make proteins. In experiments where scientists transplant the nucleus from one cell into the cytoplasm of another cell, the cell behaves according to the instructions in the nucleus. So, the nucleus is the true control center of the cell.

## Creating proteins: Ribosomes

*Ribosomes* are small structures found in the cytoplasm of cells. The instructions for proteins are copied from the DNA into a new molecule called *messenger RNA* (mRNA). The mRNA leaves the nucleus and carries the instructions to the ribosomes out in the cytoplasm of the cell. The ribosomes then organize the mRNA and other molecules that are needed to build proteins (for the full scoop on how proteins are made, flip to Chapter 8).

Thinking of ribosomes like workbenches where proteins are built is a good way to remember their function.

## Serving as the cell's factory: The endoplasmic reticulum

The *endoplasmic reticulum* (ER) is a series of canals that connects the nucleus to the cytoplasm of the cell. (*Endo* means "inside," and *reticulum* refers to the netlike appearance of the ER, so *endoplasmic reticulum* basically means "a netlike shape inside the cytoplasm.") As you can see in Figure 4-4, part of the ER is covered in dots, which are actually ribosomes that attach to it during the synthesis of certain proteins. This part is called the *rough ER,* or RER. The part of the ER without ribosomes is called the *smooth ER* (SER).

Ribosomes on the RER make proteins that either get shipped out of the cell or become part of the membrane. (Proteins that stay in the cell are put together on ribosomes that float free in the cytoplasm.) The SER is involved in the metabolism of *lipids* (fats). Proteins and lipids made at the ER get packaged up into little spheres of membrane called *transport vesicles* that carry the molecules from the ER to the nearby Golgi apparatus.

To help you remember the ER's purpose, think of the ER as the cell's internal manufacturing plant because it produces proteins and lipids and then ships them out (to the Golgi apparatus).

## Preparing products for distribution: The Golgi apparatus

The *Golgi apparatus,* which is located very close to the ER (as you can see in Figure 4-4), looks like a maze with water droplets splashing off of it. The "water droplets" are transport vesicles bringing material from the ER to the Golgi apparatus.

Inside the Golgi apparatus, products produced by the cell, such as hormones or enzymes, are chemically tagged and packaged for export either to other organelles or to the outside of the cell. After the Golgi apparatus has processed the molecules, it packages them back up into a vesicle and ships them out again. If the molecules are to be shipped out of the cell, the vesicle finds its way to the plasma membrane, where certain proteins allow a channel to be produced so that the products inside the vesicle can be secreted to the outside of the cell. Once outside the cell, the materials can enter the bloodstream and be transported through the body to where they're needed.

If it helps you remember, you can consider the Golgi apparatus the cell's post office because it receives molecular packages and tags them for shipping to their proper destination.

## Cleaning up the trash: Lysosomes

*Lysosomes* are special vesicles formed by the Golgi apparatus to clean up the cell. Lysosomes contain digestive enzymes, which are used to break down products that may be harmful to the cell and "spit" them back out into the extracellular fluid. (We fill you in on enzymes in the later "Presenting Enzymes, the Jump-Starters" section.) Lysosomes also remove dead organelles by surrounding them, breaking down their proteins, and releasing them to construct a new organelle.

Essentially, lysosomes are the waste collectors of the cell; they gather materials the cell no longer needs and break them down so some parts can be recycled.

## Destroying toxins: Peroxisomes

*Peroxisomes* are little sacs of enzymes that break down many different types of molecules and help protect the cell from toxic products. Peroxisomes help in the breakdown of lipids, making their energy available to the cell.

Some of the reactions that occur in peroxisomes produce hydrogen peroxide, which is a dangerous molecule to cells. Peroxisomes prevent your cells from being damaged by hydrogen peroxide by converting the hydrogen peroxide into plain old water plus an extra oxygen molecule, both of which are always needed by the body and can be used in any cell.

Peroxisomes are a little bit like food processors. They're involved in breaking things down, just like the blades of a food processor are used to chop up larger pieces of food.

## Providing energy, ATP-style: Mitochondria

*Mitochondria* supply cells with the energy they need to move and grow by breaking down food molecules, extracting their energy, and transferring it to an energy-storing molecule that cells can easily use. That energy-storing molecule is *ATP,* short for *adenosine triphosphate.*

Recall the role of mitochondria by thinking of them as the power plants of the cell because they produce the energy the cell needs.

The process mitochondria use to transfer the energy in foods to ATP is called *cellular respiration.* What occurs during cellular respiration is like what occurs when a campfire burns, just on a much smaller scale. In a campfire, wood burns, consuming oxygen and transferring energy (heat and light) and matter (carbon dioxide and water) to the environment. In a mitochondrion, food molecules break down, consuming oxygen and transferring energy to cells (to be stored in ATP) and the environment (as heat). For more details on cellular respiration, see Chapter 5.

## Converting energy: Chloroplasts

*Chloroplasts* are organelles found solely in plants and algae. They specialize in transferring energy from the Sun into the chemical energy in food. They often have a distinctly green color because they contain *chlorophyll,* a green pigment that can absorb sunlight. During photosynthesis, the energy from sunlight is used to combine the atoms from carbon dioxide and water to produce sugars, from which all types of food molecules can be made. (Turn to Chapter 5 for more on photosynthesis.)

Consider chloroplasts the plant equivalent of solar-powered kitchens because they use energy from the Sun and "ingredients" from the environment (carbon dioxide and water) to make food.

A very common misconception is that plants have chloroplasts rather than mitochondria. The truth is, plants have both! Think about it — it wouldn't do plants much good to make food if they couldn't also break it down. When plants make food, they store matter and energy for later. When they need the matter and energy, they use their mitochondria to break the food down into usable energy.

# Presenting Enzymes, the Jump-Starters

Chemical reactions occur whenever the molecules in cells change. They're usually part of a cycle or pathway that has separate reactions at each step. Of course, because the pace of life in cells is so fast, cells can't just wait around for chemical reactions to happen — they have to make them happen quickly. Fortunately, they have the perfect tool at their disposal in the form of proteins called enzymes.

Each reaction of a pathway or cycle requires a specific enzyme to act as a *catalyst,* something that speeds up the rate of chemical reactions. These proteins are folded in just the right way to do a specific job. Enzymes also have pockets, called *active sites,* that they use to attach to certain molecules. The molecule an enzyme binds to is called its *substrate* (see Figure 4-6).

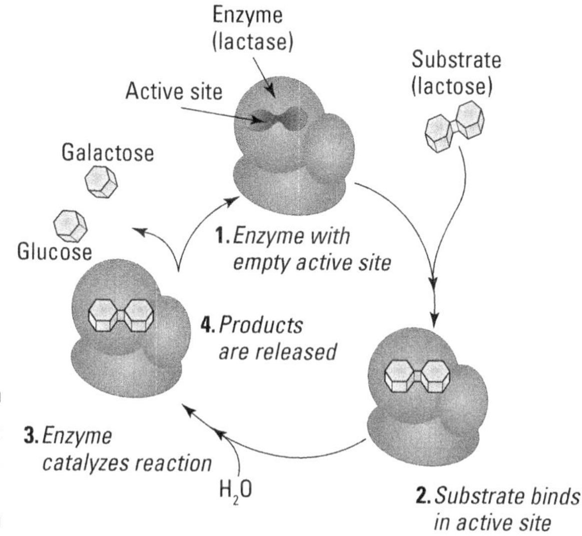


**Figure 4-6:** Enzyme catalysis.

Without the specific enzyme necessary to catalyze a particular reaction, the cycle or pathway can't be completed. The result of an uncompleted cycle or pathway is the lack of what that cycle or pathway is supposed to produce (a *product*). Without a needed product, a function can't be performed, which negatively affects the organism. For example, if people don't get enough vitamin C, the enzymes needed to make collagen can't function, resulting in a disease called scurvy. The lack of collagen in people with scurvy causes bleeding gums, loss of teeth, and abnormal bone development in children.

The sections that follow introduce you to how enzymes work, what they need to get the job done, and how cells are able to keep them under control.

## Staying the same . . .

Enzymes themselves are recycled. They're the same at the end of a reaction as they were at the beginning, and they can do their job again. For example, the first enzymatic reaction discovered was the one that breaks down urea into products that can be excreted from the body. The enzyme urease catalyzes the reaction between the reactants urea and water, yielding the products carbon dioxide and ammonia, which can be excreted easily by the body.

*Urease*

Urea + Water ↔ Carbon dioxide + Ammonia

In this reaction, the enzyme urease helps the *reactants* (molecules that enter a chemical reaction), urea and water, combine with each other. The bonds between the atoms in urea and water break and then reform between different combinations of atoms, forming the products carbon dioxide and ammonia. When the reaction is over, urease is unchanged and can catalyze another reaction between urea and water.

If you find yourself struggling to figure out which proteins are enzymes and which enzymes do what, here's a helpful hint: The names of enzymes end with *-ase* and usually have something to do with their function. For example, lipase is an enzyme that helps break down lipids (fats), and lactase is an enzyme that helps break down lactose.

## . . . while lowering activation energy

Enzymes work by reducing the amount of *activation energy* needed to start a reaction so reactions can occur more easily. On their own, reactants could occasionally collide with each other the right way to start a reaction. But they wouldn't do it nearly often enough to keep up with the fast pace of life in a cell. Without enzymes, your body wouldn't be able to, say, clear urea out of your body fast enough, leading to a toxic buildup of urea. That's where the enzyme urease comes into play. It binds the reactants in its active site and brings them together in a way that requires less energy for them to react.

Because reactions can occur more easily with enzymes, they occur more often. This increases the overall rate of the reaction in the body. One way to understand how enzymes speed up reactions is to think about reactions in terms of energy. In order for a reaction to occur, the reactants must collide with enough energy to get the reaction going. In the urea and water example, the reactants would need to collide with each other in just the right way for them to exchange partners and form into carbon dioxide and ammonia.

Whatever you do, don't fall for the idea that enzymes add energy to reactions to make them happen. They don't. In fact, they don't add *anything* to a reaction; they just help the reactants get together in the right way, lowering the "barrier" to the reaction. In other words, enzymes don't add energy; they just make it so the reactants have enough energy on their own.

## Getting some help from cofactors and coenzymes

Enzymes are proteins, but many need a nonprotein partner in order to do their job. Inorganic partners, such as iron, potassium, magnesium, and zinc ions, are called *cofactors.* Organic partners are called *coenzymes;* they're small molecules that can separate from the protein component of the enzyme and participate directly in the chemical reaction. Examples of coenzymes include many derivatives of vitamins. An important function of coenzymes is that they transfer electrons, atoms, or molecules from one enzyme to another.

## Controlling enzymes through feedback inhibition

Cells manage their activity by controlling their enzymes via *feedback inhibition,* a process in which a reaction pathway proceeds normally until the final product is produced at too high of a level. The final product then binds to the allosteric site of one of the initial enzymes in the pathway, shutting it down. (An *allosteric site* is literally an "other shape" site. When molecules bind to these "other" pockets, enzymes can be shut down.) By controlling enzymes, cells regulate their chemical reactions and, ultimately, the physiology of the entire organism.

Feedback inhibition gets its name because it uses a feedback loop. The quantity of the final product provides feedback to the beginning of the pathway. If the cell has plenty of the final product, then the cell can stop running the pathway.

By inhibiting the activity of an initial enzyme, the entire pathway is stopped. The process of feedback inhibition prevents cells not only from having to use energy creating excess products but also from having to make room to store the excess products. It's like keeping yourself from spending money on a huge quantity of food that you won't eat and would just end up storing until it rots.

Feedback inhibition is reversible because the binding of the final product to the enzyme isn't permanent. In fact, the final product is constantly binding, letting go, and then binding again. When the cell uses up its stores of the final product, the enzyme's allosteric site is empty, and the enzyme becomes active again.

# Chapter 5

# Acquiring Energy to Run the Motor

### In This Chapter

- Recognizing the important role energy plays in organisms
- Making food with photosynthesis
- Metabolizing food for energy through cellular respiration
- Counting calories

Just like you need to put gas in your car's engine so your car can move, you need to put food in your body so it can function. And you're not alone. Every person, as well as every other living thing, needs to "fill its tank" with matter and energy in the form of food. Food molecules are used to build the molecules that make up cells and are broken down to release energy to cells so they can grow and maintain themselves. Animals obtain their food by eating plants and other animals, whereas plants make their own food. In this chapter, we present some facts about the various types of energy and how they're transferred. We also demonstrate why cells need energy and take a look at how cells obtain and then store energy and matter.

## What's Energy Got to Do with It?

Whether you realize it or not, you use energy every day to cook your food, brighten your home, and run your appliances. Chances are, however, if you were asked to state what energy is, you might have some trouble. Most people have an idea of energy as something you need to do things, but they don't really know what energy is. The funny thing is that physicists, who spend a lot of time studying energy, define energy in exactly the way most people think about it: *Energy* is something that allows work to be done.

You can probably think of many kinds of energy in your life — electricity, heat, light, chemical (like gasoline). Although they may seem very different, the kinds of energy you can think of represent the two main types of energy:

- **Potential energy:** This is the energy that's stored in something because of the way it's arranged or structured. Energy in a battery, water behind a dam, and a stretched rubber band that's about to be released are all examples of potential energy. Food and gasoline also contain potential energy called *chemical potential energy* (energy that's stored in the bonds of molecules).
- **Kinetic energy:** This is the energy of motion. Light, heat, and moving objects all contain kinetic energy.

The following sections get you acquainted with the rules surrounding energy. They also explain how the cells of living things use and transfer energy, as well as how they obtain it (here's a hint: it's all about food).

## Looking at the rules regarding energy

Energy has three specific rules that are helpful to know so you can better understand how organisms use it:

- **Energy can't be created or destroyed.** The electricity that people get from hydroelectric power (or coal-burning power plants, wind turbines, or solar panels) isn't created from nothing. It's actually transferred from some other kind of energy. And when people use, say, electricity, that energy doesn't disappear. Instead, it becomes other kinds of energy, such as light or heat.

  The idea that energy can't be created or destroyed is known as the *First Law of Thermodynamics.*
- **Energy is transferred when it moves from one place to another.** To understand this rule, picture a flowing river that's used as a source of hydroelectric power. Energy from the moving river is transferred first to a spinning turbine, then to flowing electrons in power lines, and finally to the lights shining in customers' homes.
- **Energy is transformed when it changes from one form to another.** Again, think about a hydroelectric power plant. The potential energy of the water behind the plant's dam is transformed first into the kinetic energy of moving water, then the kinetic energy of a spinning turbine, and finally the kinetic energy of moving electrons.

## Metabolizing molecules

Organisms follow the rules of physics and chemistry, and the human body is no exception. The First Law of Thermodynamics (explained in the preceding section) applies to your *metabolism,* which is all the chemical reactions occurring in your cells at one time.

Two types of chemical reactions can occur as an organism metabolizes molecules:

- **Anabolic reactions:** This type of reaction builds molecules. Specifically, small molecules are combined into large molecules for repair, growth, or storage.
- **Catabolic reactions:** This type of reaction breaks down molecules to release their stored energy.

During chemical reactions, atoms receive new bonding partners, and energy may be transferred. (For more on molecules, atoms, and chemical bonds, flip to Chapter 3.)

Each type of food molecule you're familiar with — carbohydrates, proteins, and fats — are large molecules that can be broken down into smaller subunits. Complex carbohydrates, also called *polysaccharides,* break down into simple sugars called *monosaccharides;* proteins break down into *amino acids;* and fats and oils break down into *glycerol* and *fatty acids.* After cells break large food molecules down into their subunits, they can more easily reconnect the subunits and reform them into the specific molecules that they need.

## Transferring energy with ATP

Cells transfer energy between anabolic and catabolic reactions by using an energy middleman — *adenosine triphosphate* (or ATP for short). Energy from catabolic reactions is transferred to ATP, which then provides energy for anabolic reactions.

ATP has three phosphates attached to it (*tri-* means "three," so *triphosphate* means "three phosphates"). When ATP supplies energy to a process, one of its phosphates gets transferred to another molecule, turning ATP into *adenosine diphosphate* (ADP). Cells re-create ATP by using energy from catabolic reactions to reattach a phosphate group to ADP. Cells constantly build and break down ATP, creating the ATP/ADP cycle shown in Figure 5-1.

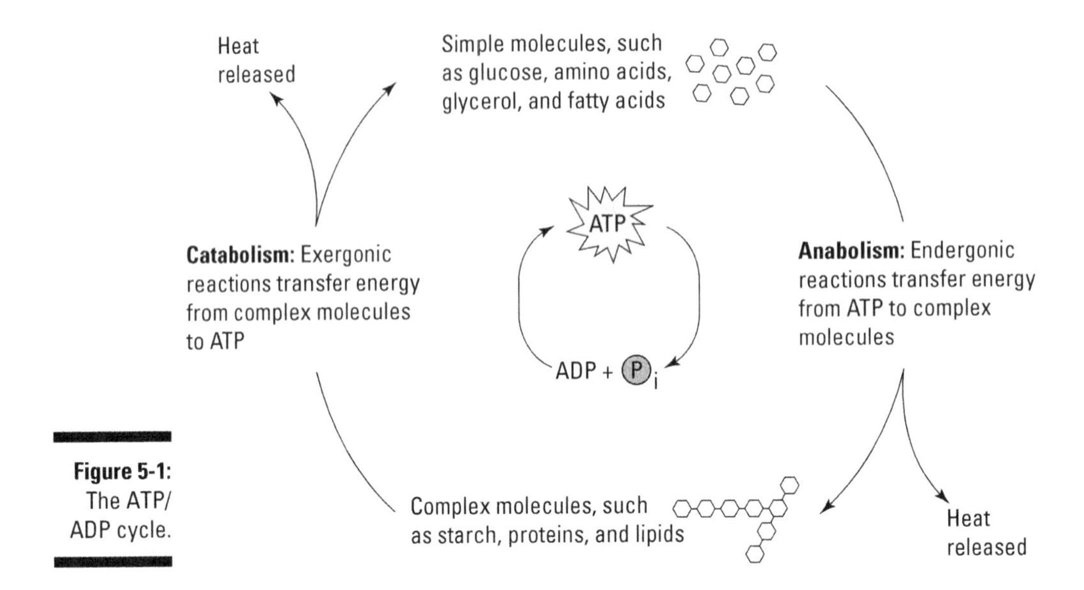


**Figure 5-1:** The ATP/ ADP cycle.

Cells have large molecules that contain stored energy, but when they're busy doing work, they need a handy source of energy. That's where ATP comes in. Cells keep ATP on hand to supply energy for all the work that they do.

Think of ATP like cash in your pocket. You may have money deposited in the bank, but that money isn't always easy to get, which is why you keep some cash in your pocket to quickly buy what you need. After you spend all of your cash, you have to go back to the bank or an ATM to get more. For living things, the energy stored in large molecules is like money in the bank. Cells break down ATP just like you spend your cash. Then, when cells need more ATP, they have to go back to the bank of large molecules and break some more down.

## Consuming food for matter and energy

Food molecules — in the form of proteins, carbohydrates, and fats — provide the matter and energy that every living thing needs to fuel anabolic and catabolic reactions and create ATP (for more on matter and molecules, see Chapter 3).

- **Organisms need matter to build their cells so they can grow, repair themselves, and reproduce.** Imagine that you scrape your knee and actually remove a fair amount of skin. Your body repairs the damage by building new skin cells to cover the scraped area. Just like a person who builds a house needs wood or bricks, your body needs molecules to build new cells (head to Chapter 4 for the full scoop on cells).

- **Organisms need energy so they can move, build new materials, and transport materials around their cells.** These activities are all examples of *cellular work,* the energy-requiring processes that occur in cells. When you walk up stairs, the muscle cells in your legs contract, and each contraction uses some energy. But the activities you decide to engage in aren't the only things that require energy. Your individual cells also need energy to do their work.

Food is a handy package that contains two things every organism needs: matter and energy.

## Finding food versus producing your own

All organisms need food, but there's one major difference in how they approach this problem: Some organisms, such as plants, can make their own food; other organisms, like you, have to eat other organisms to obtain their food. Biologists have come up with two separate categories to highlight this difference in how living things obtain their food:

- **Autotrophs can make their own food.** *Auto* means "self," and *troph* means "feed," so *autotrophs* are self-feeders. Plants, algae, and green bacteria are all examples of autotrophs.
- **Heterotrophs have to eat other organisms to get their food.** *Hetero* means "other," so *heterotrophs* are quite literally other-feeders. Animals, fungi, and most bacteria are examples of heterotrophs.

Although you may think that obtaining food is as easy as heading to the supermarket, pulling up to a drive-through window, or meeting the delivery guy at the front door, acquiring nutrients is actually a metabolic process. More specifically, food is made through one process and broken down through another. These processes are as follows:

- **Photosynthesis:** Only autotrophs such as plants, algae, and green bacteria engage in *photosynthesis,* a process that consists of using energy from the Sun, carbon dioxide from the air, and water from the soil to make sugars. (The carbon dioxide provides the matter plants need for food building.) When plants remove hydrogen atoms from water to use in the sugars, they release oxygen as waste.
- **Cellular respiration:** Both autotrophs and heterotrophs do *cellular respiration,* a process that uses oxygen to help break down food molecules such as sugars. The energy stored in the bonds of the food molecules is transferred to ATP. As the energy is transferred to the cells, the matter from the food molecules is released as carbon dioxide and water.

If you think about it, photosynthesis and cellular respiration are really the opposites of each other. Photosynthesis consumes carbon dioxide and water, producing food and oxygen. Cellular respiration consumes food and oxygen, producing carbon dioxide and water. Scientists write the big picture view of both processes as the following equations:

**Photosynthesis:**

$6\ CO_2 + 6\ H_2O + \text{Light Energy} \rightarrow C_6H_{12}O_6 + 6\ O_2$

**Cellular respiration:**

$C_6H_{12}O_6 + 6\ O_2 \rightarrow 6\ CO_2 + 6\ H_2O + \text{Usable Energy}$

Don't fall for the idea that only heterotrophs such as animals engage in cellular respiration. Autotrophs such as plants do it too. Think of it like this: Photosynthesis is a food-making pathway that autotrophs use to store matter and energy for later. So, a plant doing photosynthesis is like you packing a lunch for yourself. There wouldn't be much point in packing the lunch if you weren't going to eat it later, right? The same is true for a plant. It does photosynthesis to store matter and energy. When it needs that matter and energy, it uses cellular respiration to "unpack" its food.

# Photosynthesis: Using Sunlight, Carbon Dioxide, and Water to Make Food

Autotrophs such as plants combine matter and energy to make food in the form of sugars. With those sugars, plus some nitrogen and minerals from the soil, autotrophs can make all the types of molecules they need to build their cells. The chemical formula for *glucose,* the most common type of sugar found in cells, is $C_6H_{12}O_6$. To build glucose, autotrophs need carbon, hydrogen, and oxygen atoms, plus energy to combine them into sugar.

- The carbon and oxygen for the sugars come from carbon dioxide in the Earth's atmosphere.
- The hydrogen for the sugars comes from water found in the soil.
- The energy to build the sugars comes from the Sun (but only in autotrophs that use photosynthesis).

A common misconception is that plants get the matter they need to grow from the soil. This seems like a perfectly logical idea given that plants grow with their roots stuck in the ground. However, some very careful scientific observations by a Belgian scientist named Jean Baptiste van Helmont showed that a tree that gained 169 pounds in mass as it grew took only 2 ounces of dry material from the soil (not counting water). This experiment proved that plants don't take lots of material from the soil. Instead, they get most of the matter they need to grow from the carbon dioxide in the air. This idea may be more difficult to believe because air, including carbon dioxide, doesn't seem like much of anything, but scientists have proven that it's correct. Plants collect a lot of carbon dioxide molecules ($CO_2$) and combine them with water molecules ($H_2O$) to build sugars such as glucose ($C_6H_{12}O_6$). Plants get the water they need, plus some small amounts of minerals such as nitrogen, from the soil.

Photosynthesis occurs in two main steps (Figure 5-2 depicts both in action):

- **The light reactions of photosynthesis transform light energy into chemical energy.** The chemical energy is stored in the energy carrier ATP.
- **The light-independent reactions of photosynthesis produce food.** ATP from the light reactions supplies the energy needed to combine carbon dioxide ($CO_2$) and water ($H_2O$) to make glucose ($C_6H_{12}O_6$).

The next sections delve deeper into the process of photosynthesis.

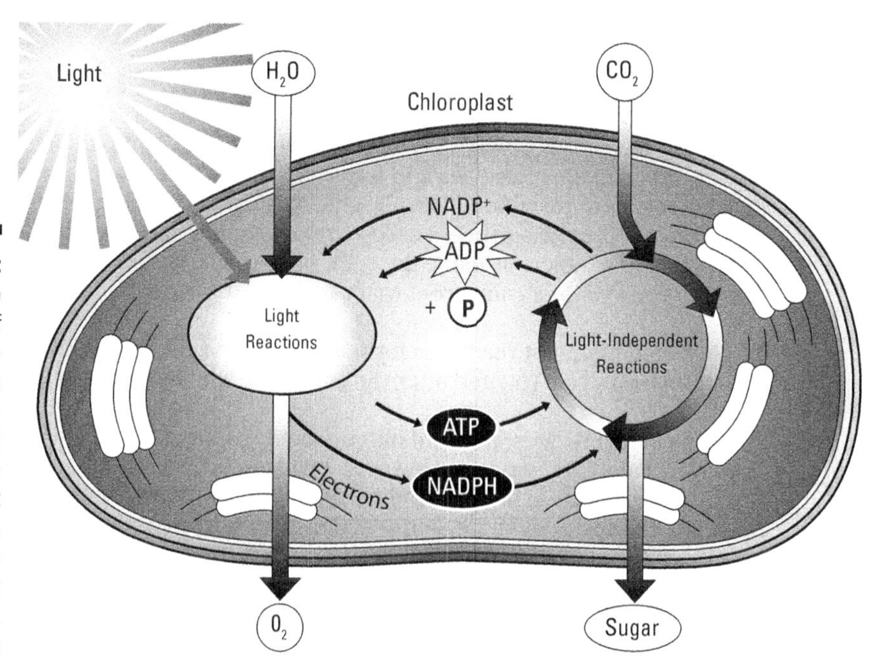


**Figure 5-2:** The two halves of photosynthesis, the light reactions and the light-independent reactions, are separate but linked.

## Transforming energy from the ultimate energy source

The Sun is a perfect energy source — a nuclear reactor positioned at a safe distance from planet Earth. It contains all the energy you could ever need . . . if only you could capture it. Well, green bacteria figured out how to do just that more than 2.5 billion years ago, showing that photosynthetic autotrophs were way ahead of humans on this one.

Plants, algae, and green bacteria use pigments to absorb light energy from the Sun. You're probably most familiar with the pigment *chlorophyll,* which colors the leaves of plants green. The chloroplasts in plant cells contain lots of chlorophyll in their membranes so they can absorb light energy (see Chapter 4 for more on chloroplasts).

During the light reactions of photosynthesis, chloroplasts absorb light energy from the Sun and then transform it into the chemical energy stored in ATP. When the light energy is absorbed, it splits water molecules. The electrons from the water molecules help with the energy transformation from light energy to chemical energy in ATP. Plants release the oxygen from the water molecules as waste, producing the oxygen ($O_2$) that you breathe.

## Putting matter and energy together

Plants use the energy in ATP (which is a product of the light reactions) to combine carbon dioxide molecules and water molecules to create glucose during the light-independent reactions. To make glucose, plants first take carbon dioxide out of the air through a process called *carbon fixation* (taking carbon dioxide and attaching it to a molecule inside the cell). They then use the energy from the ATP and the electrons that came from water to convert the carbon dioxide to sugar.

The light-independent reactions form a metabolic cycle that's known as the Calvin-Benson cycle (named after the scientists who discovered it).

As their name indicates, the light-independent reactions of photosynthesis don't need direct sunlight to occur. However, plants need the products of the light reactions to run the light-independent reactions, so really, the light-independent reactions can't happen if the light reactions can't happen.

When plants have made more glucose than they need, they store their excess matter and energy by combining glucose molecules into larger carbohydrate molecules, such as starch. When necessary, plants can break down the starch molecules to retrieve glucose for energy or to create other compounds, such as proteins and nucleic acids (with added nitrogen taken from the soil) or fats (many plants, such as olives, corn, peanuts, and avocados, store matter and energy in oils).

# Cellular Respiration: Using Oxygen to Break Down Food for Energy

Autotrophs and heterotrophs do cellular respiration to break down food to transfer the energy from food to ATP. The cells of animals, plants, and many bacteria use oxygen to help with the energy transfer during cellular respiration; in these cells, the type of cellular respiration that occurs is aerobic respiration (*aerobic* means "with air").

Three separate pathways combine to form the process of cellular respiration (you can see them all in action in Figure 5-3). The first two, glycolysis and the Krebs cycle, break down food molecules. The third pathway, oxidative phosphorylation, transfers the energy from the food molecules to ATP. Here are the basics of how cellular respiration works:

- During *glycolysis,* which occurs in the cytoplasm of the cell, cells break glucose down into *pyruvate,* a three-carbon compound. After glycolysis, pyruvate is broken down into a two-carbon molecule called acetyl-coA.
- After pyruvate is converted to acetyl-coA, cells use the *Krebs cycle* (which occurs in the matrix of the mitochondrion) to break down acetyl-coA into carbon dioxide.
- During *oxidative phosphorylation,* which occurs in the inner membrane or *cristae* of the mitochondrion), cells transfer energy from the breakdown of food to ATP.

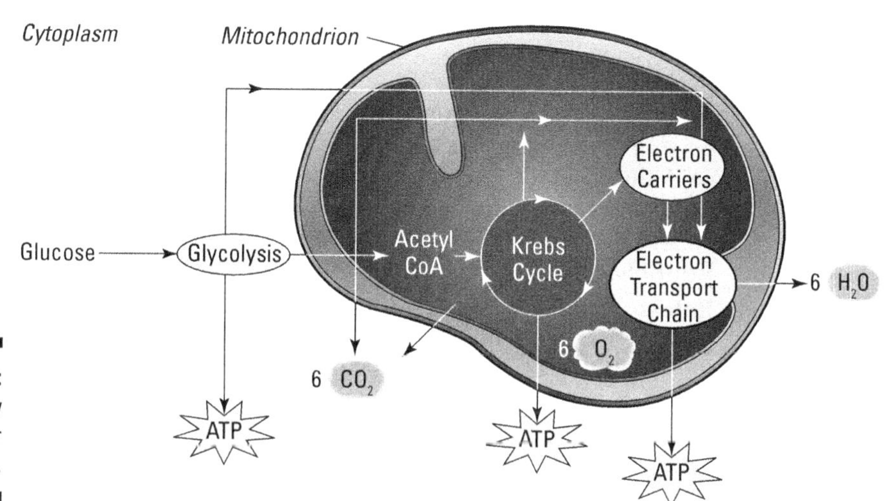


**Figure 5-3:** An overview of cellular respiration.

For a more in-depth look at cellular respiration, check out the following sections.

Cellular respiration is different from plain ol' respiration. *Respiration*, which is more commonly referred to as breathing, is the physical act of inhaling and exhaling. *Cellular respiration* is what happens inside cells when they use oxygen to transfer energy from food to ATP.

## Breaking down food

After the large molecules in food are broken down into their smaller subunits, the small molecules can be further broken down to transfer their energy to ATP. During cellular respiration, enzymes slowly rearrange the atoms in food molecules. Each rearrangement produces a new molecule in the pathway and can also produce other useful molecules for the cell. Some reactions

- **Release energy that can be transferred to ATP:** Cells quickly use this ATP for cellular work, such as building new molecules.
- **Oxidize food molecules and transfer electrons and energy to coenzymes:** *Oxidation* is the process that removes electrons from molecules; *reduction* is the process that gives electrons to molecules. During cellular respiration, enzymes remove electrons from food molecules and then transfer the electrons to the coenzymes nicotinamide adenine dinucleotide ($NAD^+$) and flavin adenine dinucleotide (FAD). $NAD^+$ and FAD receive the electrons as part of hydrogen (H) atoms, which change them to their reduced forms, NADH and $FADH_2$. Next, NADH and $FADH_2$ donate the electrons to the process of oxidative phosphorylation, which transfers energy to ATP.

  $NAD^+$ and FAD act like electron shuttle buses for the cell. The empty buses, $NAD^+$ and FAD, drive up to oxidation reactions and collect electron passengers. When the electrons get on the bus, the driver puts up the *H* sign to show that the bus is full. Then the full buses, NADH and $FADH_2$, drive over to reactions that need electrons and let the passengers off. The buses are now empty again, so they drive back to another oxidation reaction to collect new passengers. During cellular respiration, the electron shuttle buses drive a loop between the reactions of glycolysis and the Krebs cycle (where they pick up passengers) to the electron transport chain (where they drop off passengers).
- **Release carbon dioxide ($CO_2$):** Cells return $CO_2$ to the environment as waste, which is great for the autotrophs that require $CO_2$ to produce the food that heterotrophs eat. (See how it's all connected?)

Different kinds of food molecules enter cellular respiration at different points in the pathway. Cells break down simple sugars, such as glucose, in the first pathway — glycolysis. Cells use the second pathway, the Krebs cycle, for breaking down fatty acids and amino acids.

Following is a summary of how different molecules break down in the first two pathways of cellular respiration:

- During glycolysis, glucose breaks down into two molecules of pyruvate. The backbone of glucose has six carbon atoms, whereas the backbone of pyruvate has three carbon atoms. During glycolysis, energy transfers result in a net gain of two ATP and two molecules of the reduced form of the coenzyme NADH.
- Pyruvate is converted to acetyl-coA, which has two carbon atoms in its backbone. One carbon atom from pyruvate is released from the cell as $CO_2$. For every glucose molecule broken down by glycolysis and the Krebs cycle, six $CO_2$ molecules leave the cell as waste. (The conversion of pyruvate to acetyl-coA produces two molecules of carbon dioxide, and the Krebs cycle produces four.)
- During the Krebs cycle, acetyl-coA breaks down into carbon dioxide ($CO_2$). The conversion of pyruvate to acetyl-coA produces two molecules of NADH. Energy transfers during the Krebs cycle produce an additional six molecules of NADH, two molecules of $FADH_2$, and two molecules of ATP.

## Transferring energy to ATP

In the inner membranes of the mitochondria in your cells, hundreds of little cellular machines are busily working to transfer energy from food molecules to ATP. The cellular machines are called *electron transport chains,* and they're made of a team of proteins that sits in the membranes transferring energy and electrons throughout the machines.

The coenzymes NADH and $FADH_2$ carry energy and electrons from glycolysis and the Krebs cycle to the electron transport chain. The coenzymes transfer the electrons to the proteins of the electron transport chain, which pass the electrons down the chain. Oxygen collects the electrons at the end of the chain. (If you didn't have oxygen around at the end of the chain to collect the electrons, no energy transfer could occur.) When oxygen accepts the electrons, it also picks up protons ($H^+$) and becomes water ($H_2O$).

The proteins of the electron transport chain are like a bucket brigade that works by one person dumping a bucket full of water into the next person's bucket. The buckets are the proteins, or electron carriers, and the water inside the buckets represents the electrons. The electrons get passed from protein to protein until they reach the end of the chain.

While electrons are transferred along the electron transport chain, the proteins use energy to move protons ($H^+$) across the inner membranes of the mitochondria. They pile the protons up like water behind the "dam" of the inner membranes. These protons then flow back across the mitochondria's membranes through a protein called *ATP synthase* that transforms the kinetic energy from the moving protons into chemical energy in ATP by capturing the energy in chemical bonds as it adds phosphate molecules to ADP.

The entire process of how ATP is made at the electron transport chain is called the *chemiosmotic theory of oxidative phosphorylation* and is illustrated in Figure 5-4.

At the end of the entire process of cellular respiration, the energy transferred from glucose is stored in 36 to 38 molecules of ATP, which are available to be used for cellular work. (And boy do they get used quickly!)

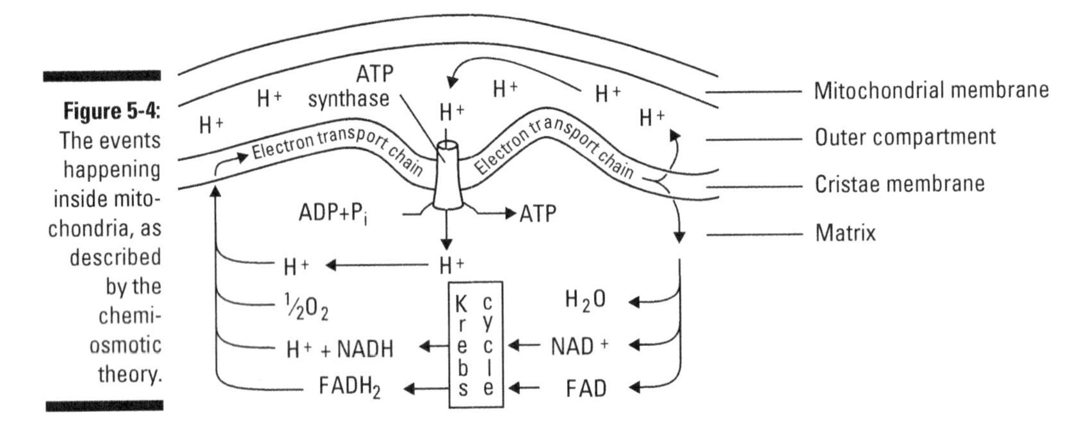


**Figure 5-4:** The events happening inside mitochondria, as described by the chemiosmotic theory.

# Energy and Your Body

Your body takes in chemical potential energy when you eat food and then transfers the energy from that food to your cells. As you use the energy to do work, that energy is eventually transformed into heat energy that you transfer to your surroundings.

Energy can be measured in many different ways, but the energy in food is measured in calories. Basically, a *calorie* is a unit of measurement for heat energy. It takes 1 calorie to raise the temperature of 1 gram of water by 1 degree Celsius (*not* Fahrenheit). The calories that you count and see written on food packages are really *kilocalories.* (*Kilo* means "1,000," so a *kilocalorie* is equal to 1,000 calories.) Kilocalories are represented by a capital *C,* whereas calories are represented by a lowercase *c.* From here on out, we use the term Calorie (with a capital *C*) to represent the kilocalories you're familiar with from nutrition facts labels.

You can get an approximate measure of your basic energy needs by performing a simple calculation to determine your *basal metabolic rate* (BMR), the approximate number of Calories you need just to maintain your body's minimum level of activity (breathing, blood pumping, digestion, and so on). Here's how to calculate BMR:

1. **Multiply your weight in pounds by 10.**
2. **Multiply your height in inches by 6.25.**
3. **Add these two numbers together.**
4. **Multiply your age by 5 and then subtract this number from the one you got in Step 3.**
5. **If you're male, add 5 to the total you found in Step 4; if you're female, subtract 161 from the total you found in Step 4.**

If you exercise, you need to consume additional Calories to supply your body with the energy it needs for increased physical activity. Use the preceding calculation and Table 5-1 to figure out how many Calories you need to consume to maintain your lifestyle.

**Table 5-1 Determining Caloric Need Based on Lifestyle**

| *If You're . . .* | *Multiply Your BMR by . . .* |
|---|---|
| Fairly sedentary (little or no exercise and desk job) | 1.2 |
| Lightly active (light exercise or sports 1 to 3 days per week) | 1.375 |
| Moderately active (moderate exercise or sports 3 to 5 days per week) | 1.55 |
| Very active (hard exercise or sports 6 to 7 days per week) | 1.725 |
| Extremely active (hard daily exercise or sports and physical job) | 1.9 |

In the past, humans had to work hard to find their food and sometimes came up empty-handed. To survive, the human body developed a mechanism for storing energy that can be used during times of low food intake. It packs energy-rich fat onto your hips, thighs, abdomen, and buttocks. So if you take in more Calories in a day than you need, the extra Calories are stored as fat in your adipose tissue. Every 3,500 extra Calories equals 1 pound of fat. And your body doesn't give up extra potential energy easily! If you continue to take in more Calories than you use, you *will* gain weight because it's much easier for your body to create fat than to use it.

# Part II

# Cell Reproduction and Genetics: Let's Talk about Sex, Baby

## In this part . . .

Living things grow and reproduce both sexually (through meiosis) and asexually (through mitosis). They also tend to pass their traits on to their offspring. The blueprints for these traits are encoded in the structure of the most famous molecule of life: deoxyribonucleic acid, or DNA. Scientists today know more about the mysteries of DNA than ever before. They've even figured out ways to harness its power for medicine, agriculture, and technology.

In this part, we explain the steps of cell division and how DNA determines the characteristics of organisms.

# Chapter 6

# Dividing to Conquer: Cell Division

### In This Chapter

- Understanding why cells reproduce and how DNA replicates itself
- Discovering how mitosis produces exact copies of cells
- Making egg and sperm cells through meiosis
- Appreciating the power of genetic diversity

All living things can reproduce their cells for growth, repair, and reproduction. Asexual reproduction by mitosis creates cells that are genetically identical to the parent cell. Sexual reproduction requires a special type of cell division called meiosis that produces cells containing half the genetic information of the parent cell. Meiosis and sexual reproduction result in greater genetic diversity in offspring and, consequently, in the populations of living things. In this chapter, we explore the reasons why cells divide and present the steps of each type of cell division. We also introduce you to the ways sexual reproduction adds variety to species of all kinds.

## Reproduction: Keep On Keepin' On

Biology is, of course, all about life. And, when you think about it, life is really all about continuation — living things keep on keepin' on from one generation to the next, passing on critical genetic information. Certainly this is one of the core differences between organisms and inanimate objects. After all, have you ever seen a chair or table replicate itself? Only living things have the ability to pass on genetic information and replicate themselves.

When cells replicate, they make copies of all of their parts, including their DNA, and then divide themselves to make new cells. If a cell makes an exact copy of itself, it's engaging in *asexual reproduction*. Single-celled prokaryotes, such as bacteria, reproduce asexually by binary fission; they're able to divide quickly and reproduce themselves in as little as 10 to 20 minutes. Some single-celled eukaryotes and individual cells within a multicellular eukaryote also reproduce asexually. However, they use a process called mitosis (which we explain in the later "Mitosis: One for you, and one for you" section) to produce new generations. If a cell produces a new cell that contains only half of its genetic information, that cell has engaged in *sexual reproduction*. A special type of cell division known as meiosis (which we explain in the later "Meiosis: It's all about sex, baby" section) is responsible for all sexual reproduction.

Cells divide for the following important reasons:

- **To make copies of cells for growth:** You started out as a single cell after mom's egg met dad's sperm, but today you have about 10 trillion cells in your body. All of those cells were produced from that first cell and its descendents by mitosis. When you watch plants grow taller or baby animals grow into adults, you're seeing mitosis at work.
- **To make copies of cells for repair:** It's a fact of life that cells wear out and need to be replaced. For instance, you constantly shed skin cells from your surface. If your body couldn't replace these cells, you'd run out of skin. And if an organism gets injured, its body uses mitosis to produce the cells necessary to repair the injury.
- **To carry on the species:** During asexual reproduction, organisms make exact copies of themselves for the purpose of creating offspring. During sexual reproduction, *gametes* (cells, specifically eggs and sperm, containing half the genetic information of their parent cells) get together to make new individuals. When the genetic information of the gametes joins together, the new individual has the correct total amount of DNA.

# Welcome to DNA Replication 101

If one cell is going to divide to produce two new cells, the first cell must copy all of its parts before it can split in half. The cell grows, makes more organelles (see Chapter 4 for the full scoop on organelles), and copies its genetic information (the DNA) so that the new cells each have a copy of everything they need. Cells use a process called *DNA replication* to copy their genetic material. In this process, the original DNA strands serve as the template for the construction of the new strands. It's particularly important that each new cell receives an accurate copy of the genetic information because this copy, whether it's accurate or faulty, directs the structure and function of the new cells.

The basic steps of DNA replication are as follows:

- First, the two parental DNA strands separate so that the rungs of the double helix ladder are split apart with one nucleotide on one side and one nucleotide on the other (see Chapter 3 for a depiction of a DNA molecule). The entire DNA strand doesn't unzip all at once, however. Only part of the original DNA strand opens up at one time. The partly open/partly closed area where the replication is actively happening is called the *replication fork* (this is the Y-shaped area in Figure 6-1).

- The enzyme DNA polymerase reads the DNA code on the parental strands and builds new partner strands that are complementary to the original strands. To build complementary strands, DNA polymerase follows the *base-pairing rules* for the DNA nucleotides: A always pairs with T, and C always pairs with G (see Chapter 3 for further details about nucleotides). If, for example, the parental strand has an A at a particular location, DNA polymerase puts a T in the new strand of complementary DNA it's building. When DNA polymerase is done creating complementary pairs, each parental strand has a brand-new partner strand.

DNA polymerase is considered *semiconservative* because each new DNA molecule is half old (the parental strand) and half new (the complementary strand).

Several enzymes help DNA polymerase with the process of DNA replication (you can see them and DNA polymerase hard at work in Figure 6-1):

- **Helicase** separates the original parental strands to open the DNA.
- **Primase** puts down short pieces of RNA, called *primers,* that are complementary to the parental DNA. DNA polymerase needs these primers in order to get started copying the DNA.
- **DNA polymerase I** removes the RNA primers and replaces them with DNA, so it's slightly different from the DNA polymerase that makes most of the new DNA (that enzyme is officially called *DNA polymerase III,* but we refer to it simply as *DNA polymerase*).
- **DNA ligase** forms covalent bonds in the backbone of the new DNA molecules to seal up the small breaks created by the starting and stopping of new strands.

The parental strands of the double helix are oriented to each other in opposite polarity: Chemically, the ends of each strand of DNA are different from each other, and the two strands of the double helix are flipped upside down relative to one another. Note in Figure 6-1 the numbers 5' and 3' (read "5 prime" and "3 prime"). These numbers indicate the chemical differences of the two ends. You can see that the 5' end of one strand lines up with the 3' end of the other strand. The two strands of DNA have to be flipped relative to each other in order for the bases that make up the rungs of the ladder to fit together the right way for hydrogen bonds to form between them. Because the two strands have opposite polarity, they're *antiparallel strands.*

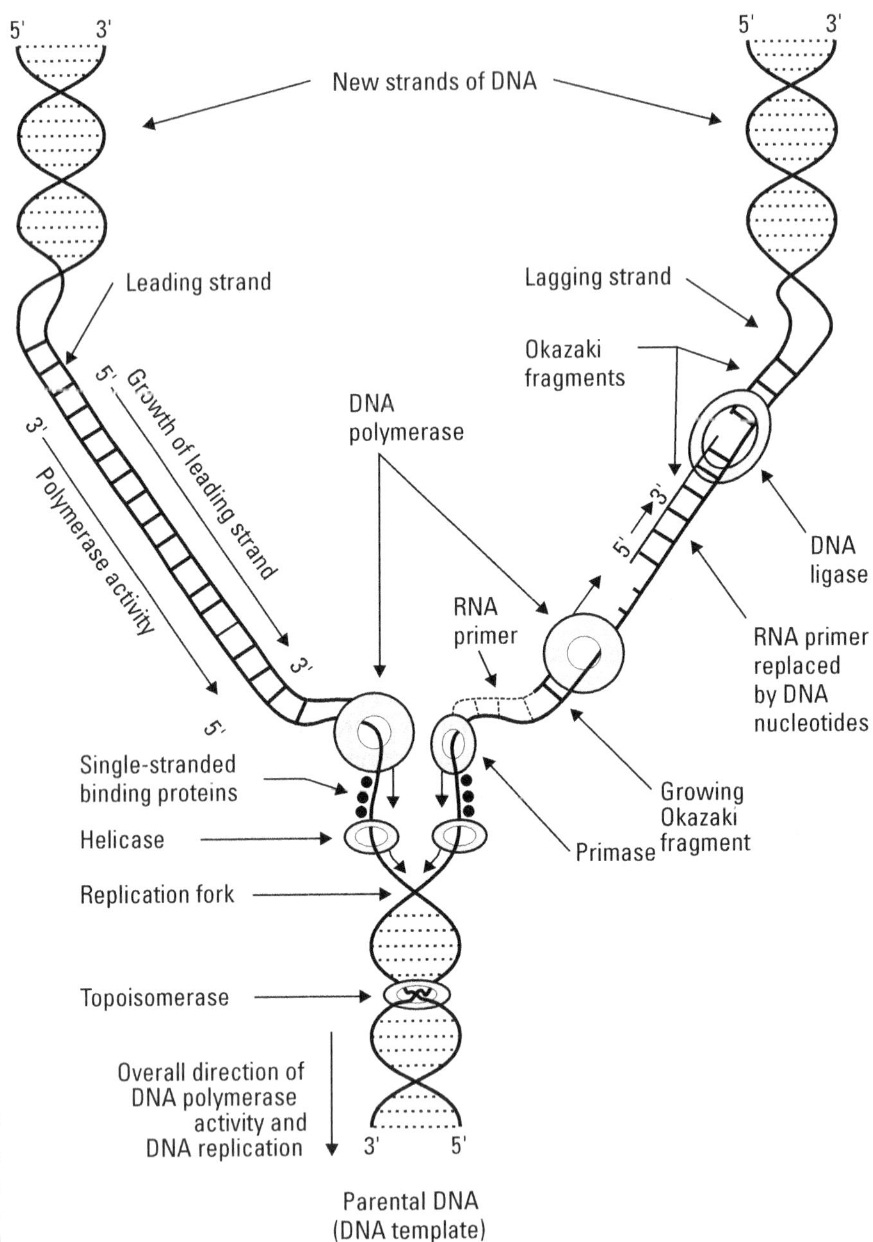


**Figure 6-1:** DNA replication.

The antiparallel strands of the parent DNA create some problems for DNA polymerase. One quirk of DNA polymerase is that it's a one-way enzyme — it can only make new strands of DNA by lining up the nucleotides a certain way. But DNA polymerase needs to use the parent DNA strands as a pattern, and they're going in opposite directions. As a result, DNA polymerase makes the two new strands of DNA a bit differently from each other, as you can see from the following:

- **One new strand of DNA, called the leading strand, grows in a continuous piece.** Refer to Figure 6-1. See how the new DNA on the left side of the replication fork is growing smoothly? The 3' end of this new strand points toward the replication fork, so after DNA polymerase starts building the new strand, it can just keep going.
- **One new strand of DNA, called the lagging strand, grows in fragments.** Look at Figure 6-1 again. Notice how the right side of the replication fork looks a little messier? That's because the replication process doesn't occur smoothly over here. The 3' end of this new strand points away from the fork. DNA polymerase starts making a piece of this new strand but has to move away from the fork to do so (because it can only work in one direction). DNA polymerase can't go too far from the rest of the enzymes that are working at the fork, however, so it has to keep backing up toward the fork and starting over. As a result, the lagging strand is made in lots of little pieces called *Okazaki fragments.* After DNA polymerase is done making the fragments, the enzyme DNA ligase comes along and forms covalent bonds between all the pieces to make one continuous new strand of complementary DNA.

# Cell Division: Out with the Old, In with the New

*Cell division* is the process by which new cells are formed to replace dead ones, repair damaged tissue, or allow organisms to grow and reproduce. Cells that can divide spend some of their time functioning and some of their time dividing. This alternation between not dividing and dividing is known as the *cell cycle,* and it has specific parts:

- The nondividing part of the cell cycle is called *interphase.* During interphase, cells are going about their regular business. If the cell is a single-celled organism, it's probably busy finding food and growing. If the cell is part of a multicellular organism, like a human, it's busy doing its job. Maybe it's a skin cell protecting you from bacteria or a fat cell storing energy for later.

- Cells that receive a signal to divide enter a division process, which is either mitosis or meiosis.
  - Cells that reproduce asexually, like a skin cell that needs to replace some of your lost skin, divide by *mitosis,* a process that produces cells that are identical to the parent cell.
  - Cells that reproduce sexually enter a special process called *meiosis* that produces special cells called *gametes* (in animals) and *spores* (in plants, fungi, and protists) that have half the genetic information of the parent cell. In you, the only cells that reproduce by meiosis are cells in your gonads. Depending on your gender, your *gonads* are your testes or your ovaries. Cells in testes produce gametes called *sperm,* and cells in ovaries produce gametes called *eggs.*

Mitosis and meiosis have many similarities, but the differences are essential. We cover both processes (as well as the interphase) in the sections that follow, but Table 6-1 can help you sort out the important differences at a glance.

**Table 6-1 A Comparison of Mitosis & Meiosis**

| *Mitosis* | *Meiosis* |
|---|---|
| One division is all that's necessary to complete the process. | Two separate divisions are necessary to complete the process. |
| Chromosomes don't get together in pairs. | Homologous chromosomes must synapse to complete the process, which occurs in prophase I. |
| Homologous chromosomes don't cross over. | Crossing-over is an important part of meiosis and one that leads to genetic variation. |
| Sister chromatids separate in anaphase. | Sister chromatids separate only in anaphase II, not anaphase I. (Homologous chromosomes separate in anaphase I.) |
| Daughter cells have the same number of chromosomes as their parent cells, meaning they're diploid. | Daughter cells have half the number of chromosomes as their parent cells, meaning they're haploid. |
| Daughter cells have genetic information that's identical to that of their parent cells. | Daughter cells are genetically different from their parent cells. |

| *Mitosis* | *Meiosis* |
|---|---|
| The function of mitosis is asexual reproduction in some organisms. In many organisms, mitosis functions as a means of growth, replacement of dead cells, and damage repair. | Meiosis creates gametes or spores, the first step in the reproductive process for sexually reproducing organisms, including plants and animals. |

## Interphase: Getting organized

During interphase, cells engage in the metabolic functions that make them unique. For instance, nerve cells send signals, glandular cells secrete hormones, and muscle cells contract. If cells get a signal to reproduce themselves, they grow, copy all of their structures and molecules, and make the structures they need to help cell division proceed in an organized fashion (*inter-* means "between," so *interphase* is literally the phase between cell divisions).

The nuclear membrane is intact throughout interphase, as you can see in Figure 6-2. The DNA is loosely spread out, and you can't see individual chromosomes. Cells that are going to divide copy their DNA during interphase.

Interphase has three subphases:

- **$G_1$ phase:** During this phase, which is typically the longest one of the entire cell cycle, the cell grows and produces cell components. Each chromosome is made up of just a single double-stranded piece of DNA (*double-stranded* is just another way of saying that the DNA is a double helix).

  Some cells actually never leave the $G_1$ phase. They never divide; instead, they just hang out and do their cellular thing. Human nerve cells are perfect examples of cells that never leave the $G_1$ phase.
- **S phase:** This phase is when the cell gets ready to divide and puts the pedal to the metal for DNA replication. Every DNA molecule is copied exactly, forming two *sister chromatids* (a pair of identical DNA molecules) that stay attached to each other in each replicated chromosome. You can see replicated chromosomes in Figure 6-2, in the cell labeled Prophase. Each replicated chromosome looks like an *X*, and each *X* represents two identical sister chromatids held together at a location on the chromosome called the *centromere*.
- **$G_2$ phase:** During this phase, the cell is packing its bags and getting ready to hit the road for cell division by making the cytoskeletal

proteins it needs to move the chromosomes around. When you look at cells that are dividing, the cytoskeletal proteins look like thin threads, hence their name — *spindle fibers.* A network of spindle fibers spreads throughout the cell during mitosis to form the mitotic spindle, which is represented by the thin curving lines drawn in the cells in Figure 6-2. The *mitotic spindle* organizes and sorts the chromosomes during mitosis.

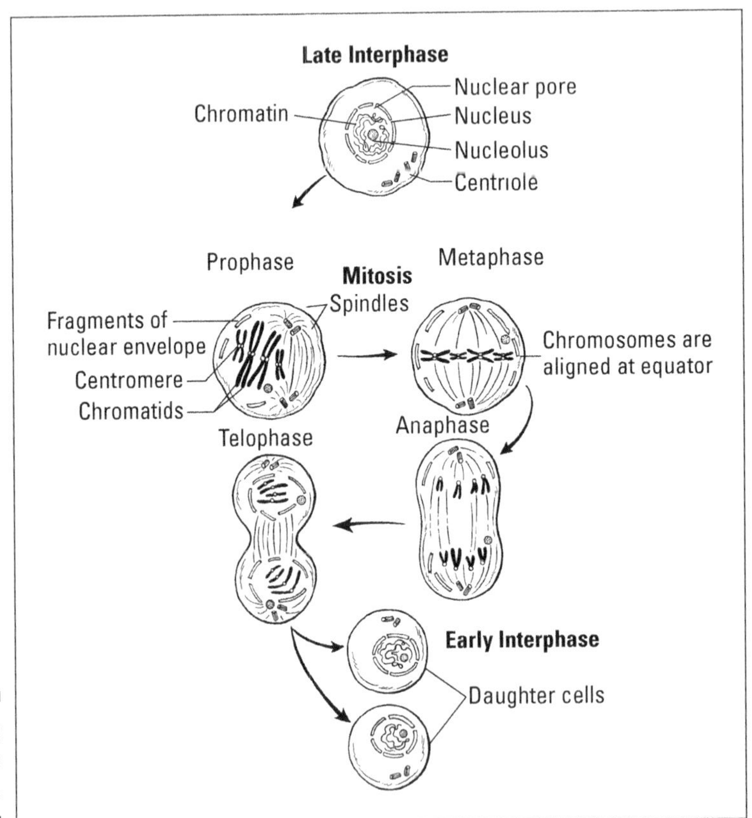


**Figure 6-2:** Interphase and mitosis.

## Mitosis: One for you, and one for you

After interphase is over (see the preceding section), cells that are going to divide to create an exact replica of a parent cell enter mitosis. During mitosis, the cell makes final preparations for its impending split. Processes during mitosis ensure that genetic material is distributed equally so each new daughter cell receives identical information (eukaryotic cells are model parents intent on avoiding bickering between their daughter cells).

The process of mitosis occurs in four phases, with the fourth phase initiating a final process called cytokinesis. We outline everything for you in the following sections.

### Examining the four phases of mitosis

Although the cell cycle is a continuous process, with one stage flowing into another, scientists divide the events of mitosis into four phases based on the major events in each stage. The four phases of mitosis are

- **Prophase:** The chromosomes of the cell get ready to be moved around by coiling themselves up into tight little packages. (During interphase, the DNA is spread throughout the nucleus of the cell in long thin strands that would be pretty hard to sort out.) As the chromosomes coil up, or *condense,* they become visible to the eye when viewed through a microscope. During prophase:
  - The chromosomes coil up and become visible.
  - The nuclear membrane breaks down.
  - The mitotic spindle forms and attaches to the chromosomes.
  - The nucleoli break down and become invisible.
- **Metaphase:** The chromosomes are tugged by the mitotic spindle until they're all lined up in the middle of the cell. (*Meta-* means "middle," so it's officially metaphase when the chromosomes are lined up in the middle; see the cell labeled Metaphase in Figure 6-2.)

- **Anaphase:** The replicated chromosomes separate so that the two sister chromatids (identical halves) from each replicated chromosome go to opposite sides (see the cell labeled Anaphase in Figure 6-2). This way each new cell has one copy of each DNA molecule from the parent cell when cell division is over.

- **Telophase:** The cell gets ready to divide into two by forming new nuclear membranes around the separated sets of chromosomes. The two daughter nuclei each have a copy of every chromosome that was in the parent cell, as you can see in Figure 6-2.

  The events of telophase are essentially the reverse of prophase.
  - New nuclear membranes form around the two sets of chromosomes.
  - The chromosomes uncoil and spread throughout the nucleus.
  - The mitotic spindle breaks down.
  - The nucleoli reform and become visible again.

### Seeing how daughter nuceli go their own way with cytokinesis

The last order of cell-division business is to give the new daughter nuclei their own cells through a process called *cytokinesis*. (*Cyto-* means "cell," and *kinesis* means "movement," so *cytokinesis* literally means "moving cells.") Cytokinesis occurs differently in animal and plant cells, as you can see from the following list and Figure 6-3:

- In animal cells, cytokinesis begins with an indentation, called a *cleavage furrow*, in the center of the cell. Cytoskeletal proteins act like a belt around the cell, contracting down and squeezing the cell in two (imagine squeezing a ball of dough at the center until it becomes two balls of dough).
- In plant cells, a new cell wall forms at the center of the cell. Because a rigid cell wall is involved, the cell can't be squeezed in two. Instead, vesicles deliver wall material to the center of the cell and then fuse together to form the cell plate. The vesicles are basically little bags made of membrane that carry the wall material, so when they fuse together, their membranes form the plasma membranes of the new cells. The wall material gets dumped between the new membranes, forming the plant's cell walls.

After cytokinesis is complete, the new cells move immediately into the $G_1$ stage of interphase. No one stops to applaud the great accomplishment of successfully completing the mitosis process, which is really too bad because it's the root of renewal and regeneration.

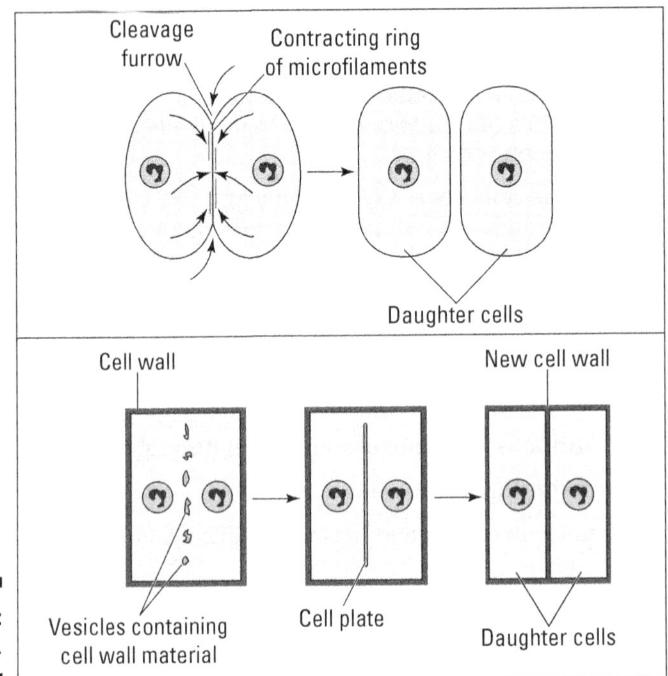


**Figure 6-3:** Cytokinesis.

# Meiosis: It's all about sex, baby

Meiosis is unique because the resulting cells have only half of their parent cells' *chromosomes,* or singular pieces of DNA. Human body cells have 46 chromosomes, 2 each of 23 different kinds. The 23 pairs of chromosomes can be sorted by their physical similarities and lined up to form a chromosome map called a *karyotype* (see Figure 6-4). The two matched chromosomes in each pair are called *homologous chromosomes* (*homo-* means "same," so these are chromosomes that have the same kind of genetic information). In each pair of your homologous chromosomes, one chromosome came from mom, and one came from dad. For every gene that your mom gave you, your dad also gave you a copy, so you have two copies of every gene (with the exception of genes on the X and Y chromosomes if you're male).

The pairs of homologous chromosomes have the same kind of genetic information. If one of the two has a gene that affects eye color, for example, the other chromosome has the same gene in the same location. The messages in each gene may be slightly different — for example, one gene could have a message for light eyes whereas the other gene could have a message for dark eyes — but both chromosomes have the same type of gene in each location.

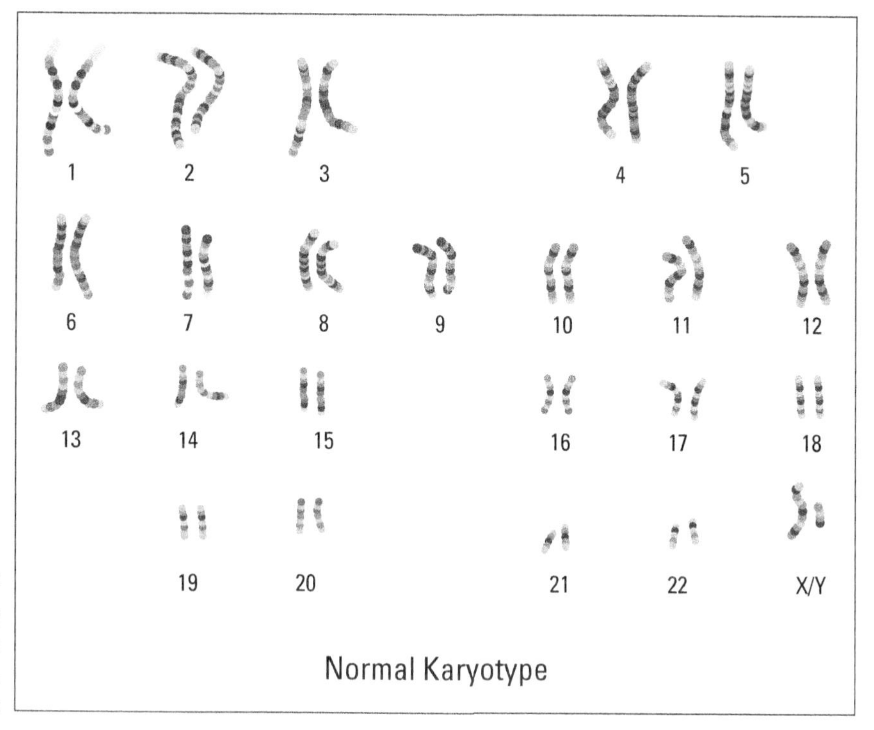


**Figure 6-4:** A human karyotype.

Human *gametes* (sperm cells and egg cells) have just 23 chromosomes. Through sexual reproduction (see Figure 6-5), a sperm and an egg join together to create a new individual, returning the chromosome number to 46. If gametes didn't have half the genetic information, then the cell they form together, called a *zygote,* would have twice the normal genetic information for a human. And when gametes are produced, they can't just get any 23 chromosomes — they have to get one of each pair of chromosomes. Otherwise, the zygote would have extras of some chromosomes and be missing others entirely. The resulting person wouldn't have the correct genetic information and probably wouldn't survive.

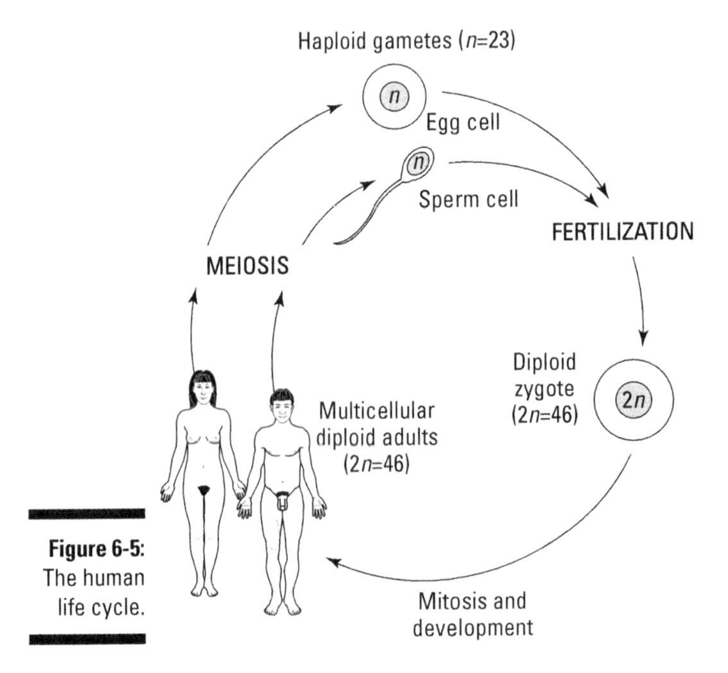


**Figure 6-5:** The human life cycle.

Meiosis is the type of cell division that separates chromosomes so gametes receive one of each type of chromosome. In humans, meiosis separates the 23 pairs of chromosomes so that each cell receives just one of each pair. Consequently, gametes have what's known as a *haploid* number of chromosomes, or a single set. When the two gametes unite, they combine their chromosomes to reach the full complement of 46 chromosomes in a normal *diploid* cell (one with a double set of chromosomes, or two of each type).

Two cell divisions occur in meiosis, and the two halves of meiosis are called *meiosis I* and *meiosis II.*

- During meiosis I, homologous chromosomes are paired up and then separated into two daughter cells. Each daughter cell receives one of each chromosome pair, but the chromosomes are still replicated. (Remember that meiosis follows interphase, so DNA replication produced a copy of each chromosome. These two copies, called sister chromatids, are held together, forming replicated chromosomes. You can see that the chromosomes still look like *X*'s after meiosis I in Figure 6-6b.)
- During meiosis II, the replicated chromosomes send one sister chromatid from each replicated chromosome to new daughter cells. After meiosis II, the four daughter cells each have one of each chromosome pair, and the chromosomes are no longer replicated. (Notice how the four daughter cells in Figure 6-6b don't have sister chromatids?)

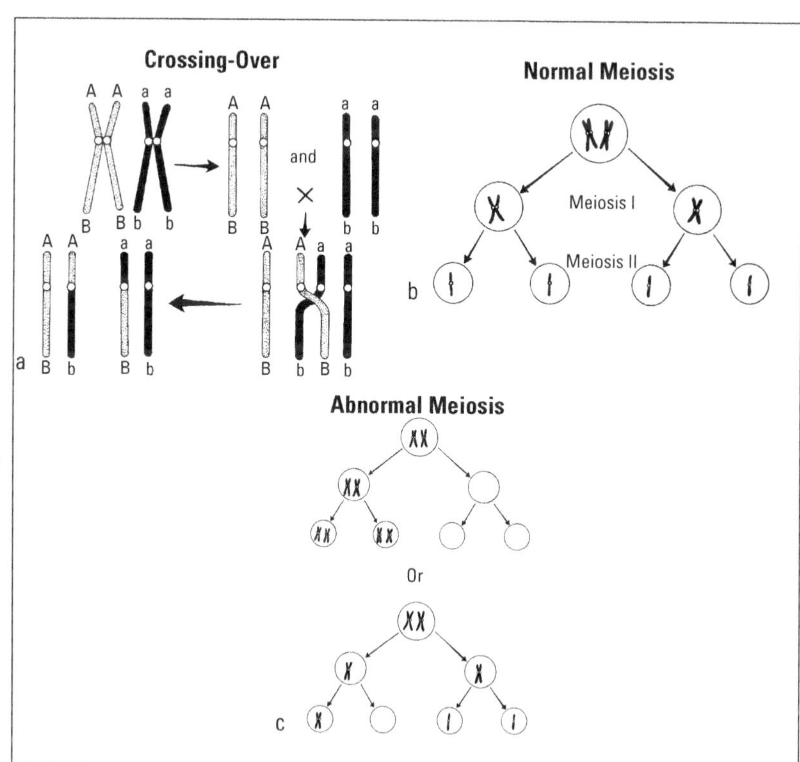


**Figure 6-6:** Crossing-over, meiosis, and nondisjunction.

In human males, meiosis takes place after puberty, when the diploid cells in the testes undergo meiosis to become haploid. In females, the process begins a lot earlier — in the fetal stage. While a little girl is dog-paddling around her mother's womb, diploid cells complete the first part of meiosis and then migrate to the ovaries, where they hang out and wait until puberty. With the onset of puberty, the cells take turns entering meiosis II. (Just one per month; no pushing or shoving, please!) Usually one single egg cell is produced per cycle, although exceptions occur, which, if fertilized, lead to fraternal twins, or triplets, or quadruplets . . . you get the idea. The other meiotic cells simply disintegrate.

When a human sperm cell and a human egg cell — each with 23 chromosomes — unite in the process of fertilization, the diploid condition of the cell is restored. Further divisions by mitosis result in a complete human being.

The phases of meiosis are very similar to the phases of mitosis; they even have the same names, which can make distinguishing between the two rather tough. Just remember that the key difference between the phases of mitosis and meiosis is what's happening to the number of chromosomes.

The next sections delve into the details of each phase of meiosis I and meiosis II.

### Meiosis I

Meiosis I is the first step in sexual reproduction. The phases are as follows:

- **Prophase I:** During this phase, the cell's nuclear membrane breaks down, the chromatids coil to form visible chromosomes, the nucleoli break down and disappear, and the spindles form and attach to the chromosomes. But that's not all. Prophase I is when something that's absolutely critical to the successful separation of homologous chromosomes occurs: synapsis.

  *Synapsis* happens when the two chromosomes of each pair find each other and stick together. The synapsis process begins when the homologous chromosomes move and lie next to each other. At this point, the two homologous chromosomes can swap equal amounts of DNA in an event called *crossing-over* (see Figure 6-6a). This swapping of materials results in four completely unique chromatids. This new arrangement of four chromatids is called a *tetrad.*

  Crossing-over between homologous chromosomes during prophase I increases the genetic variability among gametes produced by the same organism. Every time meiosis occurs, crossing-over can happen a little differently, shuffling the genetic deck as gametes are made. This is one of the reasons that siblings can be so different from each other.

- **Metaphase I:** This is when the pairs of homologous chromosomes line up in the center of the cell. The difference between metaphase I of meiosis and metaphase of mitosis is that the homologous pairs line up in the former, whereas individual chromosomes line up in the latter.
- **Anaphase I:** During this phase, the two members of each homologous pair go to opposite sides of the cell, and the chromosome number is officially reduced from diploid to haploid.
- **Telophase I:** This is when the cell takes a step back (or forward, depending on your perspective) to an interphase-like condition by reversing the events of prophase I. Specifically, the nuclear membrane reforms, the chromosomes uncoil and spread throughout the nucleus, the nucleoli reform, and the spindles break down.

### Meiosis II

During meiosis II, both daughter cells produced by meiosis I continue their dance of division so that — in most cases — four gametes are the end result. The phases of meiosis II look very similar to the phases of meiosis I with one big exception: The cells start out with half the number of chromosomes as the original parent cell.

Meiosis II separates the sister chromatids of each replicated chromosome and sends them to opposite sides of the cell. Cells going from meiosis I to meiosis II don't go through interphase again (after all, that would undo all the hard work of meiosis I).

- **Prophase II:** As in mitosis's prophase and meiosis's prophase I, the nuclear membrane disintegrates, the nucleoli disappear, and the spindles form and attach to the chromosomes.
- **Metaphase II:** Nothing too exciting here, folks. Just as in any old metaphase, the chromosomes line up at the equatorial plane. But remember that the number of chromosomes that lines up is half the number of the original parent cell (and half the number you'd see in mitosis).
- **Anaphase II:** The sister chromatids of each replicated chromosome move away from each other to opposite sides of the cell.
- **Telophase II:** The nuclear membrane and nucleoli reappear, the chromosomes stretch out for the briefest of rests, and the spindles disappear.

After meiosis II, it's time for cytokinesis, which creates four haploid cells (which is impressive considering you had just one diploid cell at the beginning of meiosis).

# How Sexual Reproduction Creates Genetic Variation

Sexual reproduction increases genetic variation in offspring, which in turn increases the genetic variability in species. You can see the effects of this genetic variability if you look at the children in a large family and note how each person is unique. Imagine this kind of variability expanded to include all the families you know (not to mention all the families of all the sexually reproducing organisms on Earth), and you begin to get a feel for the dramatic genetic impact of sexual reproduction.

The sections that follow familiarize you with some of the specific causes of genetic variation courtesy of meiosis and sexual reproduction.

## Mutations

DNA polymerase occasionally makes uncorrected mistakes when copying a cell's genetic information during DNA replication (which we walk you through earlier in this chapter). These mistakes are called *spontaneous mutations,* and they introduce changes into the genetic code. In addition, exposure of cells to *mutagens* (environmental agents, such as X-rays and certain chemicals, that cause changes in DNA) can increase the number of mutations that occur in cells. When changes occur in a cell that produces gametes, future generations are affected. (For more on the effect of mutations, see Chapter 8.)

## Crossing-over

When homologous chromosomes come together during prophase I of meiosis, they exchange bits of DNA with each other. This crossing-over (illustrated in Figure 6-6a) results in new gene combinations and new chances for variety. Crossing-over is one way of explaining how a person can have red hair from his mother's father and a prominent chin from his mother's mother. After crossing-over, these two genes from two different people wound up together on the same chromosome in the person's mother and got handed down together.

## Independent assortment

*Independent assortment* occurs when homologous chromosomes separate during anaphase I of meiosis. When the homologous pairs of chromosomes line up in metaphase I, each pair lines up independently from the other pairs. So, the way the pairs are oriented during meiosis in one cell is different from the way they're oriented in another cell. When the homologous chromosomes separate, many different combinations of homologous chromosomes can travel together toward the same side of the cell. How many different combinations of homologous chromosomes are possible in a human cell undergoing meiosis? Oh, just $2^{23}$ — that's 8,388,608 to be exact. Now you can begin to see why even large families can have many unique children.

# Fertilization

Fertilization presents yet another opportunity for genetic diversity. Imagine millions of genetically different sperm swimming toward an egg. Fertilization is random, so the sperm that wins the race in one fertilization event is going to be different than the sperm that wins the next race. And, of course, each egg is genetically different too. So, fertilization produces random combinations of genetically diverse sperm and eggs, creating virtually unlimited possibilities for variation. That's why every human being who has ever been born — and ever will be born — is genetically unique. Well, almost. Genetically identical twins can develop from the same fertilized egg, but even they can have subtle differences due to development.

# Nondisjunction

Nothing's perfect, even in the cellular world, which is why sometimes meiosis doesn't occur quite right. When chromosomes don't separate the way they're supposed to, that's called *nondisjunction.* The point of meiosis is to reduce the number of chromosomes from diploid to haploid, something that normally happens when homologous chromosomes separate from each other during anaphase I. Occasionally, however, a pair of chromosomes finds it just too hard to separate, and both members of the pair end up in the same gamete (see Figure 6-6c).

What happens next isn't pretty. Two of the final four cells resulting from the meiotic process are missing a chromosome as well as the genes that chromosome carries. This condition usually means the cells are doomed to die. Each of the other two cells has an additional chromosome, along with the genetic material it carries. Well, that should be great for these cells, shouldn't it? It should mean they'll have an increased chance for genetic variation, and that's a good thing, right?

Wrong! An extra chromosome is like an extra letter from the IRS. It's not something to hope for. Many times these overendowed cells simply die, and that's the end of the story. But sometimes they survive and go on to become sperm or egg cells. The real tragedy, then, is when an abnormal cell goes on to unite with a normal cell. When that happens, the resulting zygote (and offspring) has three of one kind of chromosome rather than the normal two. The term scientists use for this occurrence is *trisomy.*

Here's the real problem with this scenario: All the cells that develop by mitosis to create the new individual will be trisomic (meaning they'll have that extra chromosome). One possible abnormality occurring from an extra chromosome is *Down syndrome,* a condition that often results in some mental and developmental impairment and premature aging.

Scientists have now pinpointed the chromosome related to Down syndrome; it's Chromosome 21. If an egg with two number 21 chromosomes is fertilized with a normal sperm cell with just one number 21, the resulting offspring has 47 chromosomes (24 + 23 = 47), and Down syndrome occurs.

You probably already know that the mother's age is a factor in the creation of genetic abnormalities such as Down syndrome, but do you know why? The answer lies in the fact that meiosis I takes place in the fetal stage for females and then the cells rest in the ovaries until puberty, when one per month enters meiosis II in preparation for fertilization. If a cell has been waiting its turn for 40 or 45 years, it's pretty darned old — in cellular terms at least. (Aging gametes aren't such an issue for males because their sex cells don't actually enter meiosis until after puberty. Meiosis is a continuous process for them, producing new cells all the time.)

## Pink and blue chromosomes

Ever wish you could've been born the opposite sex so you wouldn't have to spend so much of your budget on makeup or shave your face every morning? Sorry, but that was never really your decision to make. Like all other genetic characteristics, gender is determined at a chromosomal level.

In many organisms — including humans and fruit flies — the gender of an individual is determined by specific *sex chromosomes,* which scientists refer to as the X and Y chromosomes. The 23 pairs of human chromosomes can be divided into 22 pairs of *autosomes,* chromosomes that aren't involved in the determination of gender, and one pair of sex chromosomes. Men and women have the same types of genes on the 22 autosomes and on the X chromosome. But only guys get a special gene, located on the Y chromosome, that jump-starts the formation of testes in boy fetuses when they're about 6 weeks old. After the testes form, they produce testosterone, and it's usually all boy from then on out. The Y chromosome is smaller than all the other chromosomes, but it packs one powerful little gene!

## Monocultures, a threat to genetic diversity

Up until about the 1960s, most American farms were smaller operations that planted diverse crops and relied fairly heavily upon human labor to get the crops in. As farms started shifting to more mechanized processes, they grew larger and looked for greater standardization in crop plants. After all, machines work best at specialized, repetitive tasks. Farms started planting huge stands of *monocultures*, crop plants that are all genetically identical. This practice worries many scientists because they recognize the power and value of genetic diversity.

If an insect pest or fungal pathogen can successfully attack a crop plant, it has the potential to wipe out an entire monoculture at once and drastically impact the human food supply. In addition, large industrial farms increasingly rely upon artificial fertilizers and pesticides to support their specialized monocultures, which can upset the balance between the farm and the natural ecology.

As people start to realize that monoculture farming may have more dangers than benefits, some people are bucking the trend. Small farms are trying to make a comeback by producing diverse, organic crops, and companies are starting to collect older varieties of crop plants, called *heirloom varieties*, and reintroduce them into the human food supply.

## Chapter 7

# Making Mendel Proud: Understanding Genetics

### In This Chapter

- Defining heritable traits and why sexually reproducing organisms are so diverse
- Exploring Mendel's discoveries about inheritance
- Mastering the lingo of genetics
- Walking through the process of studying human traits

*Genetics* is the branch of biology that looks at how parents pass traits on to their offspring. It all started more than 150 years ago when a monk named Gregor Mendel conducted breeding experiments with pea plants that led him to discover the fundamental rules of inheritance. Although Mendel worked with peas, his ideas explain a lot about why you look and function the way you do.

In this chapter, we show you one of Mendel's experiments and present some of the most important rules of inheritance.

## Why You're Unique: Heritable Traits and the Factors Affecting Them

Dogs have puppies, hens have chicks, and your parents had you. What do all three sets of parents have in common? They all passed their traits on to their offspring. Traits that are inherited from one generation to the next are called *heritable traits.*

When living things reproduce, they make copies of their DNA and pass some of that DNA on to the next generation. The *DNA* is the genetic code for the traits of the organism (see Chapter 3 for more on DNA), which means the characteristics of the next generation are built from the parents' blueprints. Of course, in sexually reproducing species, offspring aren't exactly like their parents for several reasons:

- **Offspring receive half of their genetic information from their father and half from their mother.** Parents divide their genetic information in half through the process of meiosis (described in Chapter 6), producing sperm cells and egg cells that join to form a new individual. So each offspring brings together a potentially new combination of its parents' traits. Plus, each sperm and egg are a little bit different from the others due to crossing-over and independent assortment (we cover these causes of genetic variation in greater detail in Chapter 6), which shuffle up the genetic information as it's packaged into the sperm and eggs. The result is that there really is only one you even if you have siblings, you're all a little bit different.
- **Even heritable characteristics can change slightly.** DNA changes slightly every time it's copied due to mutation (see Chapters 6 and 8 for the full scoop on mutation). If a mutation is passed from parent to offspring, the offspring may have a new trait. For instance, Queen Victoria of England had children and grandchildren who suffered from *hemophilia,* a deadly disease in which your blood doesn't clot properly, but none of her ancestors had the disease. One possibility for why Queen Victoria's children acquired the disease is that a mutation occurred in the DNA of one of Queen Victoria's parents, who then passed that mutation on to her. Queen Victoria didn't have the disease, but she passed the mutation on to several of her children. Her sons who received the mutation suffered from the disease, and her daughters passed the mutation on to their children.
- **Some traits are acquired rather than inherited.** Riding a bicycle, speaking French, and swimming are all *acquired traits,* abilities you aren't born with but that you gain during your lifetime. Even if your parents can do these things, you still have to figure out how to do them on your own. For example, you aren't born with the natural ability to swim (unless of course you're a fish).
- **Some inherited traits are affected by the environment.** The basic color of, say, your skin or hair is written in the code of your DNA, but if you spend lots of time basking in sunlight, your skin will grow darker, and your hair will become lighter. Even something like height can be affected by your environment — your DNA controls your basic height, but nutrition plays a big role in whether you reach your full potential.

# "Monk"ing Around with Peas: Mendel's Laws of Inheritance

People have probably always realized that parents pass traits on to their children. After all, as soon as a new baby is born, people start trying to decide who the baby looks like. Yet the first person who really figured out the fundamentals of how traits are passed down was an Austrian monk named Gregor Mendel.

Mendel lived in the mid-19th century. During his time, people believed in *blending inheritance,* meaning they thought that the traits of a father blended with the traits of a mother to produce children whose traits were supposed to be the averages of the parents' traits. So a tall father and a short mother were expected to have average-size kids. (If the concept of blending inheritance were correct, no one could ever have a child that was taller than his or her parents.)

Mendel, who was very interested in science and math, tested these ideas about inheritance by breeding pea plants in the abbey garden. He studied many of the heritable traits of peas, including flower color, pea color, plant height, and pea shape. Although other people had bred plants and animals for desirable characteristics before, Mendel was extremely careful in his experiments and used math to look at inheritance in a new way, revealing patterns that no one else had noticed. We walk you through the basics of Mendel's experiments and the results they produced in the next sections.

## Pure breeding the parentals

Mendel used *pure-breeding* plants (plants that always reproduce the same characteristics in their offspring) to ensure that he knew exactly what genetic message he was starting with when he chose a particular plant for his experiments. In other words, if he chose a tall plant, he wanted to know for sure that the plant had genetic messages for the tall trait only.

To make pure-breeding plants for his experiments, Mendel self-pollinated plants that had the characteristic he wanted to study, weeding out any offspring that were different until all the offspring always had his chosen characteristic. (Pea flowers have both male and female parts, so they can be self-pollinated to produce offspring.) For example, Mendel self-pollinated tall pea plants, pulling out any short offspring until he had plants that would produce only tall offspring. He did the same thing for short pea plants, self-pollinating them until they bred purely short offspring.

Pure-breeding organisms that are used as the parents in a genetic cross are called *parentals,* or the *P1 generation.*

## Analyzing the F1 and F2 generations

In one of his experiments, Mendel bred tall pea plants with short pea plants. According to the idea of blending inheritance, all the offspring should have been average in height. However, when Mendel mated his parentals and grew the offspring, called the *F1 generation,* all the offspring were tall. It almost seemed like the short characteristic had disappeared, but when Mendel mated tall pea plants from this new F1 generation and grew their offspring, he saw both tall and short pea plants, indicating that the short characteristic had merely been hidden. The second generation, called the *F2 generation,* had about three times as many tall pea plants as short pea plants.

The F1 and F2 generations get their names from the word *filial,* which means "something that relates to a son or daughter." Consequently, the F1 generation is the first generation of "sons and daughters" from the parentals, and the F2 generation is the second generation.

## Reviewing Mendel's results

The results of Mendel's pea plant experiments were very exciting because they didn't follow what was expected. In other words, they revealed something new about inheritance.

From his results, Mendel proposed several ideas that laid the foundation for the science of genetics:

- Traits are determined by factors that are passed from parents to offspring. Today, people call these factors *genes.*
- Each organism has two copies of the genes that control every trait. Parents pass one copy of each gene on to their offspring. The offspring winds up with two copies of each gene because it gets one copy from mom and one copy from dad.

- Some variations of genes can hide the effects of other variations. Variations that are hidden are *recessive,* whereas variations that hide other variations are *dominant.* In Mendel's cross between tall and short pea plants, the tall characteristic hid the short characteristic; therefore the tall gene was the dominant one.
- The genes that control traits don't blend with each other, nor do they change from one generation to the next. Mendel knew this because the short characteristic, which had disappeared in the F1 generation, reappeared in the F2 generation.

Sexually reproducing organisms have two copies of every gene, but they give only one copy of each gene to their offspring. Mendel said that this is because the two copies of genes *segregate* (separate from each other) when the organisms reproduce. Scientists now call this idea *Mendel's Law of Segregation.*

# Diving into the Pool of Genetic Terminology

Mendel's fundamental ideas have stood the test of time, but geneticists have discovered a great deal more about inheritance since Mendel's day. As the science of genetics grew and developed, so did the language used by geneticists.

A few key genetics terms are particularly useful when talking about inheritance:

- **Genes:** Defined as factors that control traits, each *gene* is a section of nucleotides along the chain of DNA in a chromosome (refer to Chapter 4 for more about chromosomes). Some genes are thousands of nucleotides long; others are less than a hundred. Your cells have about 25,000 different genes scattered among your 46 chromosomes. Each gene is the blueprint for a worker molecule in your cells, usually a protein. Genes determine protein shape and function, and the actions of proteins control your traits.

  For example, the darkness of your skin, hair, and eyes is determined by how much of the brown pigment melanin is deposited there. Proteins called enzymes help make the melanin, and other proteins help deposit the melanin in the right locations. If your genes contain the code for proteins that do these jobs well, your coloring is darker. If your genes code proteins that don't make as much melanin or don't deposit it in your skin, hair, or eyes, your coloring is lighter.

- **Alleles:** Different forms of a gene are called *alleles;* they explain why Mendel saw both tall and short pea plants. Logically, the gene that controls pea plant height has two variations, or alleles — one for tallness and one for shortness. Plants with two alleles for tallness are tall; plants with two alleles for shortness are short. As Mendel saw, plants that have one of each kind of allele are also tall, indicating that the tall allele can hide the effects of the short allele. In other words, the allele for tall is dominant to the allele for short in pea plants.

  In people, alleles for dark coloring are usually dominant to alleles for lighter coloring. So, for a gene that controls skin color, alleles for dark color are dominant to alleles for light color. The color of any individual person's skin depends on his or her unique combination of light and dark alleles.

- **Loci:** These are the locations on a chromosome where genes are found. Each gene is located at a specific place, or *locus,* on its chromosome. In organisms of the same species, the same gene is found at the same locus in every organism. In humans, for example, each gene is found on the same chromosome in all people. The gene for blood type A, B, or O is always found on the same locus on Chromosome 9, and the gene for cystic fibrosis is always found on the same locus on Chromosome 7.

Many human traits aren't controlled by just one gene. Traits such as your height, weight, and the color of your skin, hair, and eyes result from the interaction of several genes. These traits are called *polygenic traits* (*poly* means "many," and *genic* means "genes"). Polygenic traits usually show a wide range of variety in the population. Pea plants, for example, are either short or tall with nothing in between, whereas adult humans range over a wide variety of heights. This difference is because human height is polygenic and pea height is controlled by just one gene.

# Bearing Genetic Crosses

Geneticists use their own unique shorthand when analyzing the results from a *genetic cross* (a mating between two organisms with characteristics that scientists want to study). They use a letter to stand for each gene, capitalizing the letter for dominant alleles (for a definition of alleles, see the preceding section). The same letter of the alphabet is used for each allele to show that they're variations of the same gene.

For the cross Mendel did between tall and short pea plants, see the earlier "'Monk'ing Around with Peas: Mendel's Laws of Inheritance" section for more on this experiment, the letter *T* can be used to represent the gene for plant height. In Figure 7-1, the dominant allele for tallness is shown as *T,* whereas the recessive allele for shortness is shown as *t.*

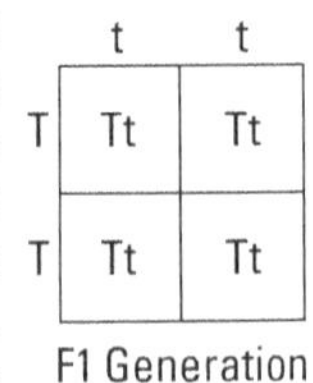


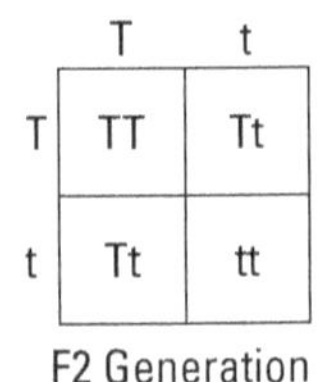


**Figure 7-1:** Mendel's cross between tall and short pea plants.

Geneticists also have special terms for describing the organisms involved in a genetic cross. Here they are:

- **Genotype:** The combination of alleles that an organism has is its *genotype*. The genotypes of the two parental plants shown in Figure 7-1 are *TT* and *tt*.
- **Phenotype:** The appearance of an organism's traits is its *phenotype*. The phenotypes of the two parental plants shown in Figure 7-1 are tall and short.

A tool called a *Punnett square* helps geneticists predict what kinds of offspring might result from a particular genetic cross. In Figure 7-1, a Punnett square shows the cross between the peas of the F1 generation. The alleles that each parent can contribute to the offspring are written along the sides of the square where it says Sperm and Eggs. All possible combinations of alleles that could result from the meeting of sperm and egg are drawn within the square.

If Mendel had used modern genetic notation and terminology, he might have analyzed his experiment like this (look to Figure 7-1 for reference if you need it):

1. The parental pea plants are purebred, so they have only one type of allele, but each individual plant has two alleles for each gene. The tall parent's alleles are shown as *TT*, and the short parent's alleles are shown as *tt*. Because both of their alleles are the same, the parental pea plants are *homozygous* for the plant height trait (*homo-* means "same," and *zygous* comes from a Greek root that means "together").
2. Each parental pea plant gives one allele to each offspring. Because the parentals are purebred, they can give only one type of allele. Tall pea plant parents always give a copy of the tall allele *(T)* to offspring, and short parents always give a copy of the short allele *(t)*. Copies of these alleles are packaged into gametes (sperm and egg cells) as the pea plants reproduce.
3. The sperm and egg of the parents combine, giving their F1 offspring two alleles for the height gene. All the F1 offspring have one copy of each allele, so their alleles are written as *Tt*. Because their alleles are different, the F1 pea plants are *heterozygous* for the plant height trait (*hetero-* means "other"). Although the F1 plants are heterozygous, they should

all look tall because the tall allele is dominant to the short allele. This is exactly what Mendel saw — the short trait from his parentals seemed to disappear in the F1 generation.

4. When F1 plants are crossed, they can each make two kinds of gametes — those that carry a dominant allele and those that carry a recessive allele. To figure out all the possible combinations of offspring the F1 plants could have, you use a Punnett square like the one shown in Figure 7-1.
5. The completed Punnett square in Figure 7-1 predicts that the F2 offspring will have three different genotypes: *TT, Tt,* and *tt.* For every one *TT* offspring, there should be two *Tt* offspring and one *tt* offspring. In other words, the *genotypic ratio* (the ratio of expected numbers of genotypes for the cross) predicted for the F2 generation is 1:2:1 for *TT:Tt:tt.*
6. The tall allele is dominant to the short allele, so F2 plants that are *TT* or *Tt* will be tall, and only F2 plants that are *tt* will be short. So, the Punnett square predicts that for every three tall plants, there'll be just one short plant. In other words, the *phenotypic ratio* (the ratio of expected numbers of phenotypes for the cross) for the F2 generation is 3:1 for tall:short. This is precisely what Mendel saw — for every one short plant he saw in his F2 generation, he saw about three tall ones.

When two organisms that are heterozygous for one trait are crossed with each other, that combination is called a *monohybrid cross.* (*Mono-* means "one," and *hybrid* means "something from two different sources," so a *monohybrid* is an organism that has two different alleles for one trait.) The cross between F1 pea plants in Figure 7-1 is an example of a monohybrid cross.

# Studying Genetic Traits in Humans

One of the reasons plants make good subjects for genetic studies is because you can control their mating. From a genetics standpoint, humans aren't nearly so cooperative. Also, humans don't produce as many offspring as plants do, and human children take a very long time to grow up so you can evaluate the appearance of traits. Consequently, when geneticists want to study human traits, they have to rely upon families that already exist.

The sections that follow show you how geneticists go about studying family trees and how they test for different inheritance scenarios. You also find out the conclusions geneticists have made about dominant and recessive traits in humans.

## Creating pedigree charts

The first steps in understanding the inheritance of a human trait are to gather information on which people in a family have the trait and record that information in a geneticist's version of a family tree, which is called a *pedigree chart.* The symbols in a pedigree chart (see Figure 7-2a) represent information about the family and the trait being studied.

- Squares indicate males; circles indicate females.
- A line drawn between two symbols represents marriage.
- A line drawn down from a marriage indicates that the marriage produced a child. Children are arranged in birth order from left to right.
- A filled-in symbol indicates people whose traits are being studied; an open symbol is used for people who don't have those traits.
- A diagonal line through a symbol represents someone who is deceased.
- Roman numerals shown to the left of each row represent the different generations. Each person in a generation is assigned an Arabic number so that any person in a pedigree chart can be identified by the combination of his or her generation number and individual number. The deceased person in Figure 7-2c, for example, is identified as Individual VI-1.

By studying a pedigree chart, geneticists can often figure out whether a trait is caused by a dominant or recessive allele. The trait shown in Figure 7-2c, for example, must be caused by a recessive allele. The symbol of Individual VI-1 is shaded, which means this person had the trait being studied. However, neither of this person's parents shows the trait. The parents must have carried the allele because children get their alleles from their parents, but the allele wasn't visible in either parent. When an allele is present but not seen in a person's phenotype, the trait must be recessive.

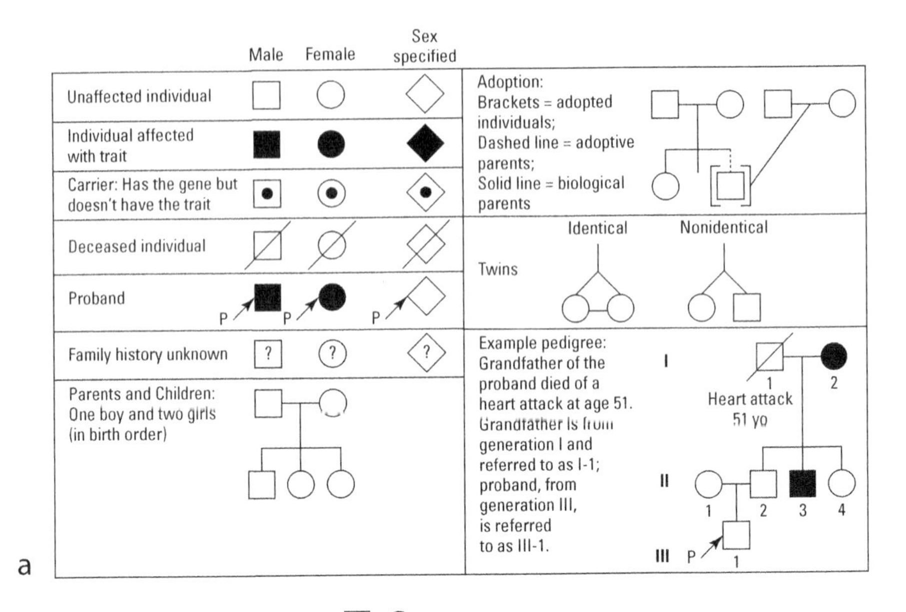


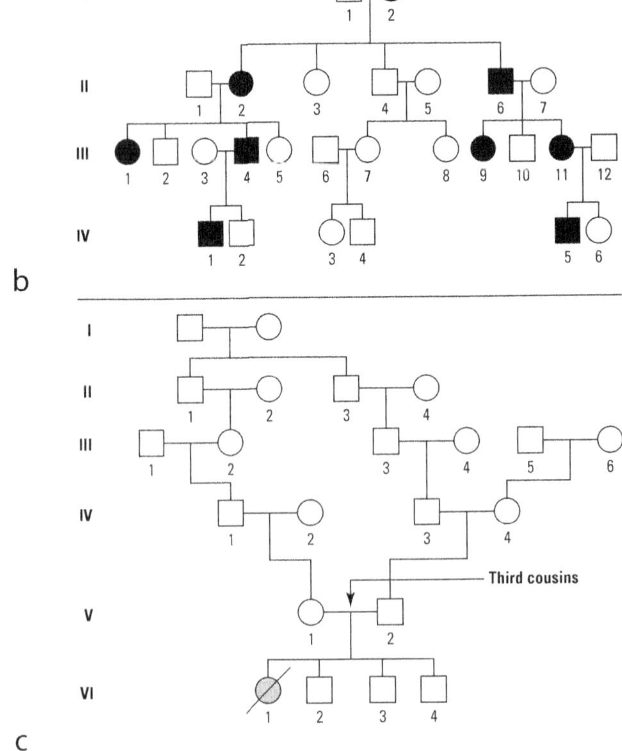


**Figure 7-2:** Human pedigree charts and symbols.

## Testing different inheritance scenarios

One pedigree chart alone doesn't always provide enough information to determine the inheritance pattern of a trait. In order to test whether a particular type of inheritance works for a pedigree, follow these steps:

1. **Decide which type of inheritance you want to test.**

   Say, for example, that you want to test the hypothesis that the trait in Figure 7-2b is caused by a dominant allele.

2. **Set up your notation for the alleles.**

   If you're testing to see whether the trait is dominant, then you can propose that *A* should represent the dominant allele and *a* should represent the recessive allele.

3. **Assign genotypes to people whose genotype is certain.**

   Because you're suggesting that the trait is dominant, then people who don't show the trait must have two copies of the recessive allele. (If they had even just one copy of the dominant allele, then you'd see the trait, and their symbols would be shaded.) In Figure 7-2b, all the individuals with open symbols must be homozygous recessive, or *aa*. Write their genotypes under their symbols.

4. **Work from the individuals whose genotypes are certain to figure out the genotypes of the other individuals.**

   Use the clues of parents and other children to figure things out. For example, Individual I-2 has the trait, which means she must have at least one copy of the dominant allele. Right away you know that her genotype is *A*– (the dash represents her second, unknown allele). To figure out her second allele, look at her children. If she was homozygous dominant, or *AA*, then she could only ever give one type of allele to her kids — the *A* allele. Because you're testing whether this trait could be dominant, then any kid who has even just one *A* is going to show the trait. However, two of I-2's children don't have the trait. Consequently, she must be able to give the little *a* allele too, which makes her genotype *Aa*.

   If you can't figure out someone's complete genotype by looking at the genotypes of people whose information is certain, you can leave a dash in that person's genotype to represent the uncertain allele.

5. **If you find a place in the pedigree chart where the genotypes of parents and children have a conflict, you must reject your hypothesis about the inheritance of the trait.**

   For example, if you proposed that the trait in Figure 7-2c was caused by a dominant allele, then Individual VI-1 would have to have had at least one dominant allele. This is impossible because neither of VI-1's parents shows the trait, and one of the parents would have to carry at least one dominant allele in order to pass that on to VI-1. Because the proposed genotypes are impossible, you must reject the hypothesis that this trait could be carried by a dominant allele.

The trait shown in Figure 7-2b looks like it's caused by a dominant allele because someone in every generation has the trait. In other words, the trait never hides like a recessive trait does. However, from just this one pedigree chart, you can't be absolutely sure. If you set up a hypothesis that the trait is recessive and test your idea using the preceding steps, you can give everyone a genotype without encountering an impossible conflict. In situations like this one, in which more than one type of inheritance is possible based on one pedigree chart, geneticists gather more information from other families. In fact, sometimes they have to study many pedigrees before they can prove the inheritance of a particular trait.

## Drawing conclusions about traits

In general, dominant and recessive inheritance show two distinct patterns in a human pedigree:

- **Traits carried by recessive alleles often skip generations.** For example, a red-headed child may have a red-headed grandparent but blonde-haired parents. Any time parents who *don't* show a certain trait have children who *do* show the trait, that's proof of a recessive trait.
- **Traits carried by dominant alleles often show up in every generation.** Brown-eyed parents who have more than one child usually have some children with brown eyes. However, people showing the dominant brown-eyed trait may have only one copy of the dominant allele, so they could potentially pass on recessive alleles to their children. This fact is why two brown-eyed people can produce some blue-eyed children. In fact, whenever two parents who show a particular trait have kids who don't show the trait, then you know for sure that the trait is dominant.

# Chapter 8

# Reading the Book of Life: DNA and Proteins

### In This Chapter

- Grasping why DNA and proteins are so important
- Producing proteins
- Surveying the various DNA mutations that can occur
- Controlling your genes

Without deoxyribonucleic acid, or DNA for short, your cells — and all the cells of every other living thing on Earth — wouldn't exist. That's because DNA controls the structure and function of organisms, largely because it's essential to the production of the proteins that determine your traits. When changes occur in the DNA of one or more cells, the effects on the organism made up of those cells can be disastrous.

We show you just how important DNA and proteins are to your everyday life in this chapter. Get ready to discover how DNA and RNA (ribonucleic acid) work together to produce proteins, the types of DNA mutations that occur and how they affect you, and more.

## Proteins Make Traits Happen, and DNA Makes the Proteins

You probably know that DNA is your genetic blueprint and that it carries the instructions for your traits. But what you may not know is exactly how your DNA causes you to look and function a certain way. DNA contains the instructions for the construction of the molecules that carry out the functions of your cells. These functional molecules are mostly proteins, and the instructions for creating these proteins are found in your *genes,* sections of DNA that lie along your chromosomes.

Think of your cells as little factories that have to carry out certain functions. Each function relies upon the actions of the robot workers in the factory. DNA is the instructions for the construction of each type of robot. If the robots are built correctly according to their instructions, they work the way they're supposed to. Based on how the robots work, the factory accomplishes specific tasks. Your cells don't have little robots running around, but they do have worker molecules that carry out the functions of the cell. If one of these molecules isn't working correctly, your cells will work differently than they're supposed to, which could affect your traits.

One gene equals one blueprint for a functional molecule. Because many of the functional molecules in your cells are proteins, genes often contain the instructions for building the polypeptide chains that make up proteins (for more on protein structure, head to Chapter 3). So, one gene often equals one polypeptide chain.

Sometimes it's hard to imagine how one little protein can affect you in an important way. After all, humans have about 25,000 genes, so you make a lot of different proteins. How can just one faulty protein make a difference? Well, if your skin cells couldn't make the protein collagen, your skin would fall off at the slightest touch. Also, if the cells in your pancreas couldn't make the protein insulin, you'd have diabetes. So you see, all the functions of your body that you probably take for granted — the way it's built, the way it looks, and the way it functions — are controlled by the actions of proteins.

# Moving from DNA to RNA to Protein: The Central Dogma of Molecular Biology

The instructions in DNA determine the structure and function of all living things, which makes DNA pretty darn important. Every time a cell reproduces, it must make a copy of these instructions for the new cell. When cells need to build a functional molecule (usually a protein), they copy the information in the genes into an RNA molecule instead of using the DNA blueprint directly (see Chapter 3 for more about RNA molecules). Here's an outline of the process:

- Cells use *transcription* to copy the information in DNA into RNA molecules.
- The information to build proteins is copied into a special type of RNA called *messenger RNA* (mRNA), which carries the blueprint for the protein from the nucleus to the cytoplasm where it can be used to build the protein.
- Cells use *translation* to build proteins from the information carried in mRNA molecules.

The idea that information is stored in DNA, copied into RNA, and then used to build proteins is considered the *central dogma of molecular biology*.

Transcription and translation are two pretty similar sounding words for two very different processes in cells. One way to remember which process is which is to think about the English meanings of these words. When you transcribe something, you copy it. Transcription in cells takes the information in DNA and uses it to make RNA. DNA and RNA are similar molecules, so it's not like you're really changing anything — you're just copying the information down. When you translate something, on the other hand, you change it from one language to another. Translation in cells takes the information in mRNA and uses it to build a protein, which is a different type of molecule. So, translation changes the language of molecules from RNA to protein.

The sections that follow give you an in-depth look at transcription, RNA processing, and translation.

## Rewriting DNA's message: Transcription

DNA molecules are long chains made from four different building blocks called *nucleotides* that biologists represent with the letters A, T, C, and G (flip to Chapter 3 for the scoop on DNA structure). These chemical units are joined together in different combinations that form the instructions for cells' functional molecules, which are mostly proteins.

When your cells need to build a particular protein, the enzyme RNA polymerase locates the gene for that protein and makes an RNA copy of it. (RNA polymerase is shown in Step 2 of Figure 8-1.) Because RNA and DNA are similar molecules, they can attach to each other just like the two strands of the DNA double helix do. RNA polymerase slides along the gene, matching RNA nucleotides to the DNA nucleotides in the gene.

The base-pairing rules for matching RNA and DNA nucleotides are almost the same as those for matching DNA with DNA (see Chapter 3). The exception is that RNA contains nucleotides with uracil (U) rather than thymine (T). During transcription, RNA polymerase pairs C with G, G with C, A with T, and U with A. (Figure 8-1 gives you the visual of this. Note that the new RNA strand CUAG pairs up with the DNA sequence GATC in the gene.)

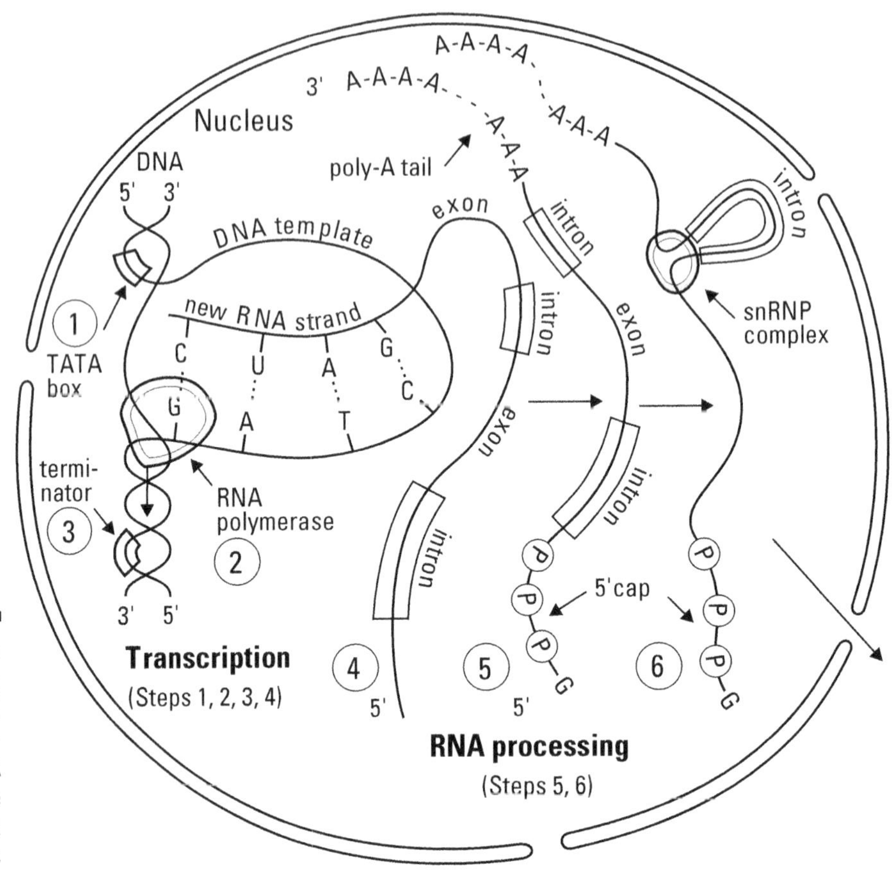


**Figure 8-1:** Transcribing DNA and processing mRNA within the nucleus of a eukaryotic cell.

It may seem strange that your cells copy the information in your genes into a sort of mirror image made of RNA, but it actually makes a lot of sense. Your genes are extremely important and need to be protected, so they're kept safe in the nucleus at all times. Your cells make copies of any information they need so the original DNA doesn't get damaged.

You can think of your chromosomes like drawers in a file cabinet. When your cells need information from the files, they open a drawer, take out a file (the gene), and make a copy of the information (the RNA molecule) that can travel out into the world (the cytoplasm). The original blueprint (the DNA) is kept safely locked away in the file cabinet.

Of course, RNA polymerase and DNA aren't the only things involved in transcription. The following sections introduce you to the other players and walk you through the process of transcription step by step.

### Finding out what else is involved

RNA polymerase locates the genes it needs to copy with the help of proteins called *transcription factors*. These proteins look for certain sequences in the DNA that mark the beginnings of genes; these sequences are called *promoters*.

Transcription factors find the genes for the proteins the cell needs to make and bind to the promoters so RNA polymerase can attach and copy the gene. Many promoters contain a particular sequence called the TATA box because it contains alternating T and A nucleotides. Transcription factors bind to the TATA box first, followed by RNA polymerase. (Step 1 in Figure 8-1 depicts a gene's promoter, along with its TATA box.)

The ends of genes are marked by a special sequence called the transcription terminator. *Transcription terminators* can work in different ways, but they all stop transcription. (Step 3 of Figure 8-1 shows a transcription terminator.)

### Walking through the process

As you can see from the following, the process of transcription is pretty straightforward:

1. **RNA polymerase binds to the promoter with the help of transcription factors.**

   By binding to the promoter, RNA polymerase gets set up on the DNA so it's pointed in the right direction to copy the gene.

2. **RNA polymerase separates the two strands of the DNA double helix in a small area.**

   By opening the DNA, RNA polymerase can use one of the DNA strands as its pattern for building the new RNA molecule. (The strand of DNA that's being used as a pattern in Figure 8-1 is labeled as the DNA template. The new RNA strand is shown as it's being built against the template strand.)

   As RNA polymerase slides along the DNA, it opens a new area, and the DNA behind it closes back up.

3. **RNA polymerase uses base-pairing rules to build an RNA strand that's complementary to the DNA in the template strand.**

   Because the base-pairing rules are specific, the new RNA molecule contains a mirror image of the code in the DNA.

4. **RNA polymerase reaches the termination sequence and releases the DNA.**

   Some terminators have a sequence that causes the new RNA to fold up on itself at the end, making a little bump that causes RNA polymerase to get knocked off of the DNA.

Your cells use transcription to make several types of RNA molecules. Some of these RNA molecules are worker molecules for the cell, others are part of cellular structures, and one type — mRNA — carries the code for proteins to the cytoplasm.

## Putting on the finishing touches: RNA processing

After RNA polymerase transcribes one of your genes and produces a molecule of mRNA, the mRNA isn't quite ready to be translated into a protein. In fact, when the mRNA is hot off the presses, it's called a *pre-mRNA* or *primary transcript* because it's not quite finished.

Before the pre-mRNA can be translated, it has to undergo a few finishing touches via RNA processing (see Steps 5 and 6 in Figure 8-1):

- **The 5' cap, a protective cap, is added to the beginning of the mRNA.** The *5' cap* (shown in Step 5 of Figure 8-1) tells the cell it should translate this piece of RNA.
- **The poly-A tail, an extra bit of sequence, is added to the end of the mRNA.** Like its name suggests, the *poly-A tail* (see Step 5 of Figure 8-1) is a chain of repeating nucleotides that contain adenine (A). It protects the finished mRNA from being broken down by the cell.
- **The pre-mRNA is spliced to remove introns (noncoding sequences).** One kind of weird thing about your genes (and ours too) is that the code for proteins is interrupted by sequences called *introns*. Your cells remove the introns before shipping the mRNA out to the cytoplasm (see Step 6 in Figure 8-1). The sections of the pre-mRNA that wind up getting translated are called *exons*. When your cells cut the introns out of pre-mRNA, the exons all come together to form the blueprint for the protein.

If you get confused about what introns and exons do, just remember that *in*trons *in*terrupt and *ex*ons get to *ex*it the nucleus.

## Converting the code to the right language: Translation

After a mature mRNA leaves the nucleus of a cell, it heads for a ribosome in the cell's cytoplasm where the code it contains can be translated to produce a protein (for more on ribosomes, see Chapter 4). As the strand of mRNA slides through the ribosome, the code is read three nucleotides at a time.

A group of three nucleotides in mRNA is called a *codon*. If you take the four kinds of nucleotides in RNA — A, G, C, and U — and make all the possible three-letter combinations you can, you'd come up with 64 possible codons. Each codon specifies one amino acid in the polypeptide chain of a protein.

To figure out the amino acid a singular codon represents, follow the labels on the edges of the table in Figure 8-2. So to find out what the codon CGU represents:

1. **Look first to the left side of the table and find the row marked by the first letter in the codon.**

   The letter C is the second letter down, so the amino acid represented by the C portion of the codon CGU is found in the second row of the table.

2. **Then look to the top of the table and find the column marked by the second letter in the codon.**

   The letter G is the last letter in the row, so the amino acid represented by the G part of the codon CGU is found at the intersection of the second row (which is marked by C) and the last column under the Second Letter heading.

3. **Look to the right side of the table and find the row marked by the third letter in the codon.**

   The letter U is listed first, so the amino acid represented by the U portion of the codon CGU is the first one listed in the intersection of the second row and the last column under the Second Letter heading. Put it all together and you find that the amino acid represented by the codon CGU is arginine.

To translate a molecule of mRNA, begin at the start codon closest to the 5' cap of the mRNA, divide the message up into codons, and look the codons up in a table of the genetic code that shows the names of the 20 different amino acids found in the proteins of living things. For example, 5'CCGCAUGCGAAAAUGA3' translates into methionine-arginine-lysine.

| First Letter ↓ | Second Letter: U | C | A | G | Third Letter ↓ |
|---|---|---|---|---|---|
| U | phenylalanine | serine | tyrosine | cysteine | U |
| | phenylalanine | serine | tyrosine | cysteine | C |
| | leucine | serine | STOP | STOP | A |
| | leucine | serine | STOP | tryptophan | G |
| C | leucine | proline | histidine | arginine | U |
| | leucine | proline | histidine | arginine | C |
| | leucine | proline | glutamine | arginine | A |
| | leucine | proline | glutamine | arginine | G |
| A | isoleucine | threonine | asparagine | serine | U |
| | isoleucine | threonine | asparagine | serine | C |
| | isoleucine | threonine | lysine | arginine | A |
| | methionine & START | threonine | lysine | arginine | G |
| G | valine | alanine | aspartate | glycine | U |
| | valine | alanine | aspartate | glycine | C |
| | valine | alanine | glutamate | glycine | A |
| | valine | alanine | glutamate | glycine | G |

**Figure 8-2:** The genetic code.

The following sections get you acquainted with specialized codons and the anticodons all codons need to pair up with for translation to occur. They also help you understand the overall process of translation.

### *Making sense of codons and anticodons*

The genetic code is amazingly similar for all the organisms on Earth, from you to *E. coli.* To read it, you need to know about the unique features of some of the codons:

- **The codon AUG is called the start codon.** AUG is called the *start codon* because translation begins here. When one of your cells starts translating mRNA into a polypeptide, the first AUG closest to the 5' cap of the mRNA is the first codon to be read. AUG also represents the amino acid methionine, so methionine is the first amino acid added to the polypeptide chain.

- **The codons UAA, UAG, and UGA are called stop codons.** Translation ends when a stop codon is read in the mRNA. A *stop codon* only indicates when translation should end; it doesn't represent an amino acid. When stop codons are read in the mRNA, translation stops without adding any new amino acids to the polypeptide chain.
- **Some amino acids are represented by more than one codon.** For instance, arginine is represented by more than one codon; CGU, CGC, CGA, and CGG all represent arginine. Because of this situation, biologists say that the genetic code is *redundant* (more than one codon represents some amino acids).

In order for your cells to decode mRNA, they need the help of an important worker: transfer RNA (tRNA). *Transfer RNA* decodes the message in the mRNA. Like all RNA molecules, tRNA is made of nucleotides that can pair up with other nucleotides according to base-pairing rules.

During translation, tRNA molecules pair up with the codons in mRNA to figure out which amino acid should be added to the chain. Each tRNA has a special group of three nucleotides, called an *anticodon,* that pairs up with the codons in mRNA. Each tRNA also carries a specific amino acid. So, the tRNA that has the right anticodon to pair with a specific codon adds its amino acid to the growing polypeptide chain.

Because the pairing of anticodon to codon is specific, only one tRNA can pair up with each codon. The specific relationship between tRNA anticodons and mRNA codons ensures that each codon always specifies a particular amino acid.

### Breaking down the translation process

Although the process of translation is fairly complicated, it's pretty easy to understand when you break it down into three main steps: the beginning *(initiation),* the middle *(elongation),* and the end *(termination*). Follow along in Figure 8-3 as we present these three steps:

1. **During initiation, the ribosome and the first tRNA attach to the mRNA (see #1 in Figure 8-3).**

   The small subunit of the ribosome binds to the mRNA. Then the first tRNA, which carries the amino acid methionine, attaches to the start codon. The start codon is AUG, so the first tRNA has the anticodon UAC (see #2 in Figure 8-3). After the first tRNA is bound to the mRNA, the large subunit of the ribosome attaches to form a complete ribosome.

2. **During elongation, tRNAs enter the ribosome and donate their amino acids to the growing polypeptide chain.**

Each tRNA enters a pocket in the ribosome called the *A site* (see #3 in Figure 8-3). An adjacent pocket, called the *P site,* holds a tRNA with the growing polypeptide chain (see #4 in Figure 8-3). When a tRNA is parked in the A site and the P site, the ribosome catalyzes the formation of a *peptide bond* between the growing polypeptide chain and the new amino acid. In Figure 8-3, a bond is forming between the amino acids cysteine (cys) and proline (pro) because they're next to each other in the ribosome.

After the new amino acid is added to the growing chain, the ribosome slides down the mRNA, moving a new codon into the A site. After a new codon is in the A site, another tRNA can enter the ribosome, and the process of elongation can continue.

3. **During termination, a stop codon in the A site causes translation to end.**

   The ribosome slides down the mRNA until a stop codon enters the A site. When a stop codon is in the A site, an enzyme called a *release factor* enters the ribosome and cuts the polypeptide chain free. Translation stops, and the ribosome and mRNA separate from each other.

Following translation, polypeptide chains may be modified a bit before they fold up and become functional proteins. Often, more than one polypeptide chain combines with another chain to form the complete protein.

## Growing wild: Cancer and aging

*Cancer,* a disease caused by uncontrolled cell division, occurs much more frequently as people age, and the reality is that the longer you live, the more likely you are to develop cancer. Why? Because throughout your life, tiny genetic changes may occur that you don't even realize are happening — a tiny mutation caused by an X-ray here or another tiny mutation caused by breathing in polluted air there. Some mutations are repaired; others just exist in your body's cells without much notice. Each individual mutation may be harmless, but after genetic changes accumulate over time, the chance that some mutation will occur to change the proteins that control cell division grows greater.

If your proteins don't do their jobs correctly, your cells may be thrown into an uncontrolled cycle of dividing and growing, causing a mass of cells called a *tumor* to form. As even greater numbers of mutations accumulate in the cells over time, their characteristics change, and the tumor may become *malignant* (more likely to spread and cause death). The spread of cancer cells around the body occurs through *metastasis,* when malignant cells leave their tissue of origin and enter the circulatory or lymphatic systems.

People living in the United States have a 1 in 3 chance of developing cancer during their lifetimes. That statistic is alarming, but you can do your part to reduce your risk of cancer by making good lifestyle choices — avoiding saturated fat, cigarettes, and obesity will all lower your cancer risk.

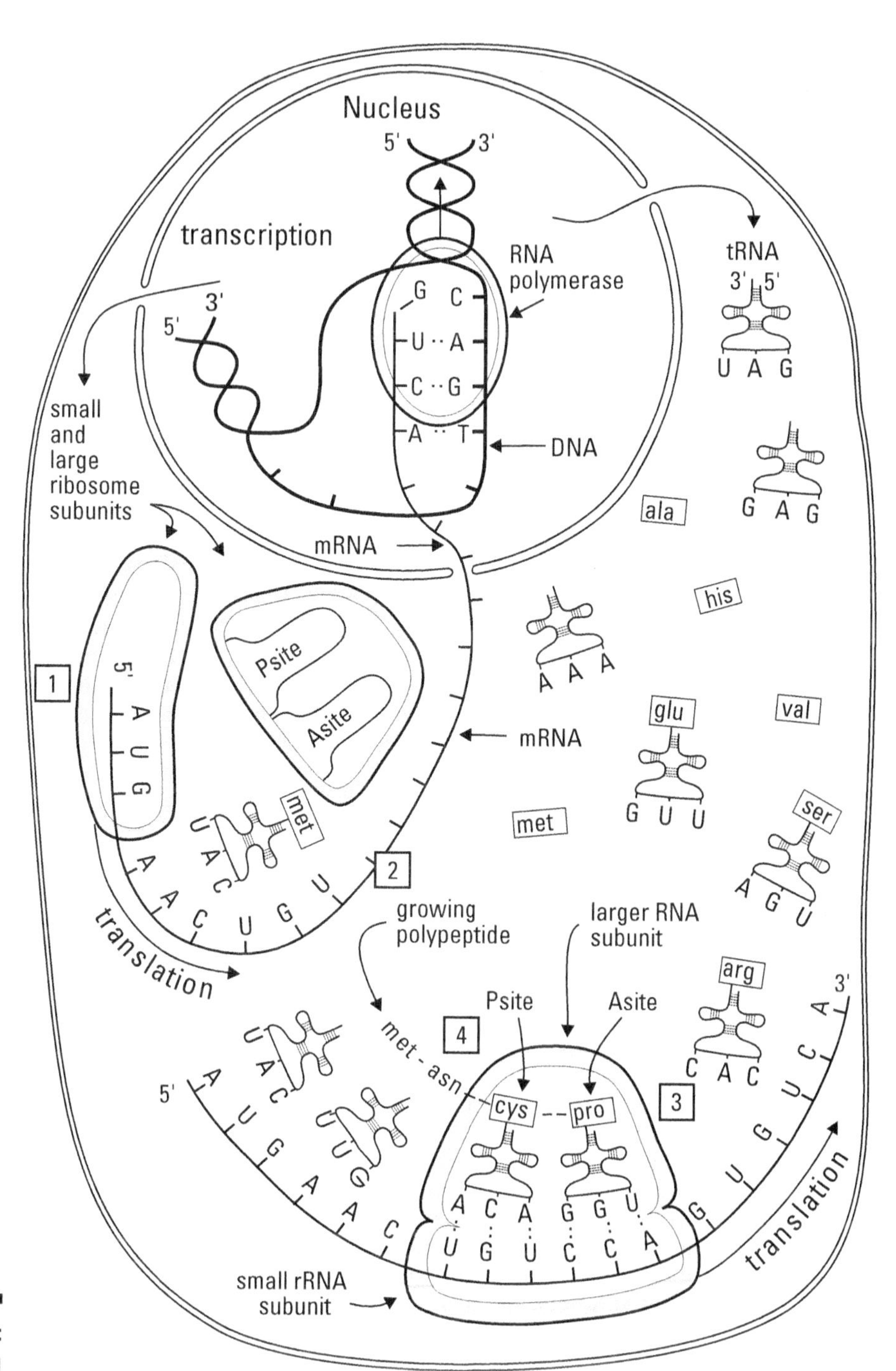


Figure 8-3: Translating mRNA into protein.

Protein Synthesis

# Mistakes Happen: The Consequences of Mutation

If a mistake in a strand of DNA goes undetected or unrepaired, the mistake becomes a mutation. A *mutation* is a change from the original DNA strand — in other words, the nucleotides aren't in the order that they should be.

Mutations in DNA lead to changes in RNA, which can lead to changes in proteins. When proteins change, the functions of cells, and the traits of organisms, can also change.

Mutations usually happen as the DNA is being copied during DNA replication (see Chapter 6 for a description of DNA replication). Two main types of mutations occur:

- **Spontaneous mutations:** These result from uncorrected mistakes by *DNA polymerase,* the enzyme that copies DNA. DNA polymerase is a very accurate enzyme, but it's not perfect. In general, DNA polymerase makes one mistake for every billion base pairs of DNA it copies. One in a billion isn't bad . . . unless you're talking about your DNA. Then, any change can eventually cause problems. Cancer, for example, usually occurs as people age because they've lived long enough to accumulate mutations in certain genes that control cell division.
- **Induced mutations:** These result from environmental agents that increase the error rate of DNA polymerase. Anything that increases the error rate of DNA polymerase is a *mutagen.* The most common mutagens are certain chemicals (such as formaldehyde and compounds in cigarette smoke) and radiation (like ultraviolet light and X-rays).

When mutations occur during DNA replication, some daughter cells formed by mitosis or meiosis inherit the genetic change (we explain how cells divide in Chapter 6). The types of mutations these cells inherit can be divided into three major categories:

- **Base substitutions:** When the wrong nucleotides are paired together in the parent DNA, a *base substitution* occurs. If the parent DNA molecule has a nucleotide containing thymine (T), then DNA polymerase should bring in a nucleotide containing adenine (A) for the new strand of DNA. However, if DNA polymerase makes a mistake and brings in a nucleotide with guanine (G) by mistake, that's a base substitution. Because just one nucleotide was changed, the mutation is called a *point mutation.* The effect of point mutations ranges from nothing to severe.

  - *Silent mutations* have no effect on the protein or organism. Because the genetic code is redundant, changes in DNA may lead to changes in mRNA that don't cause changes in the protein (see the earlier section "Making sense of codons and anticodons" for more on the redundancy of the genetic code).
  - *Missense mutations* change the amino acids in the protein. Changes in DNA can change the codons in mRNA, leading to the addition of different amino acids into a polypeptide chain. The severity of missense mutations depends on how different the original amino acid is from the new amino acid and where in the protein the change occurs.
  - *Nonsense mutations* introduce a stop codon into the mRNA, preventing the protein from being made. If the DNA changes so that a codon in the mRNA becomes a stop codon, then the polypeptide chain gets cut short. Nonsense mutations usually have severe effects and are the cause of many genetic diseases, including certain forms of cystic fibrosis, Duchenne's muscular dystrophy, and thalassemia (an inherited form of anemia).

- **Deletions:** When DNA polymerase fails to copy all the DNA in the parent strand, that's a *deletion.* If nucleotides in the parent DNA are read but the complementary bases aren't inserted, the new strand of DNA is missing nucleotides. If one or two nucleotides are deleted, then the codons in the mRNA will be skewed from their proper threes, and the resulting polypeptide chain will be very altered. Mutations that change the way in which the codons are read are called *frameshift mutations.* Deletions of three nucleotides result in the deletion of one amino acid. Serious diseases such as cystic fibrosis and Duchenne muscular dystrophy result from deletions.
- **Insertions:** When DNA polymerase slips and copies nucleotides in the parent DNA more than once, an *insertion* occurs. Just like deletions, insertions of one or two nucleotides can cause frameshift mutations that greatly alter the polypeptide chain. *Huntington's disease,* an illness that causes the nervous system to degenerate starting when a person is in his 30s or 40s, is caused by insertions of the sequence CAG into a normal gene up to 100 times. Although the sequence is a multiple of three (so technically not a frameshift mutation), the abundance of these insertions messes up the reading of the normal genetic code, causing either abnormal protein production or a lack of protein production.

## Fighting for breath

*Cystic fibrosis* (CF), a disease that clogs the lungs and smaller passageways of the body with mucus, is the most common genetic disease found among people of northern European ancestry (about 30,000 people are living with CF in the United States alone). The mucus in the lungs makes it difficult to breathe and provides a breeding ground for bacteria, causing repeated lung infections. People with CF don't usually live past young adulthood because their lungs ultimately fail from all the damaging infections. CF patients and their families struggle against this disease daily, using therapies to break up the mucus and fight the infections, trying to hold back the effects of this dreadful disease.

All of this difficulty and pain is caused by a mutation in one protein: the CFTR protein. Normal CFTR protein is found in the plasma membranes of people's cells, where it helps move chloride across the membrane. The movement of chloride affects the movement of water, which is why the mucus gets so thick on the outside of cells in people who have the disease. The most common mutation in CFTR is a deletion of three nucleotides, resulting in the loss of one amino acid. That doesn't seem like a very big change, but the loss of that *one* amino acid alters the polypeptide chain so it doesn't fold up properly. Other proteins, acting as quality-control proteins, see the abnormal folding and mark the mutated protein for destruction. So, in people with the most common form of CF, the CFTR protein never even makes it to the plasma membrane.

Scientists are currently searching for ways to correct this problem. One new and very exciting development is the discovery of a drug that helps mutant CFTR proteins make it to the plasma membrane. In lung cells grown in the laboratory, the mutant CFTR proteins were able to transport some chloride, suggesting that the drug may be able to reduce the severity of the disease. The researchers who discovered this effect are taking their discovery to the next level by working on appropriate doses of the drug and testing it on people who suffer from CF.

# Giving Cells Some Control: Gene Regulation

Even though your DNA is in control of the proteins your body makes, and those proteins are in charge of determining your traits, your cells do have some say in life. Because each one of your cells has a complete set of your chromosomes, your cells are able to practice *gene regulation,* meaning they can choose which genes to use (or not use) and when.

When a cell uses a gene to make a functional molecule, that gene is *expressed* in the cell. Gene regulation is the process cells use to choose which genes to express at any one time. (Scientists talk about gene regulation as cells turning genes "on" or "off.")

Genes are regulated by the action of proteins that bind to DNA and either help or block RNA polymerase from accessing the genes. In your cells, proteins that help RNA polymerase bind to your genes are called *transcription factors.* They bind to special sequences on the DNA near genes' promoters and make it possible for RNA polymerase to bind the promoter. Transcription and translation occur, producing the protein in the cell.

Gene regulation allows your cells to do two things: adapt to environmental changes and make it so that each cell type has a distinct role in the body. We fill you in on both in the next sections.

## Adapting to environmental changes

The world around you is always changing, which means you need to be able to respond to environmental signals in order to maintain your physiological balance. Gene regulation allows you to do just that. When your cells need to respond to environmental changes, they turn genes on or off to make the proteins needed for the response.

Suppose you're getting too much sunlight. To protect your skin, the cells on the tip of your nose need to darken a bit by making more of the skin pigment melanin. The extra sunlight on your skin triggers certain proteins to bind to the genes needed for melanin production and help RNA polymerase access the genes. RNA polymerase reads the genes, making mRNA that contains the blueprints for the necessary proteins. The mRNA is translated, and the proteins are made. The proteins do their jobs, and the skin on your nose turns a darker color. This example of how your skin gets darker illustrates how cells can access genes when they need them in order to respond to signals from the environment.

## Becoming an expert through differentiation

You have more than 200 different types of cells in your body, including skin cells, muscle cells, and kidney cells. Each of these cells does a different job for your body, and like any good craftsman, each of these cell types requires the right tools for its job. To a cell, the right tool for the job is usually a specific protein. For instance, skin cells need lots of the protein keratin, muscle cells needs lots of contractile proteins, and kidney cells need water-transport proteins.

*Cell differentiation* is the process that makes cells specialized for certain tasks. Your differentiated cells have all the blueprints for all the tools because they each have a full set of your chromosomes; what makes them different from one another is which blueprint they use.

Cells differentiate from one another due to gene regulation. For example, when a sperm met an egg to form the cell that would become the future you, that first cell had the ability to divide and form all the different cell types your body needed. As that cell and its descendents divided, however, signals caused different groups of cells to change their gene expression. Proteins in these cells bound to the DNA molecules, activating some genes and silencing others. As you grew and developed in your mother's uterus, your cells became more and more distinct from each other. Some cells became part of your nervous tissue, whereas others formed your digestive tract. Each of these changes occurred as cells transcribed and translated the genes for the proteins that they needed to do their particular job.

After a cell becomes differentiated, it usually can't go back. It's specialized for a certain task and can't access the genes for proteins that aren't in its job description.

## Stem cells: An introduction and a moral dilemma

*Stem cells* are cells that have the potential to become any type of cell in the body. When they divide, their descendants can either remain stem cells or become differentiated into a specialized cell type. Stem cells are also unique because they're immortal — unlike most human cells, which stop dividing after about 40 or 50 cell divisions, stem cells can continue to produce new cells indefinitely. They're therefore extremely important to the body because they can help replace cells and repair damage in almost every kind of tissue. They also have the potential to cure human diseases and save lives.

For many years the best source of stem cells for research was human embryos made by in vitro fertilization but no longer wanted by the people who donated the sperm and egg. Unwanted embryos are usually destroyed, but some people choose to donate their frozen embryos to scientific research. Human embryonic stem cells are special because they're *pluripotent,* meaning they have the potential to become every type of human cell. Although adult stem cells can be taken from fully developed humans, these stem cells are usually only able to become a limited number of cell types based on their tissue of origin. For example, hematopoietic stem cells from the bone marrow have the ability to divide and produce many different types of blood cells, but they typically can't produce nerve cells.

Because human embryonic stem cells are pluripotent, they're more valuable for some types of research. However, some people feel that destroying human embryos is morally wrong, no matter how early in development they are (embryos frozen after in vitro fertilization typically have about five to eight cells). This belief even led to a ban on federal funding for the development of new stem cell lines derived from human embryos. Those who support stem cell research argue that the potential benefits of stem cells for treating disease and saving lives make it imperative that scientists continue their research with embryonic stem cells.

Will this moral dilemma ever be solved? One potential solution is to make adult stem cells behave as if they're embryonic stem cells. A group of scientists has figured out a way to make this happen, but additional research is needed to determine whether these induced pluripotent stem cells are truly equivalent to embryonic stem cells.

## Chapter 9

# Engineering the Code: DNA Technology

### In This Chapter

- Figuring out how DNA technology works
- Getting to know the Human Genome Project
- Debating the benefits of genetically modified organisms

Gregor Mendel's pea plant experiments in the 1850s (which we fill you in on in Chapter 7) began a scientific exploration into the mysteries of heredity that continues to this day. After Mendel showed that traits were controlled by hereditary factors that pass from one generation to the next, scientists were determined to figure out the nature of these factors and how they were transmitted. They discovered the presence of DNA in cells, observed the movement of chromosomes during cell division, and conducted experiments demonstrating that DNA is in fact the hereditary material.

Almost 100 years after Mendel, James Watson and Francis Crick figured out that DNA was a double helix and proposed how it might be copied. Scientists deciphered the genetic code and explored how to work with it in the lab. During the last 40 years, scientists have developed an amazing array of tools to read DNA, copy it, cut it, sort it, and put it together in new combinations. The power of this DNA technology is so great that scientists have even determined the sequence of all the chromosomes in human cells as part of the Human Genome Project. A new world of human heredity is now open for exploration as scientists seek to understand the meanings hidden within human DNA — what they find out will likely change the way we see ourselves and our place in the world.

In this chapter, we get you acquainted with all that's involved in DNA technology and explore the ways scientists have mapped and manipulated the human genome and the genomes of other species.

# Understanding Just What's Involved in DNA Technology

For years the very structure of DNA made studying it rather challenging. After all, DNA is incredibly long and very tiny. Fortunately, the advent of *DNA technology,* the tools and techniques used for reading and manipulating the DNA code, has made working with DNA much easier. Scientists can even combine DNA from different organisms to artificially create materials such as human proteins or to give crop plants new characteristics. They can also compare different versions of the same gene to see exactly where disease-causing variations occur.

The following sections break down the various aspects of DNA technology so you can see how they all combine to provide a window into the very essence of existence.

## Cutting DNA with restriction enzymes

Scientists use *restriction enzymes,* essentially little molecular scissors, in the lab to cut DNA into smaller pieces so they can analyze and manipulate it more easily. Each restriction enzyme recognizes and can attach to a certain sequence on DNA called a *restriction site.* The enzymes slide along the DNA, and wherever they find their restriction site, they cut the DNA helix.

Figure 9-1 shows how a restriction enzyme can make a cut in a circular piece of DNA and turn it into a linear piece.

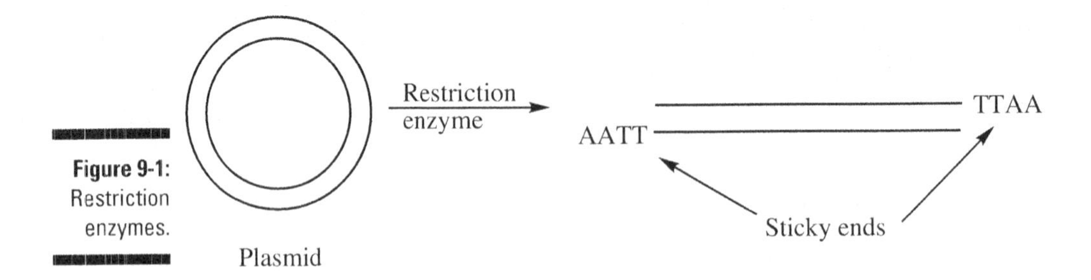


**Figure 9-1:** Restriction enzymes.

## Combining DNA from different sources

After DNA has been chopped into smaller, more workable bits (see the preceding section), scientists can combine pieces of DNA to change the characteristics of a cell. For example, they can put genes into crop plants to make them resistant to pesticides or to increase their nutritional value. This manipulation of a cell's genetic material in order to change its characteristics is called *genetic engineering*.

Because the DNA from all cells is essentially the same, scientists can even combine DNA from very different sources. For example, human DNA can be combined with bacterial DNA.

When a DNA molecule contains DNA from more than one source, it's called *recombinant DNA*.

If a recombinant DNA molecule containing bacterial and human genes is put into bacterial cells, the bacteria read the human genes like their own and begin producing human proteins that scientists can use in medicine and scientific research. Table 9-1 lists a few useful proteins that are made through genetic engineering.

**Table 9-1 Some Beneficial Genetically Engineered Proteins**

| | |
|---|---|
| Alpha-interferon | Used to shrink tumors and treat hepatitis |
| Beta-interferon | Used to treat multiple sclerosis |
| Human insulin | Used to treat people with diabetes as a safer alternative to pig insulin |
| Tissue plasminogen activator (tPA) | Given to patients who've just had a heart attack or stroke to dissolve the blockage that caused the attack |

Here's how scientists go about putting a human gene into a bacterial cell:

1. **First, they choose a restriction enzyme that forms sticky ends when it cuts DNA.**

   *Sticky ends* are pieces of single-stranded DNA that are complementary to other pieces of single-stranded DNA. Because they're complementary, the pieces of single-stranded DNA can stick to each other by forming hydrogen bonds (see Chapter 3 for more on bonds and DNA). For example, the sticky ends shown in Figure 9-1 have the sequences 5'AATT3' and 3'TTAA5'. A and T are complementary base pairs, so these ends can form hydrogen bonds and stick to each other.

2. **Next, they cut the human DNA and bacterial DNA with the same restriction enzyme.**

   When you cut bacterial DNA and human DNA with the same restriction enzyme, all the DNA fragments have the same sticky ends.

3. **Then they combine human DNA and bacterial DNA.**

   Because the two types of DNA have the same sticky ends, some of the pieces stick together.

4. **Finally, they use the enzyme DNA ligase to seal the sugar-phosphate backbone between the bacterial and human DNA.**

   DNA ligase forms covalent bonds between the pieces of DNA, sealing together any pieces that are combined.

## Using gel electrophoresis to separate molecules

Scientists separate molecules from cells such as DNA and proteins in order to study them. If a scientist wants to study just one protein, for example, she must separate that protein from the other proteins in the cell. When scientists want to separate DNA molecules from cells in order to look for relationships between DNA from two different sources, they use *gel electrophoresis,* which separates molecules based on their size and electrical charge.

Scientists conduct gel electrophoresis by inserting DNA molecules (see Figure 9-2a) into little pockets called *wells* within a slab of a gelatin-like substance (see Figure 9-2b). They then place the gel in a box, called an electrophoresis chamber, that's filled with a salty, electricity-conducting buffer solution.

The DNA molecules, which have a negative charge, are attracted to the gel box's positive electrode. When the scientists run an electrical current through the gel (see Figure 9-2c), the gel becomes like a racetrack for the DNA molecules, only instead of trying to cross the finish line, the DNA is trying to get to the positively charged end of the box.

When the power is turned off, all the DNA molecules stop where they are in the gel, and the scientists stain them. The stain sticks to the DNA, creating stripes called bands (see Figure 9-2d). Each *band* represents a collection of DNA molecules that are the same size and stopped in the same place in the gel.

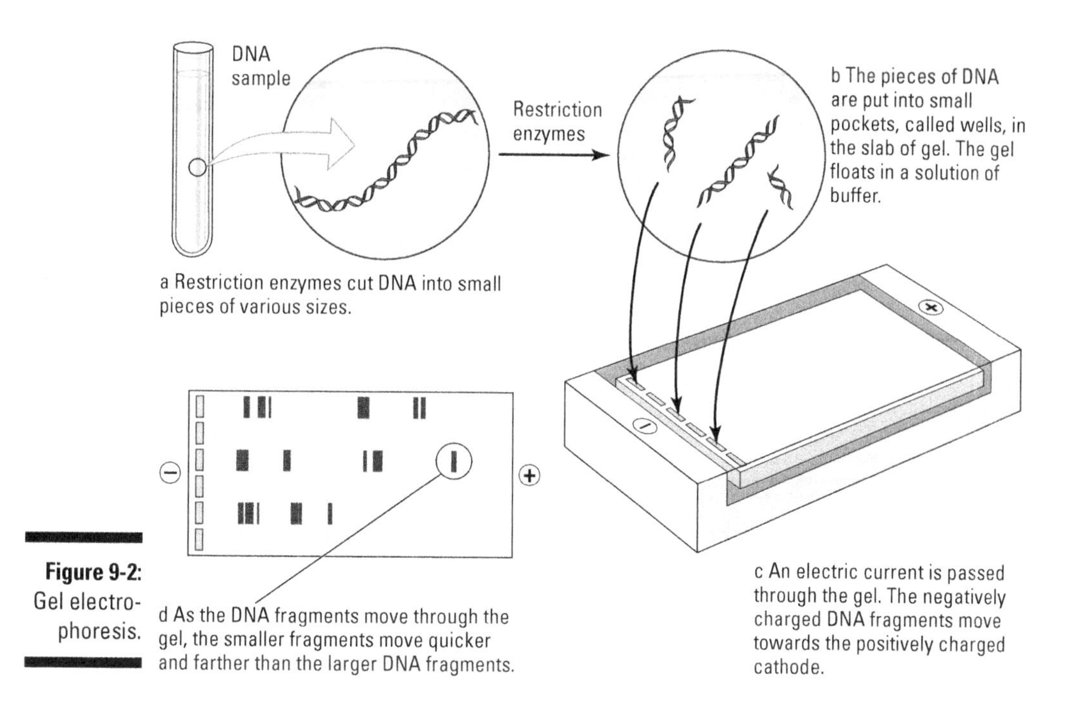


**Figure 9-2:** Gel electrophoresis.

# *Copying a gene with PCR*

The *polymerase chain reaction* (PCR) is a process that can turn a single copy of a gene into more than a billion copies in just a few hours. It gives medical researchers the ability to make many copies of a gene whenever they want to genetically engineer something (see the earlier "Combining DNA from different sources" section for more on genetic engineering).

PCR targets the gene to be copied with *primers,* single-stranded DNA sequences that are complementary to sequences next to the gene to be copied.

To begin PCR, the DNA sample that contains the gene to be copied is combined with thousands of copies of primers that frame the gene on both sides (see Figure 9-3). DNA polymerase uses the primers to begin DNA replication and copy the gene (refer to Chapter 6 for more on DNA replication). The basic steps of PCR are repeated over and over until you have billions of copies of the DNA sequence between the two primers.

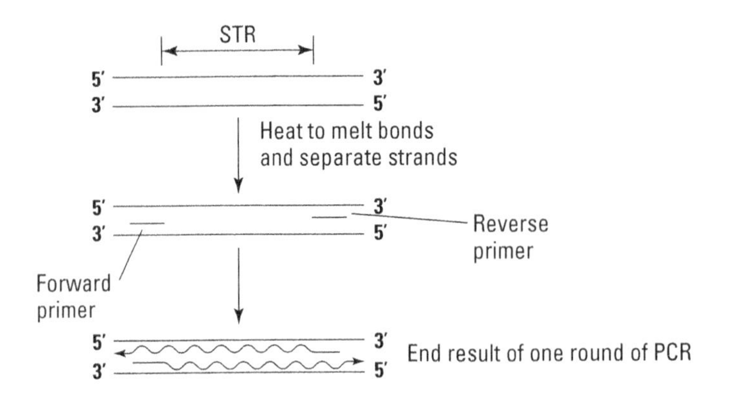


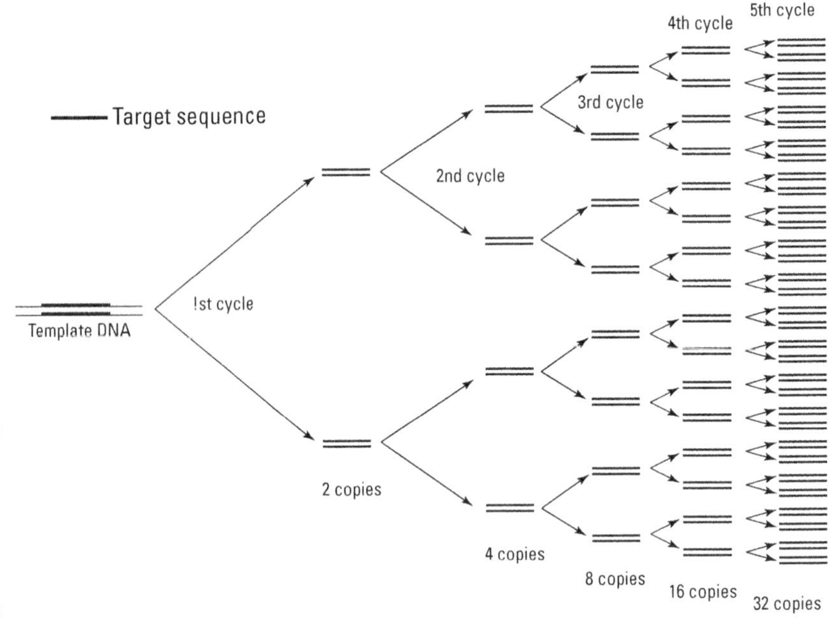


**Figure 9-3:** The polymerase chain reaction.

PCR works a little like chain e-mails. If you get a chain e-mail and send it on to two friends, who each send it on to two of their friends, and so on, pretty soon everyone has seen the same e-mail. In PCR, first a DNA molecule is copied, then the copies are copied, and so on, until you have 30 billion copies in just a few hours.

## Reading a gene with DNA sequencing

*DNA sequencing,* which determines the order of nucleotides in a DNA strand, allows scientists to read the genetic code so they can study the normal versions of genes. It also allows them to make comparisons between normal versions of a gene and disease-causing versions of a gene. After they know the order of nucleotides in both versions, they can identify which changes in the gene cause the disease.

As you can see in Figure 9-4, DNA sequencing uses a special kind of nucleotide, called ddNTP (short for dideoxyribonucleotide triphosphate). Regular DNA nucleotides and ddNTPs are somewhat similar, but the ddNTPs are different enough that they stop DNA replication. When a ddNTP is added to a growing chain of DNA, DNA polymerase can't add any more nucleotides. DNA sequencing uses this chain interruption to determine the order of nucleotides in a strand of DNA.

Most DNA sequencing done today is *cycle sequencing,* a process that creates partial copies of a DNA sequence, all of which are stopped at different points. After the partial copies are made, scientists load them into a machine that uses gel electrophoresis to put the copies into order by size. As the partial sequences pass through the machine, a laser reads a fluorescent tag on each ddNTP, noting the DNA sequence.

Although both cycle sequencing and PCR copy DNA, the two processes are quite different. Cycle sequencing uses both normal DNA nucleotides and ddNTPs and makes only partial copies of DNA that are all slightly different. PCR uses only normal DNA nucleotides and makes exact copies of a DNA sequence.

# Mapping the Genes of Humanity

The *Human Genome Project* (HGP) was a hugely ambitious task to determine the nucleotide sequence of all the DNA in a human cell. To give you an idea just how ambitious this project was, when it was first proposed in 1985, the pace of DNA sequencing was so slow that it would've taken 1,000 years to sequence the 24 unique human chromosomes (22 autosomes plus X and Y, as explained in Chapter 6). Fortunately, scientists cooperated and technology improved during the project, allowing the majority of the human genome to be sequenced by 2003. (A *genome* is the total collection of genes in a species.)

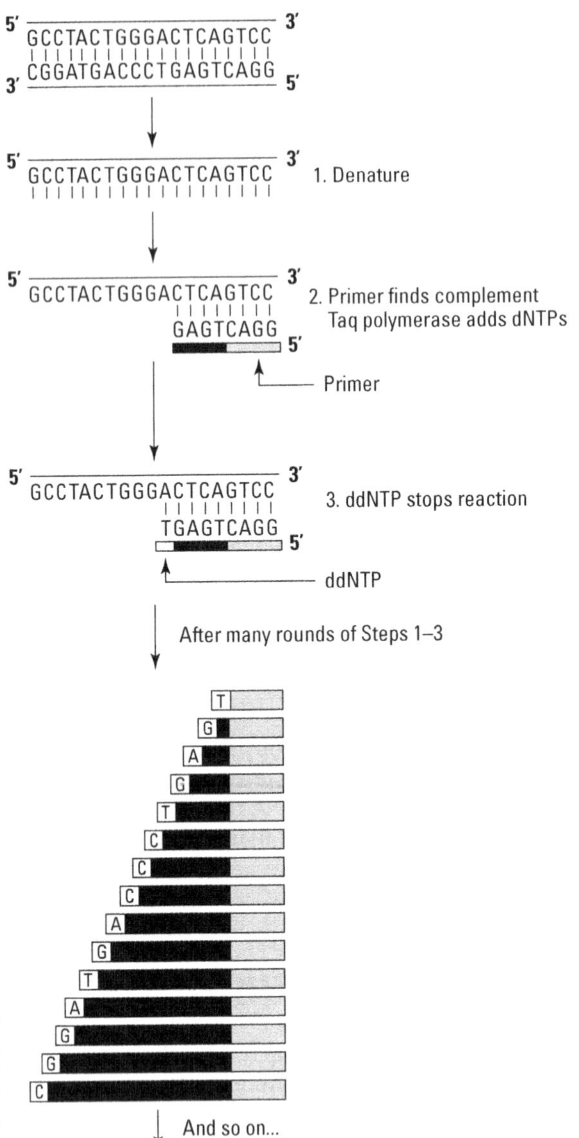


**Figure 9-4:** DNA sequencing.

If you're wondering why the HGP is a big deal, think of it this way. If you were a researcher and you wanted to study a specific human gene, first you'd have to know what chromosome it "lived" on. The map of nucleotide sequences created by the HGP is a huge step forward in providing the "address" of each human gene. Armed with a roadmap of where every gene is located, researchers can turn their attention toward making good use of that information, like seeking out the genes that cause disease.

The HGP and the technological advances that came along with it resulted in many other current and potential benefits to society, including

- Drugs designed to best treat an individual person with minimal side effects
- Earlier detection of disease
- Exploration of microbial genomes for identification of species that can be used to produce new biofuels or clean up pollution
- Comparison of DNA from crime scenes to that of suspects in order to help determine likely guilt or innocence
- Study of the evolutionary relationships of life on Earth

## Designer people?

As scientists delve deeper into the mysteries of the human genome, complex issues about the nature of humanity and the rights of individuals are generating fear and discussion. The questions people are pondering include the following:

- If a person's genome can be read, should insurance companies or employers be allowed to know about increased risks for disease?
- Should people be allowed to screen their embryos to prevent diseased children from being born or to select only the ones with desired characteristics?
- Should only people with "good" genetic stock be allowed to have children?

These questions may seem far-fetched, but history shows that they aren't. After the work of Gregor Mendel became known and people understood that genes determine human traits, a group of people founded the eugenics movement. People who joined the eugenics movement believed that certain human characteristics were more desirable than others; they wanted to control human breeding to "better" the human race. These ideas were carried to extremes by the Nazis during World War II, leading to the extermination of whole groups of people who were judged to be undesirable.

Many people are afraid that greater knowledge about the human genome will again be used to harm. The reality is that knowledge brings power, and power can be used for good or evil. Scientists seek knowledge, but they can't always control how that knowledge is used. What they can do, however, is be part of the conversation as people explore these complex moral issues. In fact, one of the primary goals of the Human Genome Project was to "address the ethical, legal, and social issues that may arise from the project." That this statement was included in the primary goals of the project is a pretty big deal because scientific proposals are usually just about the science.

# Genetically Modifying Organisms

*Genetically modified organisms* (abbreviated as GMOs and sometimes called *genetically engineered organisms* or *transgenic organisms*) contain genes from other species that were introduced using recombinant DNA technology (see the earlier "Combining DNA from different sources" section for more on this type of DNA technology). GMOs are a hot topic these days due to the controversy surrounding genetically modified crop plants and farm animals. The sections that follow take a look at both sides of the "Are GMOs good or bad?" debate.

## Why GMOs are beneficial

Genetic modification has its upsides. It not only makes growing crops easier but it can also boost the profitability of those crops. And it may even help improve human health. Here are some specific scenarios that illustrate how GMOs can be beneficial:

- **If crop plants are given genes to resist herbicides and pesticides, then a farmer can spray the fields with those chemicals, killing only the weeds and pests, not the crop plant.** This is much easier and less time-consuming than labor-intensive weeding. It can also increase crop yields and profits for the farmer.
- **If crop plants or farm animals raised for human consumption are given genes to improve their nutrition, people could be healthier.** Improved nutrition in crop plants could be a huge benefit in poor countries where malnutrition stunts the growth and development of children, making them more susceptible to disease. One of the most famous examples of improved nutrition through genetic engineering is the creation of "golden rice" — rice that has been engineered to make increased amounts of a nutrient that's necessary for vitamin A production. According to the World Health Organization, vitamin A deficiencies cause 250,000 to 500,000 children to go blind each year. The company that produced golden rice is giving the rice to poor countries for free so they can grow it for themselves and make it available to people who need it.
- **If farm animals raised for human consumption are given genes to increase their yield of meat, eggs, and milk, then more food may be available for the growing human population, and these greater yields may also increase profits for farmers.** Currently, many dairy cows are given recombinant bovine growth hormone (rBGH) to increase their

milk production. BGH is a normal growth hormone found in cows; rBGH is a slightly altered version that's produced by genetically engineered bacteria. When rBGH is given to cows, the animals' milk production increases by 10 to 15 percent.

## Why GMOs cause concern

What make GMOs so controversial are the ethical concerns. The list of concerns surrounding genetic modification is long and so serious that some countries in the European Union have banned the sale of foods containing products from GMOs. The concerns expressed include the following:

- **The use of GMOs in agriculture unfairly benefits big agricultural companies and pushes out smaller farmers.** Companies that produce seeds for genetically engineered crops retain patents on their products. The prices on these seeds can be much higher than for traditional crops, giving large agricultural companies an advantage in the marketplace. This issue is particularly worrisome when large agricultural companies from rich nations start competing in the global economy with smaller farmers from poor countries.
- **The use of GMOs in agriculture encourages unsound environmental practices and discourages best farming practices.** Farmers who plant crops engineered for pesticide or herbicide resistance use chemicals rather than manual labor to control weeds and pests. Not only do these pesticides and herbicides affect the health of plants and animals living in the area around farms but they can also get into the drinking water and possibly affect human health. Also, large-scale plantings of just a few species of plants decrease the genetic diversity in food species and put the food supply at risk for large-scale catastrophes should one of the crop species fail.
- **Animals that are engineered to produce more milk, eggs, or meat may be at greater risk for health problems.** Cows treated with rBGH to increase milk production get more infections in their milk ducts and have to be treated with antibiotics more often. Overuse of antibiotics is a human health concern because it reduces the effectiveness of antibiotics on bacteria that cause human infections.
- **Cross-pollination between genetically engineered plants and wild plants can spread resistant genes into wild plants.** Farmers can put up fences, but wind blows all over the place. If a crop plant that contains a gene for herbicide resistance can pollinate a wild plant, then the wild plant could pick up that gene, creating a weed species that can't be controlled.

- **Increased levels of bovine hormones in dairy products may have effects on humans who drink the milk.** When rBGH is injected into cows to pump up their milk production, the levels of IGF-1 (an insulin-like protein) in their bodies and milk increase. Human bodies also make IGF-1, and increased levels of this hormone have been found in patients with some types of cancer. People are worried that increased IGF-1 in milk from hormone-treated cows may put them at greater risk for cancer, but no clear link has yet been found between IGF-1 in milk and human cancer.
- **Genetic modification of foods may introduce allergens into foods, and labeling may not be sufficient to protect the consumer.** People who have food allergies have to be very careful about which foods they eat. However, if foods contain products from GMOs, it's possible that the introduced genes produced a product that's not indicated on the food label.
- **Fear of "unnatural" practices and new technologies makes people afraid of GMOs and lowers their value in the marketplace.** Some people see humans as becoming out of balance with the rest of nature and think we need to slow down and try to leave less of a footprint on the world. For some, this belief includes rejecting technology that alters organisms from their natural state.

## DNA fingerprinting

Although the genomes of human beings are extremely similar — 99.9 percent similar, to be exact — people do have some unique sequences that make them, well, unique. In fact, all people have a *DNA fingerprint,* which is their personal genetic profile. This DNA fingerprint is found in 13 areas of the human genome that tend to be very different from person to person. By looking at all 13 areas, not just 1 or 2 of them, scientists decrease the chances of getting a random match between two samples.

DNA fingerprinting is very helpful in *forensics,* the science that collects and interprets physical evidence for legal purposes. However, solving crimes isn't the only use for DNA fingerprinting. Some of its other interesting uses include

- Identifying the bodies of victims of disasters or massacres when the bodies themselves are unrecognizable
- Testing paternity, maternity, and other familial relationships
- Examining foods for evidence of genetically modified organisms (GMOs)
- Detecting harmful bacteria in food and water
- Making genetic pedigrees of crop plants and different breeds of animals
- Revealing the presence of endangered and protected species among materials taken from poachers

# Part III

# It's a Small, Interconnected World

## In this part . . .

Life on Earth comes in all forms, shapes, and sizes. It inhabits almost every environment that people have explored (and even some they haven't), from the deep, dark caves of the world's oceans to the hot springs of Yellowstone National Park. All the wonderfully diverse organisms of this planet are connected to each other in fundamental cycles of energy and matter transfer. Prepare to explore the diversity (and interconnectedness) of life on Earth in this part.

Organisms that are successful in obtaining what they need reproduce, creating offspring in their own image. As the Earth changes over time, the requirements for success change, causing shifts in the populations of organisms. In this part, we also explain the connections between living things in space and time.

# Chapter 10

# Biodiversity and Classification

### In This Chapter

- Exploring biodiversity
- Getting to know the various forms of life on Earth
- Looking at how all living things are organized into groups

As human beings, we're deeply connected to the living world around us. Yet as we transform the world in order to meet our needs, we're changing it in ways that make it less hospitable to other species.

This chapter is designed to help you understand why biodiversity is so important to the future of humanity on planet Earth. It's your chance to get acquainted with the diversity of life around you and discover how biologists organize all of those diverse life-forms into a specific classification system.

## Biodiversity: Recognizing How Our Differences Make Us Stronger

The diversity of living things on Earth is referred to as *biodiversity.* Almost everywhere biologists have looked on this planet — from the deepest, darkest caves to the lush Amazonian rain forests to the depths of the oceans — they've found life. In the deepest, darkest caves where no light ever enters, bacteria obtain energy from the metals in the rocks. In the Amazonian rain forest, plants grow attached to the tops of trees, collecting water and forming little ponds in the sky that become home to insects and tree frogs. In the deep oceans, blind fish and other animals live on the debris that drifts down to them like snow from the lit world far above. Each of these environments presents a unique set of resources and challenges, and life on Earth is incredibly diverse due to the ways in which organisms have responded to these challenges over time.

The following sections not only clue you in to the reasons why biodiversity is so important and how human actions are harming it but also how human actions can protect biodiversity moving forward.

## Valuing biodiversity

Most people choose to live with species that are a lot like them — other people, dogs, cats, and farm animals, for example. Living things that are different, such as slugs, bugs, and bacteria, may seem annoying, gross, weird, or even scary. On the other hand, some people are fascinated by the diversity of life on Earth and make a habit of watching nature shows on television; visiting zoos, aquariums, and botanical preserves; or traveling to different places on Earth to see unique organisms in their natural habitats.

Whether or not you appreciate the diversity of life on Earth, biodiversity is important — and worth valuing — for the following reasons:

- **The health of natural systems depends on biodiversity.** Scientists who study the interconnections between different types of living things and their environments (see Chapter 11) believe that biodiversity is important for maintaining balance in natural systems. Each type of living thing plays a role in its environment, and the loss of even one species can have widespread effects.
- **Many economies rely upon natural environments.** A whole industry called *ecotourism* has grown up around tour guides leading people on trips through natural habitats and explaining the local biology along the way.
- **Human medicines come from other living things.** For example, the anticancer drug taxol was originally obtained from the bark of the Pacific yew, and the heart medicine digitalin comes from the foxglove plant.
- **Biodiversity adds to the beauty of nature.** Natural systems have an aesthetic value that's pleasing to the eye and calming to the mind in today's technologically driven world.

## Surveying the threats posed by human actions

As the human population grows and uses more and more of the Earth's resources (head to Chapter 11 for the scoop on human population growth), the populations of other species are declining as a direct result. Following are the ways in which human actions pose major threats to biodiversity:

- **Development is reducing the size of natural environments.** People need places to live and farms to raise food. In order to meet these needs, they burn rain forests, drain wetlands, cut down forests, pave over valleys, and plow up grasslands. Whenever people convert land for their own use, they destroy the habitats of other species, causing habitat loss. Even if some natural habitat remains, those patches become small and scattered. This *habitat fragmentation* has the biggest impact on large animals, such as mountain gorillas and tigers, that need big habitats in which to roam.
- **Unnatural, human-produced wastes are polluting the air and water.** Automobiles and factories burn gasoline and coal, releasing pollution into the air. Metals from mining and chemicals from factories, farms, and homes get into groundwater. After pollution enters the air and water, it travels around the globe and can hurt multiple species, including humans.
- **The overharvesting of species to provide food and other materials for human consumption is driving some species to near extinction.** Because they can reproduce, living things such as trees and fish are considered renewable resources. However, if people harvest these resources faster than they can replace themselves, the numbers of individual trees and fish decline. If too few members of a species remain, then survival of that species becomes very unlikely. Case in point: Many important fisheries, like the Great Banks off the coast of Newfoundland, have actually crashed, which means the population of fish declined to a point that the area is no longer fishable and may never recover.
- **Human movements around the globe sometimes carry species into new environments.** An *introduced species* is a foreign species that's brought into a new environment. Introduced species that are very aggressive and take over habitats are called *invasive species.* Invasive species often have a large environmental impact and cause the numbers of *native species* (organisms belonging to a particular habitat) to decline; they can also attack crop plants and cause human diseases. One example of an invasive species is water hyacinth, a plant that was introduced into the American South during the 1884 exposition in New Orleans. Water hyacinth spread throughout the waterways of the southeastern

United States where it choked rivers and lakes with huge masses of floating plants, slowing water flow, blocking the light for aquatic species, and reducing biodiversity. Maintenance crews in modern-day Florida work constantly to try and weed out water hyacinth in order to keep the state's rivers and lakes usable for recreation and other species.

## Exploring the extinction of species

The combined effects of all the various human actions in Earth's ecosystems are reducing the planet's biodiversity. In fact, the rate of extinctions is increasing along with the size of the human population. No one knows for certain how extensive the loss of species due to human impacts will ultimately be, but there's no question that human practices such as hunting and farming have already caused numerous species to become extinct.

Many scientists believe Earth is experiencing its sixth *mass extinction,* a certain time period in geologic history that shows dramatic losses of many species. (The most famous mass extinction event is the one that occurred about 65 million years ago and included the extinction of the dinosaurs.) Scientists theorize that most of the past mass extinctions were caused by major changes in Earth's climate and that the current extinctions (most recently including black rhinos, Zanzibar leopards, and golden toads) began as a result of human land use but may increase as a result of global warming.

The loss in biodiversity that's currently happening on Earth could have effects beyond just the loss of individual species. Living things are connected to each other and their environment in how they obtain food and other resources necessary for survival. If one species depends on another for food, for example, then the loss of a prey species can cause a decline in the predator species.

The sections that follow introduce you to two classifications of species that biologists are keeping an eye on when it comes to questions of extinction.

### Keystone species

Some species are so connected with other organisms in their environment that their extinction changes the entire composition of species in the area. Species that have such great effects on the balance of other species in their environment are called *keystone species.* As biodiversity decreases, keystone species may die out, causing a ripple effect that leads to the loss of many more species. If biodiversity gets too low, then the future of life itself becomes threatened.

An example of a keystone species is the purple seastar, which lives on the northwest Pacific coast of the United States. Purple seastars prey on mussels in the intertidal zone. When the seastars are present, they keep the mussel population in check, allowing a great diversity of other marine animals to live in the intertidal zone. If the seastars are removed from the intertidal zone, however, the mussels take over, and many species of marine animals disappear from the environment.

### Indicator species

One way biologists can monitor the health of particular environments and the organisms that live in them is by measuring the success of *indicator species,* species whose presence or absence in an environment gives information about that environment.

In the Pacific Northwest region of the United States, the health of old-growth forests is measured by the success of the northern spotted owl, a creature that can only make its home and find food in mature forests that are hundreds of years old. As logging decreases the number and size of these old forests, the number of spotted owls has declined, thereby making the number of spotted owls an indicator of the health, or even the existence, of old-growth forests in the Pacific Northwest. Of course, old-growth forests aren't just home to spotted owls — they shelter a rich diversity of living things including plants, such as sitka spruce and Western hemlock, and animals, such as elk, bald eagles, and flying squirrels. Old-growth forests also perform important environmental functions such as preventing erosion, floods, and landslides; improving water quality; and providing places for salmon to spawn. If old-growth forests become extinct in the Pacific Northwest, the effects will be far reaching and have many negative impacts on the people and other species in the area.

## Protecting biodiversity

Biodiversity increases the chance that at least some living things will survive in the face of large changes in the environment, which is why protecting it is crucial. So what can people do to protect biodiversity and the health of the environment in the face of the increasing demands of the human population? No one has all the answers, but here are a few ideas worth trying:

- Keep wild habitats as large as possible and connect smaller ones with *wildlife corridors* (stretches of land or water that wild animals travel as they migrate or search for food) so organisms that need a big habitat to thrive can move between smaller ones.

- Use existing technologies and develop new ones to decrease human pollution and clean up damaged habitats. Technologies that have minimal effects on the environment are called *clean* or *green technologies.* Some businesses are trying to use these technologies in order to reduce their impact on the environment.
- Strive for sustainability in human practices including manufacturing, fishing, logging, and agriculture. Something that's *sustainable* meets current human needs without decreasing the ability of future generations to meet their needs.

The Great Law of the Iroquois says that "people must consider the impact of their actions not just on the current generation, but on future generations that aren't yet born." This law is often quoted as: "In our every deliberation, we must consider the impact of our decisions on the next seven generations." A generation spans roughly 25 years, so if people follow the rule of the Iroquois, they must consider the effects their actions will have 175 years from now.

- Regulate the transport of species around the world so that species aren't introduced into foreign habitats. This includes being careful about the transport of not-so-obvious species. For example, ships traveling from one port to another are often asked to empty their ballast water offshore so they don't accidentally release organisms from other waters into their destination harbors.

# Meet Your Neighbors: Looking at Life on Earth

Life on Earth is incredibly diverse, beautiful, and complex. Heck, you could spend a lifetime exploring the microbial universe alone. The deeper you delve into the living world around you, the more you can appreciate the similarities between all life on Earth — and be fascinated by the differences. The following sections give you a brief introduction to the major categories of life on Earth; we encourage you to check them out for yourself.

## Unsung heroes: Bacteria

Most people are familiar with disease-causing bacteria such as *Streptococcus pyogenes, Mycobacterium tuberculosis,* and *Staphylococcus aureus.* Yet the vast majority of bacteria on Earth don't cause human diseases. Instead, they play important roles in the environment and health of living things, including humans. Photosynthetic bacteria make significant contributions to planetary food and oxygen production (see Chapter 5 for more on photosynthesis), and *E. coli* living in your intestines make vitamins that you need to stay healthy. So when you get down to it, plants and animals couldn't survive on Earth without bacteria.

Generally speaking, bacteria range in size from 1 to 10 micrometers in length and are invisible to the naked eye. Along with being nucleus-free, they have a genome that's a single circle of DNA. They reproduce asexually by a process called binary fission. Some bacteria move about by secreting a slime that glides over the cell's surface, allowing it to slide through its environment. Others have *flagella* (little whiplike appendages made of protein) that they swish around to swim through their watery homes.

Bacteria have many ways of getting the energy they need for growth and various strategies for surviving in extreme environments. Their great metabolic diversity has allowed them to colonize just about every environment on Earth.

## A bacteria impersonator: Archaeans

*Archaeans* are prokaryotes, just like bacteria. In fact, you can't tell the difference between the two just by looking, even if you look very closely using an electron microscope, because they're about the same size and shape, have similar cell structures, and divide by binary fission.

Until the 1970s, no one even knew that archaeans existed; up to that point, all prokaryotic cells were assumed to be bacteria. Then, in the 1970s, a scientist named Carl Woese started doing genetic comparisons between prokaryotes. Woese startled the entire scientific world when he revealed that prokaryotes actually separated into two distinct groups — bacteria and archaea — based on sequences in their genetic material.

The first archaeans were discovered in extreme environments (think salt lakes and hot springs), so they have a reputation for being *extremophiles* (*-phile* means "love," so *extremophiles* means "extreme-loving"). Since their initial discovery, however, archaeans have been found everywhere scientists have looked for them. They're happily living in the dirt outside your home right now, and they're abundant in the ocean.

Because archaeans were discovered fairly recently, scientists are still learning about their role on planet Earth, but so far it looks like they're as abundant and successful as bacteria.

## A taste of the familiar: Eukaryotes

Unless you're a closet biologist, you're probably most familiar with life in eukaryotic form because you encounter it every day. As soon as you step outside, you can find a wealth of plants and animals (and maybe even a mushroom or two if you look around a little).

On the most fundamental level, all eukaryotes are quite similar. They share a common cell structure with nuclei and organelles (we cover eukaryotic cells in greater detail in Chapter 4), use many of the same metabolic strategies (explained in Chapter 5), and reproduce either asexually by mitosis or sexually by meiosis (both of which are covered in Chapter 6).

Despite these similarities, we bet you still feel that you're pretty different from a carrot. You're right to feel that way. The differences between you and a carrot are what separate you into two different kingdoms. In fact, enough differences exist between eukaryotes to separate them into four different kingdoms:

- **Animalia:** Animals are organisms that begin life as a cell called a *zygote* that results from the fusion of a sperm and an egg and then divides to form a hollow ball of cells called a *blastula.* If you're wondering when the fur, scales, and claws come into play, these familiar animal characteristics get factored in much later, at the point when animals get divided up into phyla, families, and orders (see the "Organizing life into smaller and smaller groups" section later in this chapter for more on these groupings). Although the animal kingdom contains familiar animals such as dogs, cats, lizards, birds, and fish, the defining characteristic of animals must be true for all members of the kingdom — including slugs, worms, and sea sponges.
- **Plantae:** Plants are photosynthetic organisms that start life as embryos supported by maternal tissue. This definition of plants includes all the plants you're familiar with: pine trees, flowering plants (including carrots), grasses, ferns, and mosses. All plants have cells with cell walls made of cellulose. They reproduce asexually by mitosis (described in Chapter 6), but they can also reproduce sexually. (Flip to Chapter 20 for more on plant structures and life cycles.)

  The definition of plants, which specifies a stage where an embryo is supported by maternal tissue, excludes most of the algae, like seaweed, found on Earth. Algae and plants are so closely related that many people include algae in the plant kingdom, but many biologists draw the line at including algae in the plant kingdom.

- **Fungi:** Fungi may look a bit like plants, but they aren't photosynthetic. They get their nutrition by breaking down and digesting dead matter. Their cells have walls made of *chitin* (a strong, nitrogen-containing polysaccharide), and they don't produce swimming cells during their life cycle. Kingdom Fungi includes mushrooms, molds that you see on your bread and cheese, and many rusts that attack plants. Yeast is also a member of kingdom Fungi even though it grows differently (most fungi grow as filaments, but yeast grows as little oval cells).
- **Protista:** Kingdom Protista is defined as everything else that's eukaryotic. Seriously. Biologists have studied animals, plants, and fungi for a long time and defined them as distinct groups long ago. But many, many, eukaryotes don't fit into these three kingdoms. A whole world of microscopic protists exists in a drop of pond water. The protists are so diverse that some biologists think they should be separated into as many as 11 kingdoms of their own. But so far no one has pushed to make that happen, which is certainly good news for you because we bet you don't want to memorize the names of 15 kingdoms of eukaryotes.

## What about viruses?

Here's a riddle for you: What has genetic material, exists by the billions, functions like a living parasite, but isn't truly a living thing? A virus! That's right, those nasty little bugs that cause diseases, ranging from the human immunodeficiency virus (HIV) and food poisoning to the common cold and even some forms of cancer, may be the world's most efficient parasites. But, in the strictest sense of the word, viruses aren't really alive because they can't reproduce outside of a host cell.

Unlike living things, viruses aren't made of cells. They're just very tiny pieces of DNA or RNA covered with protein as protection. Because they're so small — a fraction of the size of bacteria — you can't see them, even with the aid of a light microscope.

No one really knows how viruses evolved. Some biologists think they were originally intracellular parasites that got so good at what they did — being parasitic, that is — that they were able to survive with nucleic acid alone. Others think viruses are cellular escapees, genes that ran away from home but can't replicate until they return to a specific kind of host cell. Still others think viruses may represent an offshoot of life from its very beginning, before cells evolved. For the details on how viruses attack cells and reproduce, head to Chapter 17.

# Climbing the Tree of Life: The Classification System of Living Things

Much like you'd draw a family tree to show the relationships between your parents, grandparents, and other members of your family, biologists use a *phylogenetic tree* (a drawing that shows the relationships between a group of organisms) to represent the relationships between living things. This "tree of life" allows them to categorize all the diverse organisms that call planet Earth home and organize them into manageable classifications.

Although you probably know how your family members are related to each other, biologists have to use clues to figure out the relationships between living things. The types of clues they use to figure out these relationships include

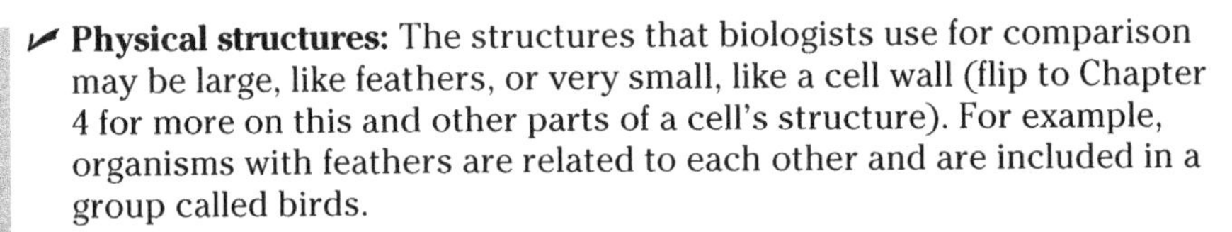

- **Physical structures:** The structures that biologists use for comparison may be large, like feathers, or very small, like a cell wall (flip to Chapter 4 for more on this and other parts of a cell's structure). For example, organisms with feathers are related to each other and are included in a group called birds.

  Biologists consider reproductive structures to be especially important for determining relationships.

- **Chemical components:** Some organisms produce unique chemicals. Bacteria, for example, are the only cells that make a hybrid sugar-protein molecule called peptidoglycan. If biologists encounter an organism that produces this molecule, they can safely group it in with other bacteria.
- **Genetic information:** An organism's genetic code determines its traits, so by reading the genetic code in DNA, biologists can go right to the source of differences between species. Even organisms that seem incredibly different, like you and the bacteria *E. coli,* have some traits in common. For example, all cells on Earth contain ribosomes for making proteins, which means biologists can read a gene that has the code for one of your ribosomal proteins and compare it to the gene that has the code for one of *E. coli*'s ribosomal proteins.

The more characteristics two organisms have in common with each other, the more closely related they are. Characteristics that organisms have in common are called *shared characteristics.*

Based on structural, cellular, biochemical, and genetic characteristics, biologists can classify life on Earth into groups that reflect the evolutionary history of the planet. That history indicates that all life on Earth began from one original universal ancestor after the Earth formed 4.5 billion years ago. All the diversity of life that exists today is related because it's descended from that original ancestor.

The next sections break down the various classifications of living things and explain how each one gets its own unique scientific name.

## Mastering the domains

You can interpret the degree of relationship between two organisms by looking at their positions on a phylogenetic tree.

- **Each organism or group being compared has a branch on the tree.** One branch represents all the animals, and another branch represents all the plants.
- **The smaller the distance between two groups, the closer the relationship between them.** The distance between animals and plants is a great deal less than the distance between animals and any of the bacteria, so animals are much more closely related to plants than they are to bacteria.
- **If two branches meet at a common point, then the groups represented by those branches evolved from a common ancestor.** Groups with a common point that's far away separated further back in time than groups with a common point that's near.

When biologists used genetic information to compare all life on Earth, they discovered that living things fall into three main groups called *domains.* The three domains of life are

- **Bacteria:** Consisting mostly of single-celled organisms, bacteria are prokaryotic, meaning they lack a nuclear membrane around their DNA (refer to Chapter 4 for the details on prokaryotic cells). Most bacteria have a cell wall made of peptidoglycan.
- **Archaea:** These are single-celled, prokaryotic organisms. In addition to their genetic differences from bacteria, archaeans have some chemical differences, including the fact that their cell walls are never made of peptidoglycan.
- **Eukarya:** Organisms in the Eukarya domain may be single-celled or multicellular; either way, their cells are eukaryotic, meaning they have a nuclear membrane around their DNA (see Chapter 4 for more on eukaryotic cells). This domain contains familiar organisms such as animals, plants, mushrooms, and seaweed.

## Organizing life into smaller and smaller groups

Being able to categorize the three largest, and most distantly related, groups of living things on Earth into domains (as explained in the preceding section) is great, but biologists need smaller groups to work with in order to determine how similar different types of organisms are. Hence the creation of the *taxonomic hierarchy,* a naming system that ranks organisms by their evolutionary relationships. Within this hierarchy, living things are organized into the largest, most-inclusive group down to the smallest, least-inclusive group.

The taxonomic hierarchy is as follows, from largest to smallest. (Note that organisms are placed into each category based on similarities within that particular group of organisms. Whatever characteristics are used to define a category must be shared by all organisms placed into that category.)

- **Domain:** Domains group organisms by fundamental characteristics such as cell structure and chemistry. For example, organisms in domain Eukarya are separated from those in the Bacteria and Archaea domains based on whether their cells have a nucleus, the types of molecules found in the cell wall and membrane, and how they go about protein synthesis.
- **Kingdom:** Kingdoms group organisms based on developmental characteristics and nutritional strategy. For example, organisms in the animal kingdom (Animalia) are separated from those in the plant kingdom (Plantae) because of differences in the early development of these organisms and the fact that plants make their own food by photosynthesis whereas animals ingest their food. (Kingdoms are most useful in domain Eukarya because they're not well defined for the prokaryotic domains.)
- **Phylum:** Phyla separate organisms based on key characteristics that define the major groups within the kingdom. For example, within kingdom Plantae, flowering plants (Angiophyta) are in a different phylum than cone-bearing plants (Coniferophyta).
- **Class:** Classes separate organisms based on key characteristics that define the major groups within the phylum. For example, within phylum Angiophyta, plants that have two seed leaves (dicots, class Magnoliopsida) are in a separate class than plants with one seed leaf (monocots, class Liliopsida).
- **Order:** Orders separate organisms based on key characteristics that define the major groups within the class. For example, within class

Magnoliopsida, nutmeg plants (Magnoliales) are put in a different order than black pepper plants (Piperales) due to differences in their flower and pollen structure.

- **Family:** Families separate organisms based on key characteristics that define the major groups within the order. For example, within order Magnoliales, buttercups (Ranunculaceae) are in a different family than roses (Rosaceae) due to differences in their flower structure.
- **Genus:** Genera separate organisms based on key characteristics that define the major groups within the family. For example, within family Rosaceae, roses *(Rosa)* are in a different genus than cherries *(Prunus)* thanks to differences in their flower structure.
- **Species:** Species separate eukaryotic organisms based on whether they can successfully reproduce with each other. You can walk through a rose garden and see many different colors of China roses *(Rosa chinensis)* that are all considered one species because they can reproduce with each other.

Think of how biologists organize living things like how you might organize your clothing. In your first round of organizing, you might make groups of pants, shirts, socks, and shoes. From there, you might go into the shirt group and organize your shirts into smaller groups, such as short-sleeved versus long-sleeved shirts. Then perhaps you'd organize them by type of fabric, then color, and so on. At some point, you'd have very small groups with very similar articles of clothing — perhaps a group of two short-sleeved, button-down, blue shirts, for example. All of your clothing would be organized in a hierarchy, from the big category of clothing all the way down to the small category of short-sleeved, button-down, blue shirts.

People have lots of funny sayings to help them remember the taxonomic hierarchy. Our favorite, because it's so hard to forget, is *Dumb Kids Playing Chase On Freeways Get Squished.* The first letter of each word in the sentence represents the first letter of a category in the taxonomic hierarchy. If you don't like this particular saying, just search the Internet for "taxonomic hierarchy mnemonic," and you'll find many more.

All life on Earth is related, but relative position within the taxonomic hierarchy demonstrates the degree of that relationship. For instance, you and a carrot are both in domain Eukarya, so you definitely have some things in common, but you have more characteristics in common with organisms that are part of the animal kingdom.

Table 10-1 compares the classification, or *taxonomy,* of you, a dog, a carrot, and *E. coli.*

**Table 10-1 Comparing the Taxonomy of Several Species**

| *Taxonomic Group* | *Human* | *Dog* | *Carrot* | *E. coli* |
|---|---|---|---|---|
| **Domain** | Eukarya | Eukarya | Eukarya | Bacteria |
| **Kingdom** | Animalia | Animalia | Plantae | Eubacteria |
| **Phylum** | Chordata | Chordata | Angiophyta | Proteobacteria |
| **Class** | Mammalia | Mammalia | Magnoliopsida | Gammaproteo-bacteria |
| **Order** | Primates | Carnivora | Apiales | Enterobacteriales |
| **Family** | Hominidae | Canidae | Apiaceae (Umbelliferae) | Enterobacteriaceae |
| **Genus** | *Homo* | *Canus* | *Daucus* | *Escherichia* |
| **Species** | *H. sapiens* | *C. familiaris* | *D. carota* | *E. coli* |

Of the organisms listed in Table 10-1, you have the most in common with a dog. You're both animals possessing a central nervous chord (phylum Chordata), and you're both mammals (class Mammalia), which means you have hair and the females of your species make milk. However, you also have many differences, including the tooth structure that separates you into the order Primates and a dog into the order Carnivora. If you compare yourself to a plant, you can see that you have certain features of cell structure that place you together in domain Eukarya, but little else in common.

Two organisms that belong to the same species are the most similar of all. For most eukaryotic organisms, members of the same species can successfully sexually reproduce together, producing live offspring that can also reproduce. Bacteria and archaea don't reproduce sexually, so their species are defined by chemical and genetic similarities.

## Playing the name game

When biologists discover a new organism, they give it a scientific name. They've been doing this for hundreds of years according to a system developed by Swedish naturalist Carl Linnaeus in the 1750s. Linnaeus created a classification system that included some of the categories, such as kingdom and class, that are still used today in the taxonomic hierarchy. Linnaeus also proposed the system of *binomial nomenclature* that modern biologists use to give every type of living thing a unique name that has two parts.

In binomial nomenclature, the first part of an organism's name is the genus, and the second part is the species name, or *specific epithet.* The rules for using binomial nomenclature are as follows:

- The genus is always capitalized.
- The species name is never written without the genus, although the genus can be abbreviated by just the first letter.
- Both the genus and species should be italicized or underlined to indicate that the name is the official scientific name.

According to these rules, humans may be correctly identified as *Homo sapiens* or *H. sapiens.*

# Chapter 11

# Observing How Organisms Get Along

### In This Chapter

- Discovering how organisms interact with each other and their environment
- Analyzing populations and their growth (or lack thereof)
- Tracing the path of energy and matter on planet Earth

One of the amazing things about this planet is that even though different parts of the world have different climates, the organisms living there somehow manage to get what they need to survive from each other and the world around them. This chapter explores Earth's various ecosystems and details how the interactions between organisms work to keep life on Earth in balance. It also covers how scientists study groups of organisms to stay on top of how their populations are growing (or declining).

## Ecosystems Bring It All Together

Life thrives in every environment on Earth, and each of those environments is its own *ecosystem,* a group of living and nonliving things that interact with each other in a particular environment. An ecosystem is essentially a little machine made up of living and nonliving parts. The living parts, called *biotic factors,* are all the organisms that live in the area. The nonliving parts, called *abiotic factors,* are the nonliving things in the area (think air, sunlight, and soil).

Ecosystems exist in the world's oceans, rivers, forests — they even exist in your backyard and local park. They can be as huge as the Amazon rain forest or as small as a rotting log. The catch is that the larger an ecosystem is, the greater the number of smaller ecosystems existing within it. For example, the ecosystem of the Amazon rain forest also consists of the soil ecosystem and the cloud forest ecosystem (found at the tops of the trees).

A particular branch of science called *ecology* is devoted to the study of ecosystems, specifically how organisms interact with each other and their environment. Scientists who work in this branch are called *ecologists,* and they look at the interactions between living things and their environment on many different scales, from large to small.

The sections that follow explain how ecologists classify Earth's various ecosystems and how they describe the interactions between the planet's many species. Before you check them out, take a look at Figure 11-1 to get an idea of how living things are organized.

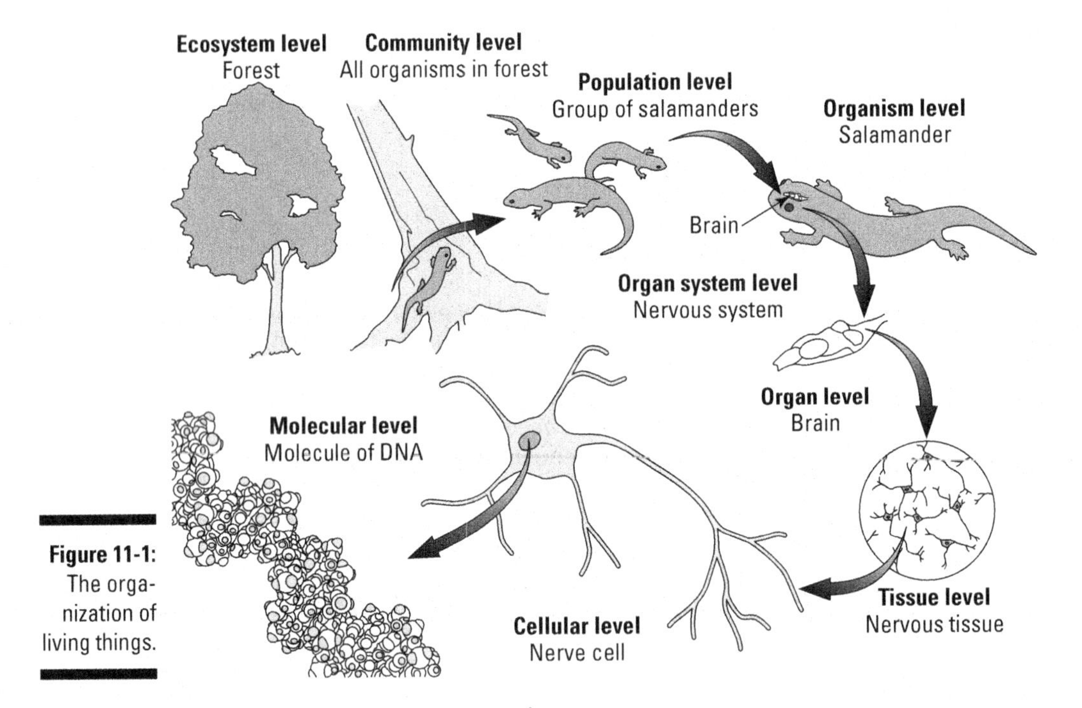


**Figure 11-1:** The organization of living things.

## Biomes: Communities of life

All the living things together in an ecosystem form a *community.* For example, a forest community may contain trees, shrubs, wildflowers, squirrels, birds, bats, insects, mushrooms, bacteria, and much more. The different types of communities found on Earth are called *biomes.* Six major types of biomes exist:

- **Freshwater biomes** include ponds, rivers, streams, lakes, and wetlands. Only about 3 percent of the Earth's surface is made up of freshwater, but freshwater biomes are home to many different species, including plants, algae, fish, and insects. Wetlands, in particular, have the greatest amount of diversity of any of the biomes.
- **Marine biomes** contain saltwater and include the oceans, coral reefs, and estuaries. They cover 75 percent of the Earth's surface and are very important to the planet's oxygen and food supply — more than half the photosynthesis that occurs on Earth occurs in the ocean (we describe the process of photosynthesis in Chapter 5). Marine biomes are home to many marine creatures such as algae, fish, octopuses, dolphins, and whales.

  *Estuaries* are areas where saltwater mingles with freshwater. They include familiar places such as bays, sounds, lagoons, salt marshes, and beaches. Estuaries are an important habitat for many different species, including birds, fish, and shellfish. Because they provide a habitat for young fish, estuaries are vital to the health of commercial fisheries. Unfortunately, estuaries are typically found on the coast, which is also prime real estate for people. As a result, estuaries are being heavily impacted by human development.

- **Desert biomes** receive minimal amounts of rainfall and cover approximately 20 percent of the Earth's surface. Plants and animals that live in deserts have special adaptations, such as the ability to store water or only grow during the rainy season, to help them survive in the low-water environment. Some familiar desert inhabitants are cacti, reptiles, birds, camels, rabbits, and dingoes.
- **Forest biomes** contain many trees or other woody vegetation; cover about 30 percent of the Earth's surface; and are home to many different plants and animals, including trees, skunks, squirrels, wolves, bears, birds, and wildcats. They're important for global carbon balance because they pull carbon dioxide out of the atmosphere through the process of photosynthesis. Forests are being heavily impacted by human development as humans seek additional land for homes and agriculture and cut down forests for their wood.

  *Rain forests* are evergreen forests that receive lots of rainfall and are incredibly rich in species diversity. As many as half the world's animals live in rain forests, including gorillas, tree frogs, butterflies, tigers, parrots, and boa constrictors.

- **Grassland biomes** are dominated by grasses, but they're also home to many other species, such as birds, zebras, giraffes, lions, buffaloes, termites, and hyenas. Grasslands cover about 30 percent of the Earth's surface and are typically flat, have few trees, and possess rich soil. Because of these features, people converted many natural grasslands for agricultural purposes.

- **Tundra biomes** are very cold and have very little liquid water. Tundras cover about 15 percent of the planet's surface and are found at the poles of the Earth as well as at high elevations. Arctic tundras are home to organisms such as arctic foxes, caribou, and polar bears, whereas mountain tundras are home to mountain goats, elk, and birds. In both types of tundra, nutrients are typically scarce, and the growing seasons are quite short.

## Why can't we be friends: Interactions between species

Not all the organisms in a given community are the same. In fact, they're often members of different species (meaning they can't sexually reproduce together). Yet these organisms must interact with each other as they go about their daily business of finding what they need to survive. Of course, just like relationships between people, relationships between other species can be good, bad, or just so-so.

Ecologists use a few specific terms to describe the types of interactions between different species:

- **Mutualism:** Both organisms benefit in a mutualistic relationship. Case in point: You give the bacteria in your small intestines a nice place to live complete with lots of food, and they make vitamins for you. Another example is when fungi in the soil form partnerships with plant roots. The fungi, called mycorrhizae, grow on the plant roots and help the roots absorb water and minerals from a wider area within the soil. In return, the mycorrhizae get some sugar from the plant.
- **Competition:** Both organisms suffer in a competitive relationship. If a resource such as food, space, or water, is in limited supply, species struggle with each other to obtain enough to survive. Just think of a vegetable garden that's overrun with weeds. In this scenario, the vegetable plants can't do well because they're competing with the weeds for water, minerals, and space. As a result, all the plants grow smaller and weaker in the crowded space than they would if they were growing by themselves.
- **Predation and parasitism:** One organism benefits at the expense of the other in predatory and parasitic relationships. When a lion eats a gazelle, the benefits are purely the lion's. Likewise, when a dog gets worms, the worms have a happy home and lots of food, but the dog doesn't get enough nutrition. The only real difference between these two situations is the speed of the interaction. In predation, one organism kills and eats the other right away; in parasitism, one organism slowly feeds off the other.

# Studying Populations Is Popular in Ecology

Each group of organisms of the same species living in the same area forms a *population*. For instance, the forests in the Pacific Northwest consist of Douglas fir trees and Western red cedars. Because Doug firs and Western red cedars are two different kinds of trees, ecologists consider two groups of these trees in the same forest to be two different populations.

*Population ecology* is the branch of ecology that studies the structures of populations and how they change. (*Population biology* is a very similar field that also includes the study of the genetics of populations.)

The following sections introduce you to some of the basic concepts of population ecology. They also help you understand the ways in which populations grow and change, as well as how scientists measure and study their growth, and give you some insight into the massive growth of the human population.

## Reviewing the basic concepts of population ecology

Like all ecologists, population ecologists are interested in the interactions of organisms with each other and with their environment. The unique thing about population ecologists, though, is that they study these relationships by examining the properties of populations rather than individuals.

The next few sections walk you through some of the basic properties of populations and show you why they're important.

### Population density

One way of looking at the structure of a population is in terms of its *population density* (how many organisms occupy a specific area).

Say you want to get an idea of how the human population is distributed in the state of New York. About 19.5 million people live in the 47,214 square miles that make up the state. If you divide the number of people by the area, you get a population density of about 413 people per square mile. However, the human population of New York isn't evenly distributed. In order to really understand how the human population is distributed in the state of New York, you need to compare the population density of the state to the population density of New York City.

The New York City metropolitan area has 8,214,426 people living in just 303 square miles, creating a population density of 27,110 people per square mile. If all the people in New York City were evenly distributed, each person would have 1½ acres of space all to herself (there are 640 acres in 1 square mile). However, people living in New York City actually have only 3/10 of an acre, which is only 1,300 square feet, to themselves.

All of these numbers just go to show that the human population in New York is heavily concentrated in New York City and much less concentrated in other areas of the state.

### Dispersion

Population ecologists use the term *dispersion* to describe the distribution of a population throughout a certain area. Populations disperse in three main ways:

- **Clumped dispersion:** In this type of dispersion, most organisms are clustered together with few organisms in between. Examples include people in New York City, bees in a hive, and ants in a hill.
- **Uniform dispersion:** Uniformly dispersed organisms are spread evenly throughout an area. Grapevines in a vineyard and rows of corn plants in a field are examples of uniform dispersion.
- **Random dispersion:** In this type of dispersion, one place in the area is as good as any other for finding the organism. (***Note:*** Random dispersion is rare in nature but may result when seeds or larvae are scattered by wind or water.) Examples of random dispersion include barnacles scattered on the surfaces of rocks and plants with wind-blown seeds settling down on bare ground.

### Population dynamics

*Population dynamics* are changes in population density over time or in a particular area. Population ecologists typically use age-structure diagrams to study these changes and note trends.

*Age-structure diagrams,* sometimes called *population pyramids,* show the numbers of people in each age group in a population at a particular time. The shape of an age-structure diagram can tell you how fast a population is growing.

- **A pyramid-shaped age-structure diagram indicates the population is growing rapidly.** Take a look at Figure 11-2a. In Mexico, more people are below reproductive age than above reproductive age, giving the age-structure diagram a wide base and a narrow top. The newest generations are larger than the generations before them, so the population size is increasing.

- **An evenly shaped age-structure diagram indicates the population is relatively stable.** According to Figure 11-2b, the number of people above and below reproductive age in Iceland is about equal, with a decrease in the population as the older group ages. The newest generations are about the same size as the generations before them, so the population is staying roughly the same size.
- **An age-structure diagram that has a smaller base than middle portion indicates the population is decreasing in size.** When you refer to Figure 11-2c, you notice that more people are above reproductive age in Japan than below it. The newest generations are smaller than the older generations, so the population is decreasing.

### Survivorship

Scientists interested in *demography* — the study of birth, death, and movement rates that cause change in populations — noticed that different types of organisms have distinct patterns in how long offspring survive after birth. The scientists followed groups of organisms that were all born at the same time and looked at their *survivorship,* which is the number of organisms in the group that are still alive at different times after birth. They then plotted out *survivorship curves,* graphs that plot survival after birth over time (like the one in Figure 11-3), to depict how long individuals typically survive in a population.

Three types of survivorship exist:

- **Type I survivorship:** Most offspring survive, and organisms live out most of their life span, dying in old age. Humans have a Type I survivorship because most individuals survive to middle age (about 40 years) and beyond.
- **Type II survivorship:** Death occurs randomly throughout the life span, usually due to predation or disease. Mice have Type II survivorship — they never know when the cat or mousetrap will strike.
- **Type III survivorship:** Most organisms die young, and few members of the population survive to reproductive age. However, individuals that do survive to reproductive age often live out the rest of their life span and die in old age. In other words, Type III organisms die young. Species such as frogs that produce offspring that must swim on their own as larvae fit into this category. Other animals eat many of the larvae before they reach the adult stage (which is when they can reproduce).

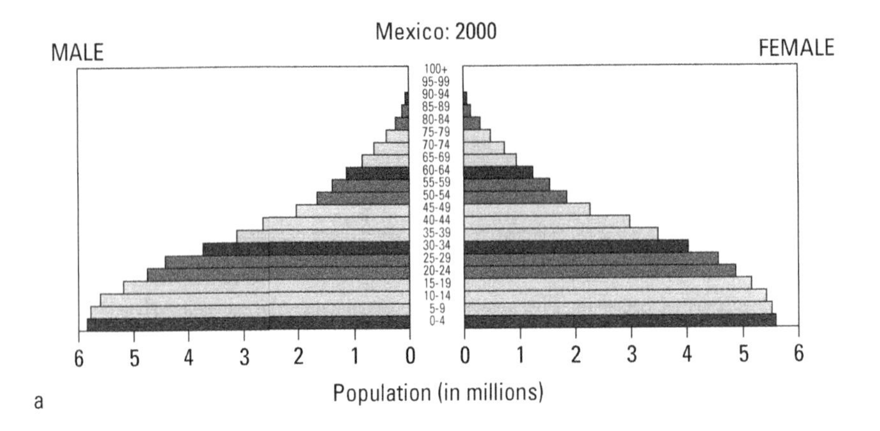


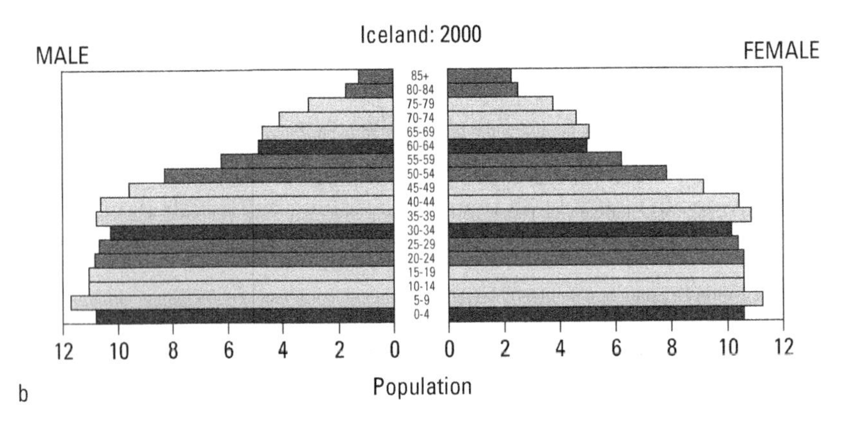


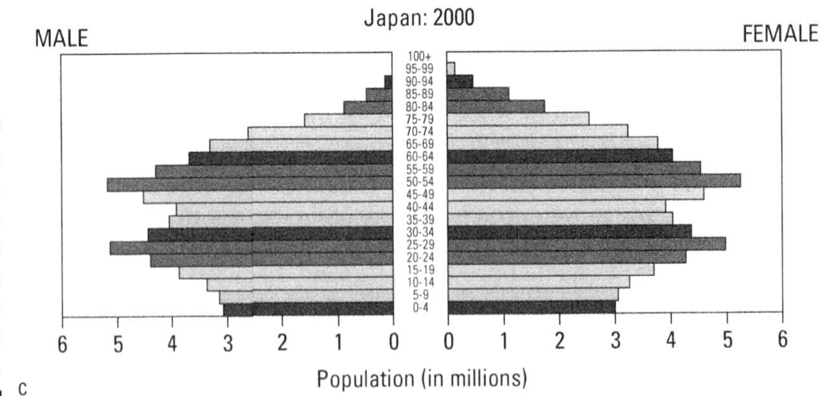


**Figure 11-2:** Age-structure diagrams break down age groups in populations.

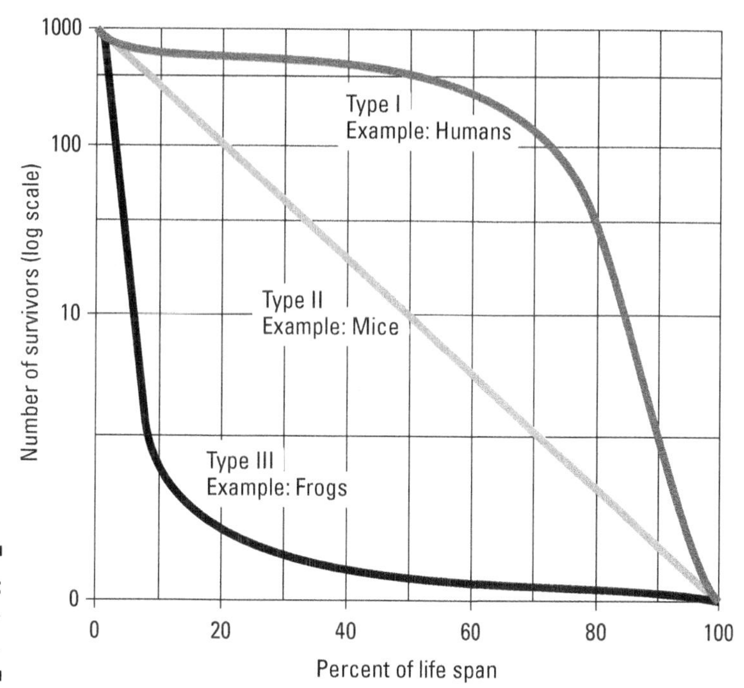


**Figure 11-3:** A survivorship curve.

## Discovering how populations grow

Populations have the potential to grow exponentially when organisms have more than one offspring. Why? Because those offspring have offspring, and the population gets even bigger.

For example, assume 1 organism has 3 offspring, creating a population of 4 organisms. Say each of the 3 original offspring has 3 offspring, adding 9 and bringing the total population to 13 organisms. If the 9 newest offspring have 3 offspring each, that adds 27 new individuals and brings the total population to 40. Although the rate of reproduction per individual, called the *per capita reproduction rate,* remains the same, the population grows larger and larger.

The next sections fill you in on the factors that affect population growth, how scientists track a population's growth, and more.

### Understanding biotic potential

The maximum growth rate of a population under ideal conditions is referred to as *biotic potential.* Ideal conditions occur when species don't have to compete for resources, such as food or water, and when no predators or diseases affect the growth of the organisms. Other factors involved in determining biotic potential include

- The age of the organisms when they're able to reproduce
- The number of offspring typically produced from one successful mating
- How often the organisms reproduce
- How long a period of time they're capable of reproducing
- The number of offspring that survive to adulthood

Bacteria, for example, have a very high biotic potential. Many bacteria can reproduce in less than an hour, and their offspring are ready to begin reproduction as soon as they're made. If one cell of *E. coli* could grow without limits for just 48 hours, the population of bacteria produced would weigh the same amount as the Earth!

### Looking at the factors affecting population growth

Population growth can be limited by a number of environmental factors, which population ecologists group into two categories:

- **Density-dependent factors** are more likely to limit growth as population density increases. For example, large populations may not have enough food, water, or nest sites, causing fewer organisms to survive and reproduce. This lower birth rate combines with a higher death rate to slow population growth.
- **Density-independent factors** limit growth but aren't affected by population density. Changes in weather patterns that cause droughts or natural disasters such as earthquakes kill individuals in populations regardless of that population's size.

Some populations can remain very steady in the face of these factors, whereas others fluctuate quite a bit.

- **Populations that depend on limited resources fluctuate more than populations that have ample resources.** If a population depends heavily on one type of food, for example, and that food becomes unavailable, the death rate will increase rapidly.
- **Populations with low reproductive rates are more stable than populations with high reproductive rates.** Organisms with high reproductive rates may have sudden booms in population as conditions change. Organisms with low reproductive rates don't experience these booms; their population growth rate is fairly steady.
- **Populations may rise and fall because of interactions between predators and prey.** When prey is abundant, predator populations grow until the increased numbers of predators eat up all the prey. When that happens, the prey population crashes, followed by the predator population as the predators starve. After the predator population crashes, the prey has a chance to recover, and its numbers increase again, starting the cycle anew.

### Reaching carrying capacity

When a population hits *carrying capacity,* it has reached the maximum amount of organisms of a single population that can survive in one *habitat* (the scientific name for a home).

As populations approach the carrying capacity of a particular environment, density-dependent factors have a greater effect, and population growth slows dramatically. If carrying capacity is exceeded even temporarily, the habitat may be damaged, further reducing the amount of resources available and leading to increased deaths. This situation decreases the population so that the carrying capacity can be met once again. However, if the habitat is damaged, the carrying capacity may be lowered even further, necessitating even more deaths to restore balance.

### Graphing growth rates

Scientists often use graphs to make sense of population data. *J-shaped growth curves,* like the one in Figure 11-4a, depict *exponential growth.* In other words, they show that a population is increasing at a steady rate (the birth and death rates are constant). In nature, populations may show exponential growth for short periods of time, but then environmental factors act to limit their growth rate.

*S-shaped growth curves,* like the one in Figure 11-4b, show *logistic growth,* meaning the population size is affected by environmental factors. In logistic growth, the growth rate is high when population density is low and then slows as population density increases.

## Painting with numbers: Using statistics to get a picture of population growth

Population ecologists use statistics to model the growth of populations. The natural growth rate of a population *(r)* is equal to the per capita birth rate *(b)* minus the per capita death rate *(d)*. In other words: $r = b - d$.

If *migration* is occurring, meaning organisms are moving from one place to another, immigration is added to the birth rate, and emigration is added to the death rate. When conditions are optimal, the growth rate reaches its maximum and is called the *intrinsic rate of increase,* or $r_{max}$. Different species have characteristic values for $r_{max}$. For example, even if conditions are optimal, the $r_{max}$ of elephants is never going to be close to the $r_{max}$ for *E. coli.*

The growth rate of any population at a particular time can be calculated by multiplying the intrinsic rate of increase ($r_{max}$) by the number of individuals in the population *(N)*. In other words: $\Delta N \div \Delta t = Nr_{max}$.

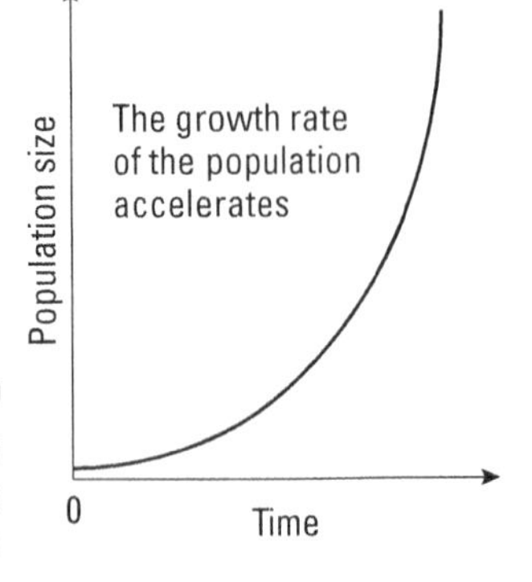


a Exponential (unrestricted) growth

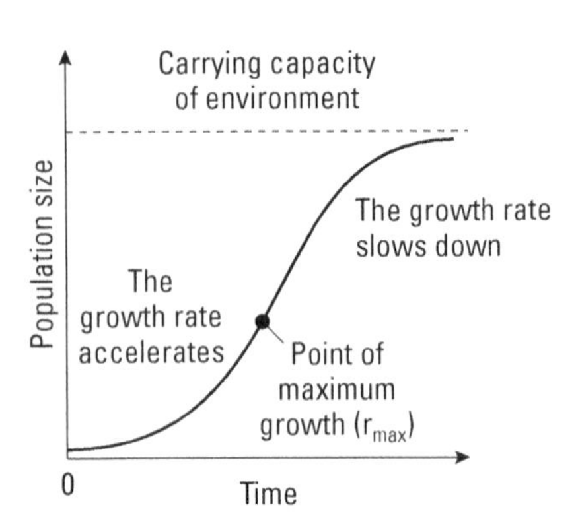


b Logistic (restricted) growth

**Figure 11-4:** Population growth curves.

## Taking a closer look at the human population

There's no doubt about it: Humans are the dominant population on Earth, and our numbers keep on rising. It's important to have an understanding of how our population is growing because of the impact humans have on the planet and all the other species on it. The following sections provide some insight into this and introduce you to the special tool population ecologists have derived to study human population growth.

### The human population explosion

Up until about a thousand years ago, human population growth was very stable. Food wasn't as readily available as it is now. Nor were there antibiotics to fend off invading bacteria, vaccines to fight against deadly diseases, and sewage treatment plants to ensure that water was safe to drink. People didn't shower or wash their hands as often, so they spread diseases more easily. All of these factors, and more, increased the death rate and decreased the birth rate of the human population.

Yet in the last 100 to 200 years, the food supply has increased and hygiene and medicines have reduced deaths due to common illnesses and diseases. So not only are more people born, but more of these people are surviving well past middle age. As you can see in Figure 11-5, the human population has grown exponentially in relatively recent history.

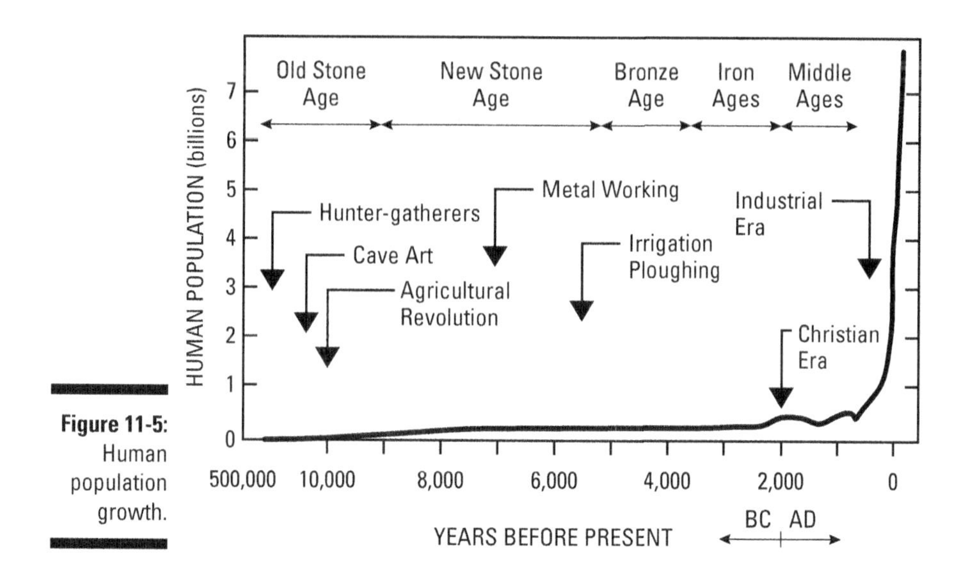


**Figure 11-5:** Human population growth.

If Figure 11-5 doesn't impress you, here are a few statistics that might:

- The human population doubled in the 40 years between 1950 and 1990.
- Every second, about three new people are born.
- The global human population passed the 6 billion mark at the end of the 20th century.

At current growth rates, the human population is projected to reach 8 to 12 billion by the end of the 21st century. Imagine that for a minute. What would your life be like if twice as many people lived on Earth as do today? There'd be twice as many people dining at restaurants, driving around, going to the movies, hiking in the woods . . . you get the idea.

What's scary is that scientists question whether the Earth can even support that many humans. The exact carrying capacity of the Earth for humans isn't known because, unlike other species, humans can use technology to increase the Earth's carrying capacity for the species. Currently, scientists estimate that humans are using about 19 percent of the Earth's *primary productivity,* which is the ability of living things like plants to make food. Humans also use about half of the world's freshwater. If humans continue to use more and more of the Earth's resources, the increased competition will drive many other species to extinction. (This pressure on other species from human impacts is already being seen, endangering species such as gorillas, cheetahs, lions, tigers, sharks, and killer whales.)

### The demographic transition model

Comprehending human population growth is a bit tougher than following the population growth of other organisms. Technology, education, and other factors affect how different human populations grow. Richer, industrialized nations (such as European countries, the United States, and Iceland), have reached a stage of *zero population growth,* meaning birth and death rates are equal in these countries. On the other hand, poorer, less industrialized nations (such as many countries in Africa) have very high birth rates relative to their death rates and have rapidly growing populations.

The major difference between these groups of nations is fertility. In less industrialized nations, families that have more children gain an economic benefit because the children are needed for labor-intensive tasks. In more industrialized nations, fewer children are needed to work for the family, and raising children becomes more expensive, so people choose to have fewer children.

Population ecologists have developed a special *demographic transition model* to note the stages of development the human population goes through in any given country on its way to stabilization. Based on human history over the past century or so, the process to reach full demographic transition has four stages (see Figure 11-6 for the visual):

- **Stage 1:** Birth and death rates are both high. Basic sanitation and modern medicine aren't yet available to lower the death rate and extend the life span. Demographic transition has not yet occurred.
- **Stage 2:** Sanitation and medicine lower the death rate, but the economy still encourages a high birth rate. Farming remains a large part of the economy in Mexico, for example, so many people there still have large families.
- **Stage 3:** Increased urbanization reduces the need for large families, and the cost of raising and educating children encourages fewer births. The birth rate drops and becomes close to that of the death rate. The population still grows, however, as earlier generations reach reproductive age. As countries pass from Stage 2 to Stage 3, they make a partial demographic transition.
- **Stage 4:** The population becomes stable, and birth rates equal death rates. When countries reach Stage 4, they've made a full demographic transition.

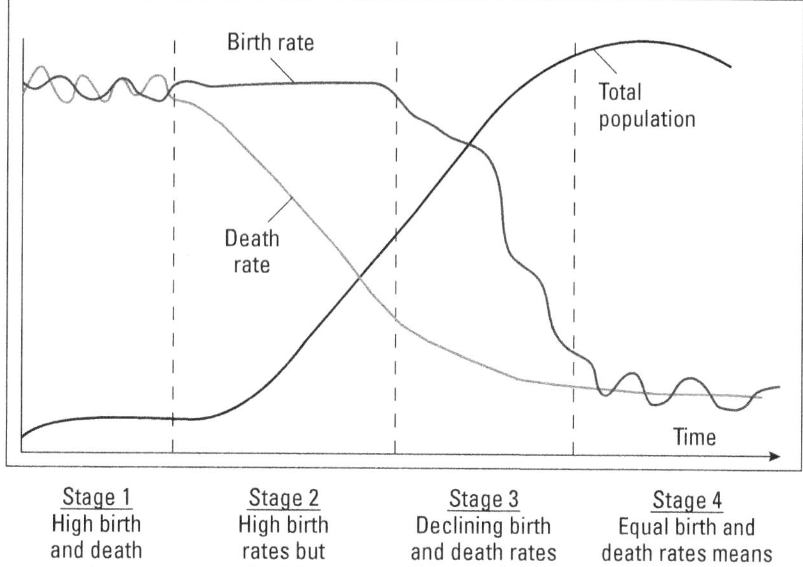


**Figure 11-6:** The demographic transition model.

# Moving Energy and Matter around within Ecosystems

Organisms interact with their environment and with other organisms to acquire energy and matter for growth. The interactions between organisms influence behavior and help the organisms establish complex relationships.

One of the most fundamental ways that organisms interact with each other is eating each other. In fact, all the various organisms in an ecosystem can be divided into four categories called *trophic levels* based on how they get their food:

- **Producers** make their own food. Plants, algae, and green bacteria are all producers that use energy from the Sun to combine carbon dioxide and water and form carbohydrates via photosynthesis. Producers can also be called *autotrophs* (see Chapter 5 for more on autotrophs and the process of photosynthesis).

- **Primary consumers** eat producers. Because producers are mainly plants, primary consumers are also called *herbivores* (plant-eating animals).
- **Secondary consumers** eat primary consumers. Because primary consumers are animals, secondary consumers are also called *carnivores* (meat-eating animals).
- **Tertiary consumers** eat secondary consumers, so they're also considered carnivores.

Organisms in the different trophic levels are linked together in a *food chain,* a sequence of organisms in a community in which each organism feeds on the one below it in the chain. Figure 11-7 shows a depiction of a simple food chain.

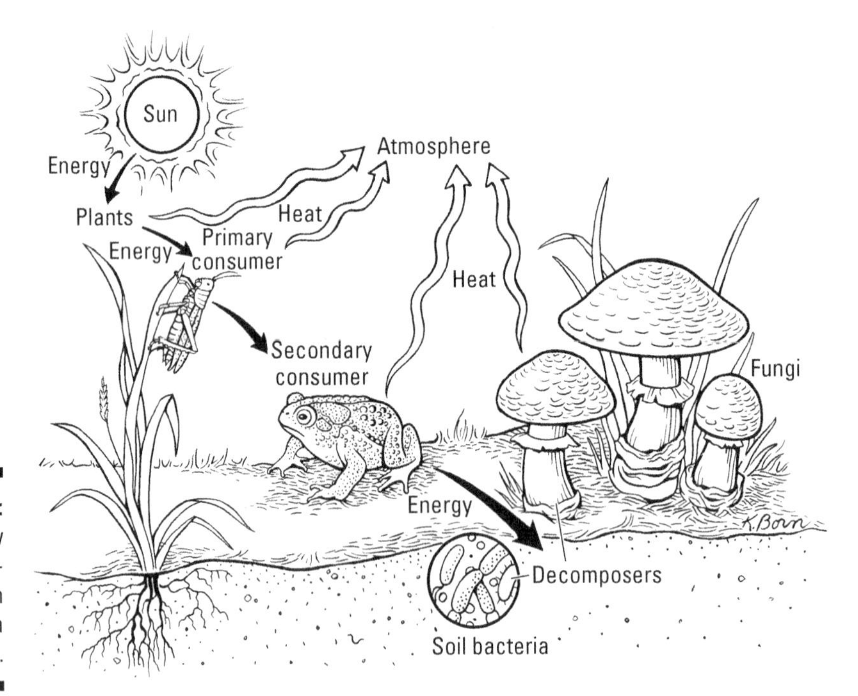


**Figure 11-7:** Energy flow in ecosystems shown through a food chain.

Interactions in ecosystems go way beyond those shown in a simple food chain because

- **Some organisms eat at more than one trophic level.** You, for example, may eat a slice of pizza with pepperoni. The grain that made the crust came from a plant, so when you eat the crust you're acting as a primary consumer. The pepperoni, however, came from an animal, so when you eat the pepperoni, you're acting as a secondary consumer.

- **Some organisms eat more than one type of food.** When you eat pepperoni pizza, you're eating food from both plants and animals. Organisms such as humans that eat both plants and animals are called *omnivores.* Also, organisms that eat more than one type of food belong to more than one food chain. When all the food chains from an ecosystem are put together, they form an interconnected *food web.*
- **Some organisms get their food by breaking down dead things.** *Decomposers,* like bacteria and fungi, release enzymes onto dead organisms, breaking them down into smaller components for absorption. *Detritivores,* such as worms, small insects, crabs, and vultures, also eat the dead.

The sections that follow delve into the details of how energy and matter move from one organism to another in a never-ending cycle that's essential to the survival of life on this planet.

## Going with the (energy) flow

The energy living things need to grow flows from one organism to another through food. Sounds simple, we know, but that energy is governed by a few key principles, perhaps most important of which is that an organism never gets to use the full amount of energy it receives from the thing it's "eating."

In the sections that follow, we reveal the principles that govern energy as well as the way in which scientists measure the flow of energy from organisms at different levels of the food chain.

### Energy principles

Some really important energy principles form the foundation of organism interactions in ecosystems:

- **Energy can't be created or destroyed.** This statement represents a fundamental law of the universe called the *First Law of Thermodynamics.* The consequence of this law is that every living thing has to get its energy from somewhere. No living thing can make the energy it needs all by itself. Even producers, who make their own food, can't make their own energy — they capture energy from the environment and store it in the food they make.

- **When energy is moved from one place to another, it's transferred.** When a primary consumer eats a producer, the energy that was stored in the body of the producer is transferred to the primary consumer.

  When describing energy transfers, be sure to say where the energy is coming from and where it's going to.

- **When energy is changed from one form to another, it's transformed.** When plants do photosynthesis, they absorb light energy from the Sun and convert it into the chemical energy stored in carbohydrates. So, during photosynthesis, light energy is transformed into chemical energy.

  When describing energy transformations, be sure to state the form of energy both before and after the transformation.

- **When energy is transferred in living systems, some of the energy is transformed into heat energy.** This statement is one way of representing another law of the universe, called the *Second Law of Thermodynamics.* The impact of this law on ecosystems is that no energy transfer is 100 percent efficient. After energy is transferred to heat, it's no longer useful as a source of energy to living things. In fact, only about 10 percent of the energy available at one trophic level is usable to the next trophic level.

  The Second Law of Thermodynamics has many impacts on energy and can be stated in several different ways. In biology, it's usually stated as "all chemical reactions spontaneously occur in the direction that increases disorder (called *entropy*) in the universe." What this means is that any process that makes things more random — like breaking down molecules or spreading molecules randomly over an area — can occur without the input of energy. This tendency of the universe to become more random also applies to the distribution of energy. Food molecules represent a very concentrated form of energy — that's why living things like them so much. Heat, on the other hand, is a much more dispersed, or random, form of energy. So, according to the Second Law of Thermodynamics, if energy from food molecules is involved in an energy transfer, some of that energy is going to become more randomly dispersed, meaning it transforms into heat energy.

You may hear people, even scientists, say that "energy is lost" or "energy is lost as heat." These statements can be confusing because they make it sound like energy disappears somehow. But you know from the First Law of Thermodynamics that energy can't be destroyed or disappear. The correct interpretation of statements like these is that useful energy is lost from the system as it's transformed into heat. In other words, after energy is transformed into heat, organisms in an ecosystem can't use it as a source of energy for growth.

Never use the words *lost, disappear, destroyed,* or *created* when you're talking about energy. Use the words *transfer* and *transformed* instead, and you'll avoid a great deal of confusion.

### The energy pyramid

Scientists use an *energy pyramid* (also called a *trophic pyramid;* see Figure 11-8) to illustrate the flow of energy from one trophic level to the next. Energy pyramids show the amount of energy at each trophic level in proportion to the next trophic level — what ecologists refer to as *ecological efficiency.* To estimate ecological efficiency, ecologists use a rule of thumb called the *10-percent rule,* which says that only about 10 percent of the energy available to one trophic level gets transferred to the next trophic level.

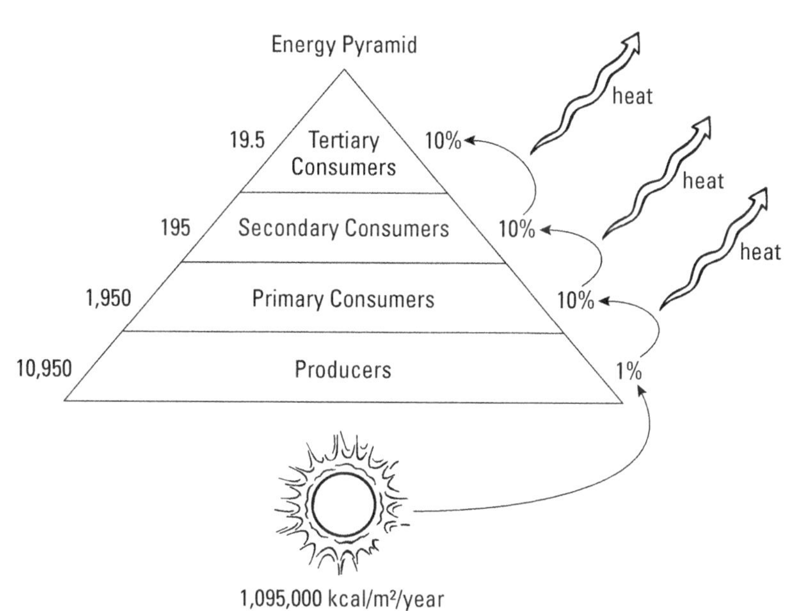


**Figure 11-8:** The energy pyramid.

Following along in Figure 11-8, you see that energy travels to the Earth from the Sun. About 1 percent of the energy available to producers is captured and stored in food. Producers grow, transferring much of their stored energy into ATP for cellular work (as explained in Chapter 5) and the molecules that make up their bodies. As producers transfer energy for growth, some energy is also transformed into heat that's transferred to the environment.

About 10 percent of the energy that was originally stored in producers is transferred to primary consumers when they eat the producers. Just like producers, the primary consumers grow, transferring energy from food into ATP for cellular work and into the molecules that make up their bodies. As primary consumers transfer energy for growth, some energy is also transformed into heat that's transferred to the environment. This process repeats itself when secondary consumers eat primary consumers and when tertiary consumers eat secondary consumers. Each level of consumer receives about 10 percent of the energy originally captured by the organism it consumed, and as the consumer uses energy for growth, it transfers some of that energy back to the environment as heat.

But the energy pyramid doesn't end there. As organisms die, some of their remains become part of the environment. Decomposers and detritivores use this dead matter as their source of food, transferring energy from food into ATP and molecules and giving off some energy as heat.

***Note:*** Food chains and energy pyramids usually don't go beyond tertiary consumers because the amount of energy available to each level is reduced as you move up the pyramid. Past the tertiary consumer level, too much energy has been depleted from the system.

## Cycling matter through ecosystems

Not only does food provide energy to living things (as explained earlier in this chapter) but it also provides the matter organisms need to grow, repair themselves, and reproduce. For example, you eat food that contains proteins, carbohydrates, and fats, and your body is made of proteins, carbohydrates, and fats (see Chapter 3 for more on these molecules). When you eat food, you break it down through digestion (covered in Chapter 16) and then transfer small food molecules around your body via circulation (covered in Chapter 15) so that all of your cells receive food. Your cells then have two options:

- They can use the food for energy by breaking it down into carbon dioxide and water through cellular respiration (see Chapter 5).
- They can rebuild the small food molecules into the larger food molecules that make up your body.

Yes, that second option means you are what you eat — well, almost. You don't use food molecules directly to build your cells; you break them down first and use the pieces to build what you need. In other words, your cells are made of human molecules that are rebuilt from the parts of molecules taken from the plants and animals you've eaten. So really you're made of molecules that you recycled from your food. (Likewise, the living things your food used to be got their molecules by recycling them from somewhere else.)

Think about what all goes into a slice of pepperoni pizza. The crust came from the grains of plants, and the pepperoni (for the sake of argument) came from a pig. Plants make their own food from carbon dioxide and water and then use that food to build their bodies, which means the plant that went into your pizza crust got the parts it needed to build its body from carbon dioxide in the air and water in the soil. Pigs get their molecules by eating whatever food the farmer gives them, which is likely some type of plant. After you eat a slice of pepperoni pizza, you can trace some of the atoms that make up your body back to carbon dioxide from the air, water from the soil, and plants that were fed to pigs.

One of the most fascinating facts about the Earth is that almost all the matter on this planet today has been here since the Earth first formed. That means all the carbon, hydrogen, oxygen, nitrogen, and other elements that make up the molecules of living things have been recycled over and over throughout time. Consequently, ecologists say that matter cycles through ecosystems.

Scientists track the recycling of atoms through cycles called *biogeochemical cycles* (*bio* because the recycling involves living things, *geo* because it involves the Earth, and *chemical* because it involves chemical processes). Four biogeochemical cycles that are particularly important to living things are the hydrologic cycle, the carbon cycle, the phosphorous cycle, and the nitrogen cycle.

### The hydrologic cycle

The *hydrologic cycle* (also known as the *water cycle*) refers to plants obtaining water by absorbing it from the soil and animals obtaining water by drinking it or eating other animals that are made mostly of water. Water returns to the environment when plants transpire (as explained in Chapter 21) and animals perspire. Water evaporates into the air and is carried around the Earth by wind. As moist air rises and cools, water condenses again and returns to the Earth's surface as precipitation (think rain, snow, sleet, and hail). Water moves over the Earth's surface in bodies of water such as lakes, rivers, oceans, and even glaciers; it also moves through the groundwater below the soil.

### The carbon cycle

The carbon cycle (depicted in Figure 11-9) may be the most important biogeochemical cycle to living things because the proteins, carbohydrates, and fats that make up their bodies all have a carbon backbone (see Chapter 3 for more on these molecules). In the carbon cycle, plants take in carbon dioxide from the atmosphere, using it to build carbohydrates via photosynthesis (covered in Chapter 5). Animals consume plants or other animals, incorporating the carbon that was in their food molecules into the molecules that make up their own bodies. Decomposers break down dead material, incorporating the carbon from the dead matter into their bodies.

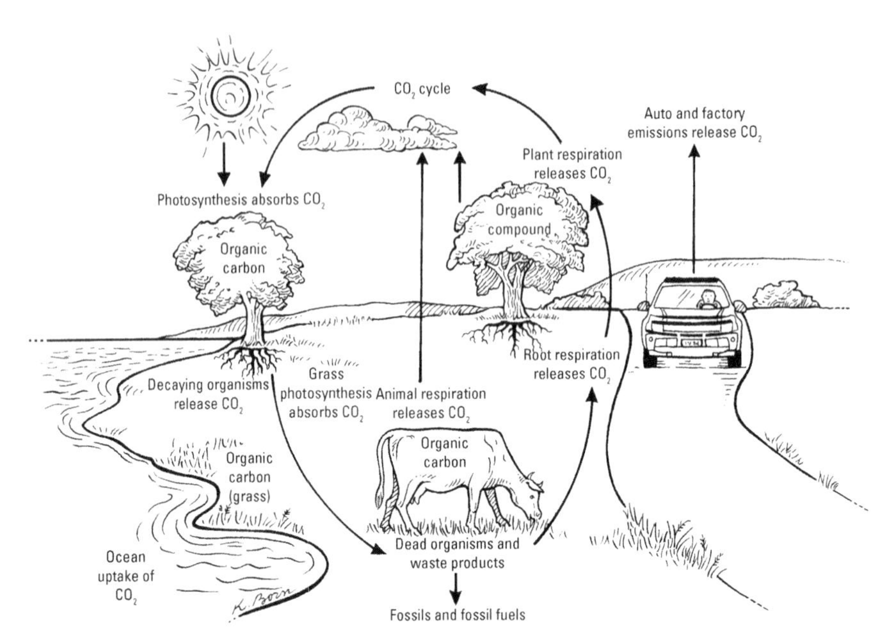


**Figure 11-9:** The carbon cycle.

All of these living things — producers, consumers, and decomposers — also use food molecules as a source of energy, breaking the food molecules back down into carbon dioxide and water in the process of cellular respiration (see Chapter 5). Cellular respiration releases the carbon atoms back into the environment as carbon dioxide, where it's again available to producers for photosynthesis.

Carbon storage in living things (in the form of proteins, carbohydrates, and fats) is purely temporary. Carbon can actually be stored in the environment for longer periods of time.

- Large forests represent significant storage of carbon, which can be suddenly released back to the environment as carbon dioxide when forests are cut and wood is burned.
- Fossil fuels contain carbon that was stored in the bodies of living things long ago and then trapped in a way that the proteins, carbohydrates, and fats were converted to coal, oil, and natural gas deposits. As people burn fossil fuels, this stored carbon is being rapidly released back into the atmosphere as carbon dioxide, causing the concentration of carbon dioxide in the atmosphere to rise to its highest levels in recorded history.
- Carbon is also stored in the world's oceans, in the form of dissolved carbon dioxide. Warm water holds less carbon dioxide than cold water, so some of this carbon may be released back to the atmosphere if ocean temperatures rise as a result of global warming.

### The phosphorus cycle

Phosphorous is an important component of the molecules that make up living things. It's found in adenosine triphosphate (ATP), the energy-storing molecule produced by every living thing, as well as the backbones of DNA and RNA molecules. The phosphorous cycle involves plants obtaining phosphorous when they absorb inorganic phosphate and water from the soil and animals obtaining phosphorous when they eat plants or other animals. Phosphorus is excreted through the waste products created by animals, and it's released by decomposers back into the soil as they break down dead materials. When phosphorus gets returned to the soil, it's either absorbed again by plants or it becomes part of the sediment layers that eventually form rocks. As rocks erode by the action of water, phosphorus is returned to water and soil.

### The nitrogen cycle

Not only is nitrogen part of the amino acids that make up proteins but it's also found in DNA and RNA (see Chapter 3 for more on these molecules). Nitrogen also exists in several inorganic forms in the environment, such as nitrogen gas (in the atmosphere) and ammonia or nitrates (in the soil).

Because nitrogen exists in so many forms, the nitrogen cycle (shown in Figure 11-10) is pretty complex.

- **Nitrogen fixation occurs when atmospheric nitrogen is changed into a form that's usable by living things.** Nitrogen gas in the atmosphere can't be incorporated into the molecules of living things, so all the organisms on Earth depend upon the activity of bacteria that live in the soil and in the roots of plants. These nitrogen-fixing bacteria convert nitrogen gas ($N_2$) into forms such as ammonium ion or nitrate ($NH_4^+$ or $NO_3^-$) that organisms can use. Plants obtain nitrogen by absorbing ammonia and nitrate along with water from the soil; animals get their nitrogen by eating plants or other animals.

  Some nitrogen fixation also occurs via lightning strikes and processes in factories that produce chemical fertilizers for plants. However, the nitrogen fixation that occurs from lightning strikes isn't enough to supply ecosystems with all the nitrogen they need, and industrial nitrogen fixation requires a lot of energy.

- **Ammonification releases ammonia into the soil.** As decomposers break down the proteins in dead things, they may not need all the nitrogen from those proteins for themselves. If the decomposers have excess nitrogen, they release some of it into the soil as ammonia ($NH_3$). In the soil, ammonia converts into ammonium ion ($NH_4^+$). The waste products of animals also contain nitrogen in the form of urea or uric acid that can be converted to ammonia by bacteria living in the soil.

- **Nitrification converts ammonia to nitrite and nitrate.** Certain bacteria get their energy by converting ammonia ($NH_3$) into nitrite ($NO_2^-$). Other bacteria get their energy by converting the $NO_2^-$ into nitrate ($NO_3^-$).
- **Denitrification converts nitrate to nitrite and nitrogen gas.** Some bacteria in the soil use nitrate ($NO_3^-$) rather than oxygen for cellular respiration (see Chapter 5 for more on cellular respiration). When these bacteria use nitrate, they convert it into nitrite that's released into the soil or nitrogen gas that's released into the atmosphere.

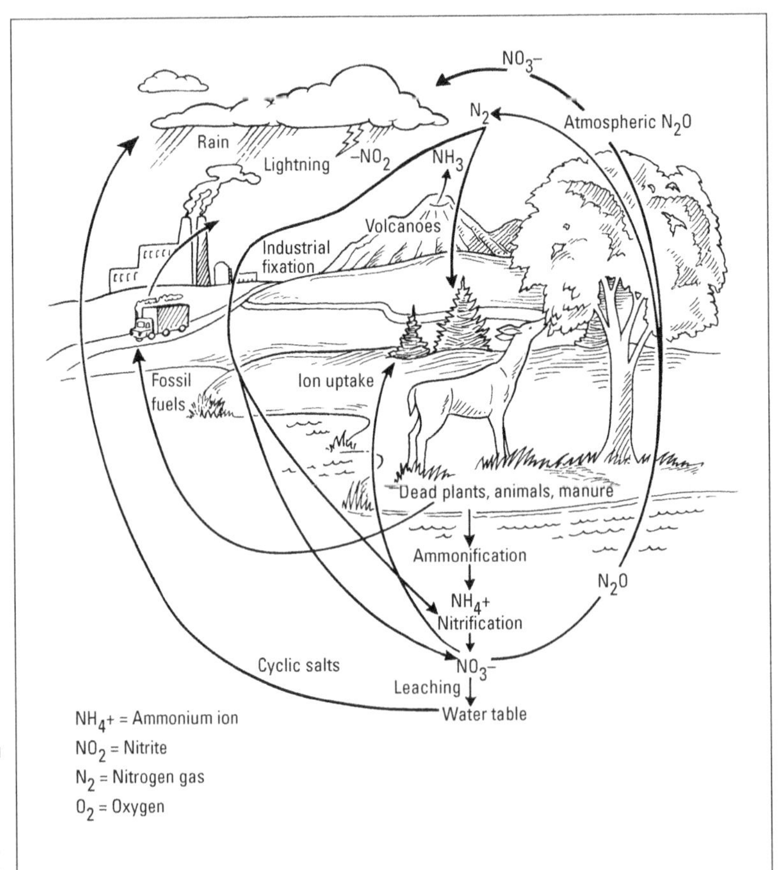


**Figure 11-10:** The nitrogen cycle.

# Chapter 12

# Evolving Species in an Ever-Changing World

### In This Chapter

- Checking out your ancestors' ideas of how organisms got here
- Diving into Darwin's controversial theories of biological evolution and natural selection
- Examining the proof of biological evolution
- Looking at both sides of the creationism-versus-evolution debate
- Finding out what species preceded the current human species

If you've ever been to a museum, then you've probably seen fossilized bones or tools from ancient ancestors. These objects are evidence of how humans have changed and expanded their knowledge over the millenia. In other words, they provide perspective on how the human species has evolved. But what was the starting point of evolution, and from what did the earliest humans evolve?

This chapter tells you about the beliefs people had regarding evolution; how Charles Darwin came up with his theory of biological evolution; and what the current thoughts are on the origin of species, how humans have evolved, and how life on Earth began. Prepare to be amazed by the proof researchers have found for biological evolution and find out some pretty fascinating facts about how things used to be and how they've changed.

## What People Used to Believe

From the time when ancient Greece was the world's cultural hotspot until the early 1800s, philosophers, scientists, and the general public believed that plants and animals were specially created at one time and that new species hadn't been introduced since then. (You could call this way of thinking *fundamentalism.*) In this view, every living thing was created in its ideal form by the hand of God for a special purpose. Aristotle classified all living things into a "great chain of being" from simple to complex, placing human beings at the top of the chain, just under the angels and very close to God.

People also thought the Earth and the universe it occupied were unchanging, or *static,* throughout time. They believed that God created the Earth, the stars, and the other planets all at once and that nothing had changed since the dawn of creation. These ideas held through much of human history, extending virtually unquestioned through the Middle Ages, when people even accepted that their place in society was unchanging and predestined by their birth.

Beginning in the 15th century and continuing into the 18th century, explorers, scientists, and naturalists made new discoveries that challenged the old ideas of a static universe.

- Various explorers fell upon the New World (the Western Hemisphere of the Earth). The New World revealed many different species of living things, including new races of people, that were previously unknown. The New World and the people that lived there weren't mentioned in the Bible, causing Europeans to debate whether the New World was created at the same time as the Old World and whether the people who lived there were descendants of Adam. These puzzles raised questions about a literal interpretation of the creation story in the Bible's book of Genesis.
- William Smith, a British surveyor, classified the types of ground material in Britain in preparation for the excavation of a canal system across the island. Smith discovered that the ground consisted of layers of different material and that different types of fossils could be found in each layer. He also found that the deeper he went into the layers, the more different the fossils appeared from the plants and animals that lived in Britain at the time. Smith used the fossils as a practical means of identifying the different layers of sediments, but he also made estimates of the age of the layers based on rates of erosion and the uplift of mountains.
- Georges Cuvier, a French anatomist, demonstrated that fossil bones found in Europe, such as those of wooly mammoths, could be recognized as very similar to existing species, such as elephants, but were clearly not from anything currently living.
- James Hutton, a Scottish geologist, proposed that the Earth was very ancient and that its surface was constantly changing due to erosion, the depositing of sediment, the uplift of mountains, and flooding. His idea, called *uniformitarianism,* was that the processes he observed on the Earth in the 1700s were the same processes that had occurred on the Earth since its creation.

# How Charles Darwin Challenged Age-Old Beliefs about Life on Earth

Charles Darwin was a gentleman from the English countryside who set out on a seafaring journey on the HMS *Beagle* in 1831 as the ship's naturalist. He was no less religious than others of his day, but he had a very active, curious mind and was acquainted with much of the scientific thinking of the time. That inquisitive mind combined with the scientific knowledge he'd gained over his short 22 years led Darwin to notice something rather unique about the finch population on the Galapagos Islands. These observations led to the creation of two of the most important biology-related theories of all time: biological evolution and natural selection. We fill you in on these theories and Darwin's inspiration for them in the following sections.

## Owing it all to the birds

While traveling on the HMS *Beagle,* Darwin visited the Galapagos Islands, which lie nearly 600 miles off the western coast of South America. He was amazed to find a variety of species that were similar to those in South America yet different in ways that seemed to make them exactly suited to the unique environment of the isolated islands.

Characteristics of organisms that make them suited to their environment are called *adaptations.*

Darwin chose to focus his attention on the Galapagos Islands' finches (a type of bird). Each island had its own unique species of finch that was distinct from the other species and from the finches on the mainland. In South America, finches ate only seeds. On the islands, some finches ate seeds, others ate insects, and some even ate cactuses. The beak of each type of finch seemed exactly suited to its food source.

Darwin thought that all the finches had a common ancestor from mainland South America that either flew or floated to the newly formed islands, perhaps during occasional storms. The islands are far enough apart from each other that finches can't really travel between them, so the different populations are geographically isolated from each other. Geographic isolation means they also can't mate with each other and combine their *genes* (sequences in DNA that control the traits of living things; see Chapter 8).

Darwin proposed that each type of island had unique conditions and that these unique conditions favored certain traits over others. Birds whose traits made them more successful at obtaining food were more likely to survive and reproduce, passing their genes and traits on to their offspring. Over time, the characteristics of the island birds shifted away from those of their mainland ancestor toward characteristics that better suited their new home. Eventually, the island birds became so different from their ancestors, and from each other, that they were unique species.

What Darwin observed in the finch population in the Galapagos Islands is a type of biological evolution referred to as adaptive radiation. *Adaptive radiation* happens when members of one species get into environmental niches and have very little competition for resources at the outset. The lack of competition allows the species to become rooted in a new environment and increase its population. As the population increases, competition for resources begins, and the original species breaks off into several new species that adapt to different environmental conditions.

## Darwin's theory of biological evolution

*Biological evolution* refers to the change of living things over time (it's not the same as *evolution,* which simply means change). Darwin introduced the world to this concept in his 1854 work, *On the Origin of Species.* In this book, Darwin proposed that living things descend from their ancestors but that they can change over time. In other words, Darwin believed in *descent with modification.*

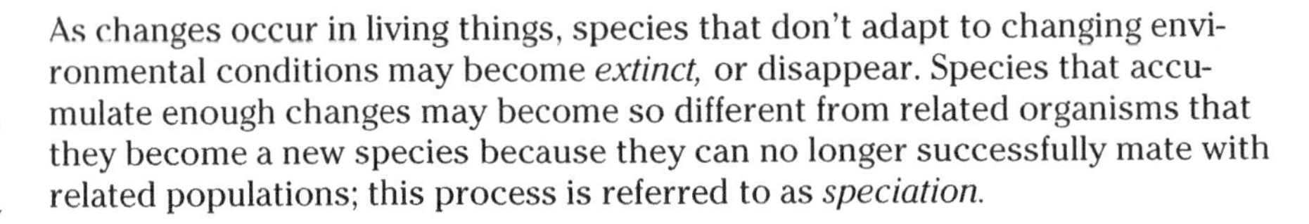
As changes occur in living things, species that don't adapt to changing environmental conditions may become *extinct,* or disappear. Species that accumulate enough changes may become so different from related organisms that they become a new species because they can no longer successfully mate with related populations; this process is referred to as *speciation.*

If you're ever interested in reading Darwin's *On the Origin of Species* (or any of his other works), check out `darwin-online.org.uk`.

## The idea of natural selection

Darwin concluded that biological evolution occurred as a result of *natural selection,* which is the theory that in any given generation, some individuals are more likely to survive and reproduce than others. When a particular trait improves the survivability of an organism, the environment is said to favor that trait or naturally select for it. Natural selection therefore acts against unfavorable traits.

The theory of natural selection is often referred to as "survival of the fittest." In biological terms, fitness doesn't have anything to do with your BMI or how often you work out. Biological fitness is basically your ability to produce offspring. So, survival of the fittest really refers to the passing on of those traits that enable individuals to survive and successfully reproduce.

In the next sections, we help you understand the difference between natural and artificial selection, why natural selection can occur in the first place, and what the different types of natural selection are.

### Comparing natural selection with artificial selection

Darwin compared his theory of natural selection with the artificial selection that results from selective breeding in agriculture.

- *Artificial selection* occurs when people choose plants or animals and breed them for certain desired characteristics. Farmers in Darwin's day bred the cows that gave the most milk, the chickens that laid the most eggs, and the pigs that got the biggest, creating many different breeds of each species. People's preferences have dramatically shaped the breeds of domestic animals and plants in a relatively short amount of time.
- *Natural selection* occurs when environmental factors "choose" which plants or animals will survive and reproduce. If a visual predator, such as an eagle, is cruising for its lunch, the individuals that it can see most easily are likely to be eaten. If the eagle's prey is mice, which can be white or dark colored (see Figure 12-1a), and the mice live in the forest against dark-colored soil, then the eagle is going to be able to see the white mice more easily. Over time, if the eagles in the area keep eating more white mice than dark mice (see Figure 12-1b), then more dark mice are going to reproduce. Dark mice have genes that specify dark-colored fur, so their offspring will also have dark fur. If the eagle continues to prey upon mice in the area, the population of mice in the forest will gradually begin to have more dark-colored individuals than white individuals (see Figure 12-1c).

  In this example, the eagle is the *selection pressure* — an environmental factor that causes some organisms to survive (the dark-colored mice) and others not to survive (the white-colored mice). A selection pressure gets its name from putting "pressure" or stress on the individuals of the population.

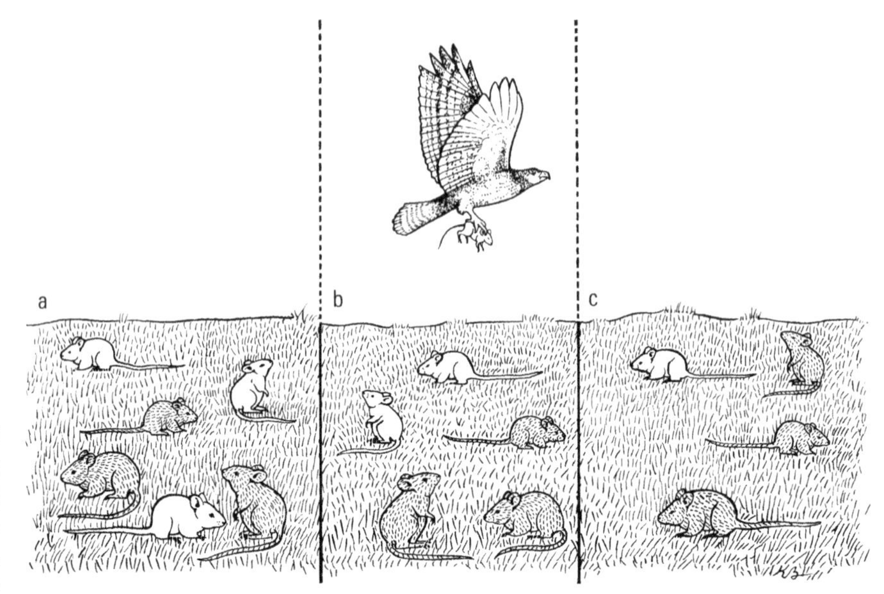


**Figure 12-1:** Natural selection in action.

### *Reviewing the conditions under which natural selection occurs*

In order for natural selection to occur in a population, several conditions must be met:

- **Individuals in the population must produce more offspring than can survive.** Human beings are somewhat unique among living things in that we can make conscious choices about how many offspring we have. Most other organisms, however, produce as many offspring as they can.
- **Those individuals must have different characteristics.** During Darwin's time, no one knew where these differences came from. Now scientists know that differences in organisms arise due to mutations in DNA combined with the mixing of genetic information during sexual reproduction (for more information on genetic variation due to sexual reproduction, see Chapter 6).
- **Offspring must inherit some characteristics from their parents.** During Darwin's time, the laws of inheritance were just beginning to be figured out, so Darwin didn't know exactly how parents passed on their traits. Modern scientists know that traits are inherited when parents pass genes on to their offspring (head to Chapter 7 for more on inheritance and Chapter 8 for more on genes).

- **Organisms with the best-suited characteristics for their environment are more likely to survive and reproduce.** This is the heart of natural selection. If there's competition for survival and not all the organisms are the same, then the ones with the advantageous traits are more likely to survive. If these traits can be inherited, then the next generation will show more of these advantageous traits.

If these four conditions are met, then the new generation of individuals will be different from the original generation in the frequency and distribution of traits, which is pretty much the definition of biological evolution.

### *Checking out the four types of natural selection*

Natural selection can cause several different types of changes in a population. How the population changes depends upon the particular selection pressure the population is under and which traits are favored in that circumstance. Individuals within a population may evolve to be more similar to or more different from each other depending on the specific circumstances and selection pressures.

The four types of natural selection are as follows:

- **Stabilizing selection:** This type eliminates extreme or unusual traits. Individuals with the most common traits are considered best adapted, which maintains the frequency of common traits in the population. Over time, nature selects against extreme variations of the trait. The size of human babies, for example, remains within a certain range due to stabilizing selection. Extremely small or extremely large babies are less likely to survive, so alleles that cause these extremes don't last in the population.
- **Directional selection:** In this type, traits at one end of a spectrum of traits are selected for, whereas traits at the other end of the spectrum are selected against. Over generations, the selected traits become common, and the other traits become more and more extreme until they're eventually phased out. The biological evolution of horses is a good example of directional selection. Ancestral horse species were built for moving through wooded areas and were much smaller than modern day horses. Over time, as horses moved onto open grasslands, they evolved into much larger, long-legged animals.
- **Disruptive selection:** In this type, the environment favors extreme or unusual traits and selects against the common traits. One example is the height of weeds in lawn grass compared with in the wild. In the wild, natural state, tall weeds compete for the resource of light better than short weeds. But in lawns, weeds have a better chance of surviving if they remain short because grass is kept short.
- **Sexual selection:** Females increase the fitness of their offspring by choosing males with superior fitness; females are therefore concerned with quality. Males contribute most to the fitness of a species by maximizing the quantity of offspring they produce. Because males are concerned with quantity, competition between males for opportunities to mate

exists in contests of strength. Therefore, structures and other traits that give a male an advantage in a contest of strength have evolved, including antlers, horns, and larger muscles. Because females choose their mates, males have also developed traits to attract females, such as certain mating behaviors and bright coloring.

Biological evolution happens to populations, not individuals. Individuals live or die and reproduce or don't reproduce depending on their circumstances. But individuals themselves can't evolve in response to a selection pressure. Imagine a giraffe whose neck isn't quite long enough to reach the tastiest leaves at the top of the tree. That individual giraffe can't suddenly grow its neck longer to reach the leaves. However, if another giraffe in the herd has a longer neck, gets more leaves, grows better, and makes more calves that inherit his long neck, then future generations of giraffes in that area may have longer necks.

# The Evidence of Biological Evolution

Since Darwin first proposed his ideas about biological evolution and natural selection, many different lines of research from many different branches of science have produced evidence supporting his belief that biological evolution occurs in part due to natural selection.

Because a great amount of data supports the idea of biological evolution through natural selection, and because no scientific evidence has yet been found to prove this idea false, this idea is considered a scientific theory. (For more on the importance of theories in science, see Chapter 2.)

The following sections describe some of the evidence, both old and new, that supports Darwin's theory and the tools modern scientists have used to obtain it.

## Biochemistry

The fundamental *biochemistry,* the basic chemistry and processes that occur in cells, of all living things on Earth is incredibly similar, showing that all of Earth's organisms share a common ancestry.

Case in point: All living things store their genetic material in DNA and build proteins out of the same 20 amino acids. Regardless of whether the organisms are flowers taking in carbon dioxide from the air, water from the soil, and light from the Sun; lions chomping down a wildebeest; or humans consuming a gourmet meal cooked by Wolfgang Puck himself, all organisms convert food sources to energy and store that energy in ATP. That stored energy is then used to power cellular processes such as the production of proteins, which is directed by the genes on strands of DNA.

## Comparative anatomy

*Comparative anatomy* — which looks at the structures of different living things to determine relationships — has revealed that the various species on Earth evolved from common ancestors. Just like you have structural characteristics that are similar to those of your family members (think small ears, a large nose, and so on), structural similarities also exist between more distantly related groups.

As you can see in Figure 12-2, the skeletons of humans, cats, whales, and bats, for example, are amazingly similar even though these animals live unique lifestyles in very different environments. From the outside, the arm of a human, the front leg of a cat, the flipper of a whale, and the wing of a bat seem very different, but when you look at the bones within them, you see that they all contain the same ones — an upper "arm," an elbow, a lower "arm," and five "fingers." The only differences in these bones are their size and shape. Scientists call similar structures such as these *homologous structures* (*homo-* means "same"). The best explanation for these homologous structures is that all four mammals are descended from the same ancestor — an idea that's supported by the fossil record.

The homologous structures of mammals are particularly interesting in the case of whales because they reveal whales' close relationship to land-dwelling animals. In fact, this evidence from comparative anatomy supports the idea that whales evolved from land-dwelling mammals into sea creatures.

## Geographic distribution of species

How populations of species are distributed around the globe helps solidify Darwin's theory of biological evolution. In fact, the science of *biogeography,* the study of living things around the globe, allows scientists to make testable predictions about biological evolution. Basically, if biological evolution is real, then you'd expect groups of organisms that are related to each other to be clustered near each other because related organisms come from the same common ancestor. (An exception to this prediction is that migratory animals could travel far from their relatives.) On the other hand, if biological evolution isn't real, then there's no reason for related groups of organisms to be found near each other. For example, a creator could scatter organisms randomly all over the planet, or groups of organisms could arise independently of other groups in whatever environments suited them best. When biogeographers compare the distribution of organisms living today, they find that species are distributed around the Earth in a pattern that reflects their genetic relationships to one another.

When Darwin compared the finches on the Galapagos Islands with those on mainland South America, the unique types of finches on the Galapagos Islands led him to hypothesize that the islands had been colonized by finches from the mainland. This hypothesis was later supported when modern scientists performed a genetic analysis of the Galapagos Islands' finches and were able to demonstrate their relationship to each other and to their mainland ancestors.

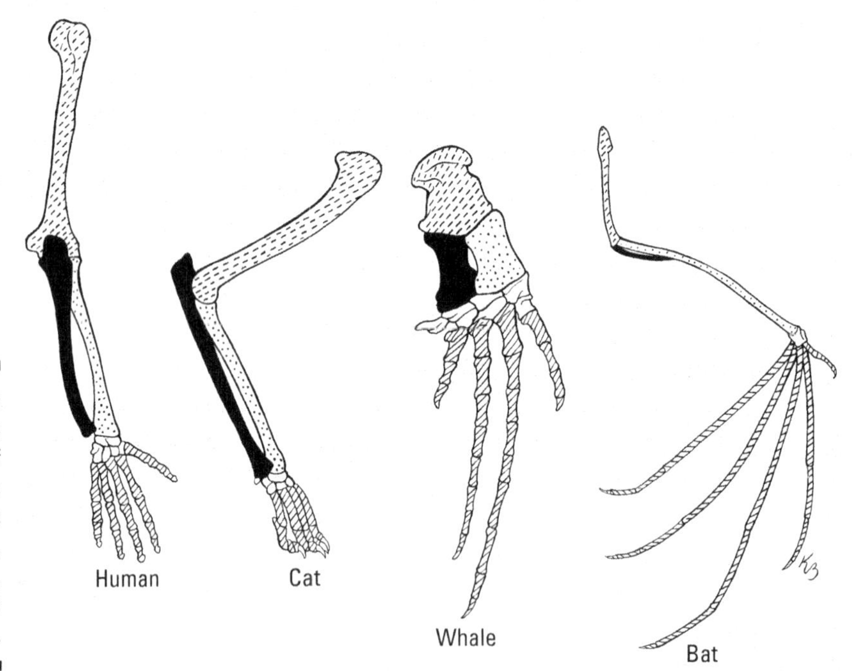


**Figure 12-2:** Comparative anatomy of the bones in front limbs of humans, cats, whales, and bats.

Since Darwin's time, many other examples have been found that illustrate how geographic distribution has influenced the biological evolution of organisms. The distribution of organisms on the Hawaiian Islands, for example, tells a very similar story to that of the Galapagos. Hawaii has types of living things that exist only on those islands but are related to living things found on the North and South American continents. The best explanation for the unusual life-forms found in Hawaii is that organisms arrived on the islands due to unusual events such as storms and then evolved separately from their mainland relatives.

Similarly, North and South America were separate continents before the Isthmus of Panama formed. Distinct groups of mammals lived in each area. Armadillos, porcupines, and opossums called South America home, whereas mountain lions, raccoons, and sloths lived in North America. The fossil record shows that these groups of mammals evolved separately until the Isthmus of Panama joined the two continents and the mammals were able to migrate back and forth.

## Molecular biology

*Molecular biology* is the branch of biology that focuses on the structure and function of the molecules that make up cells. With it, biochemists have been able to compare the structures of proteins from many different species and use the similarities to create *phylogenetic trees* (they're essentially family trees; see Chapter 10 for details) that show the proposed relationships between organisms based on similarities between their proteins.

With the development of DNA technology that allows for reading of the actual gene sequence in DNA (see Chapter 8 for more on DNA), modern scientists have also been able to compare gene sequences among species. Some proteins and gene sequences are similar between very distantly related organisms, indicating that they haven't changed in millions of years; these sequences are called *highly conserved sequences.*

One of these highly conserved sequences produces a protein called cytochrome c, which is part of the electron transport chain that occurs in mitochondria. Humans and chimpanzees have exactly the same amino acid sequences in their cytochrome c proteins, which indicates that humans and chimpanzees branched off the trunk of the evolutionary tree very recently ("recently" in evolutionary terms is still quite a long time, about 6 million years in this case). The cytochrome c protein in rhesus monkeys differs from humans and chimpanzees by just 1 amino acid (out of a total of 104), indicating that rhesus monkeys are slightly more distantly related to humans.

## Fossil record

The *fossil record* (all the fossils ever found and the information gained from them) shows detailed evidence of the changes in living things over time. During Darwin's day, the science of *paleontology,* which studies prehistoric life through fossil evidence, was just being born. Since Darwin's time, paleontologists have been busy filling in gaps left in the fossil record in order to explain the evolutionary history of organisms.

Hundreds of thousands of fossils have been found, showing the changing forms of organisms. For some types of living things, such as fish, amphibians, reptiles, and primates, the fossil record depiction of the changes from one form of the organism to another is so complete that it's hard to say where one species ends and the next one begins.

Based on the fossil record, paleontologists have established a solid timeline of the appearance of different types of living things, beginning with the appearance of prokaryotic cells (see Chapter 4) and continuing through modern humans.

## Observable data

Biological evolution can be measured by studying the results of scientific experiments that measure evolutionary changes in the populations of organisms that are alive today. In fact, you need only look in the newspaper or hop online to see evidence of biological evolution in action in the form of antibiotic-resistant bacteria.

In the 1940s, when people first started using antibiotics to treat infections, most strains of the bacterium *Staphylococcus aureus (S. aureus)* could be killed by penicillin. By using antibiotics, people applied a strong selection pressure to the populations of the *S. aureus* bacteria. The fittest *S. aureus* bacteria were those that could best withstand the penicillin. The bacteria that couldn't withstand the penicillin died, and the resistant bacteria multiplied. Today, most populations of *S. aureus* are resistant to natural penicillin. Another strain of *S. aureus* called *MRSA* has evolved that's not only resistant to natural penicillin but also to the semisynthetic methicillin, which used to be a great weapon in a doctor's staph-infection-fighting arsenal. For most strains of *MRSA*, vancomycin is the last effective treatment, but for some new, highly dangerous strains, vancomycin is beginning to fail. In the late 1990s, the first strains of *VRSA — vancomycin-resistant S. aureus* — were reported. Doctors don't currently have anything that can fight off *VRSA*; if a person gets a dangerous *VRSA* infection, chances are he'll die.

Using antibiotics is a double-edged sword. By using them to fight infection, people get healthy but they also speed up the process of natural selection. The potentially good news is that because doctors and scientists understand biological evolution, they're able to recognize what's happening and can take action to counteract these trends (such as prescribing fewer antibiotics).

## Radioisotope dating

Radioisotope dating indicates that the Earth is 4.5 billion years old — that's plenty old enough to allow for the many changes in Earth's species due to biological evolution. *Isotopes* are different forms of the atoms that make up matter on Earth (see Chapter 3 for more on isotopes). Some isotopes, called *radioactive isotopes,* discard particles over time and change into other elements. Scientists know the rate at which this radioactive decay occurs, so they can take rocks and analyze the elements within them. Using the known rates of radioactive decay and the types of elements that were originally present in the rocks, scientists can calculate how long the elements in a particular rock have been discarding particles — in other words, they can figure out the age of the rock (including rocks with fossils).

# Why So Controversial? Evolution versus Creationism

Virtually all scientists today agree that biological evolution happens and that it explains many important observations about living things, but many nonscientists don't believe in biological evolution and are often violently opposed to it. They prefer to take the Bible's creation story literally. These wildly differing viewpoints have led to one of the great debates of all time: Which is correct, evolution or creationism? (*Creationism* is the idea that God created the world and all the life on it out of nothing. Most creationists believe the creation story that's told in the Bible's book of Genesis.)

The idea of biological evolution has inspired so much controversy over the years in large part because many people think it contradicts the Christian view of humanity's place in God's design. According to the ancient Greek philosopher Aristotle, no accidents occur in nature; therefore, everything in nature is created for a purpose. A 17th-century thinker, William Paley, built on this idea with his theory of *intelligent design:* Beautiful designs don't arise by chance; if a beautiful design exists, the designer must also exist.

At the root of the controversy about biological evolution, then, seems to be this question: If living things developed in all their wonderful complexity due to natural processes and without the direct involvement of God, what does that do to man's place in the world? Is mankind not "special" to God?

If you have strong religious beliefs and you think that accepting biological evolution as a fact would somehow make you less special to God, then it's easy to see how belief in biological evolution creates conflict. But are biological evolution and religious faith necessarily in conflict? Many religious figures and scientists don't think so. In fact, many scientists have strong religious beliefs, and many religious leaders have come forward to say that they believe in biological evolution.

Ultimately, each person's beliefs are under his or her own control. But scientists stress the difference between beliefs, or faith, and science.

- Science is an attempt to explain the natural world based on observations made with the five senses. Scientific ideas, or hypotheses, must be testable — able to be proven false — by observation and experimentation (see Chapter 2 for more on the nature of science).
- The existence of God isn't within the scientific realm. God is widely believed to be a supernatural being, outside the workings of the natural world. Belief in the existence of God is therefore a matter of faith.
- Because intelligent design and creationism invoke the existence of a supernatural designer or creator, they're neither scientific ideas nor scientific theories and can't be tested or observed by scientific means. People who support intelligent design often support their arguments with observations of the natural world, but the explanations they propose for their observations aren't based in the natural world, nor do they conduct experiments of their ideas based in the natural world.

Creationism and intelligent design don't follow the fundamental rules of science and can't be considered scientific ideas.

Table 12-1 puts the scientific and creationist arguments about biological evolution side by side so you can compare them and come to your own conclusions about what you believe.

**Table 12-1 Faith-Based versus Scientific Views on Evolution**

| *What Creationists & Believers in Intelligent Design Say* | *What Scientists Say* |
|---|---|
| Nature is beautiful and complex. Many living things are perfectly suited to their role in nature. These wonderful designs couldn't have arisen by random chance; an intelligent designer must exist. | Biological evolution isn't random. Change is random, but biological evolution is based on change and natural selection. Natural selection causes populations to shift in particular directions, specifically those that are best suited to environmental conditions. If particular organisms and structures seem perfectly suited to their environment, that's because natural selection has made them that way. |

| What Creationists & Believers in Intelligent Design Say | What Scientists Say |
|---|---|
| The complexity of living things, from the many metabolic reactions in the cell to the incredible vertebrate eye, couldn't have been suddenly created through the accumulation of random changes. | Complex processes and structures aren't suddenly created out of nothing. Biological evolution works by adapting existing structures. By accumulating several changes that remake existing structures, new processes and structures are created. |
| The fossil record doesn't support biological evolution — too many gaps exist between species. Also, the missing link between humans and apes has never been found. | The fossil record in Darwin's time was incomplete, but today many evolutionary lines are well documented, including that of primates. Two particularly important fossils that show transitions between species are those of *Archaeopteryx,* a feathered reptile that appears to mark the transition between dinosaurs and birds, and *Tiktaalik,* an animal that appears intermediate between fish and four-legged animals. Tiktaalik had lungs and gills and was able to support itself on four legs. |
| Biological evolution is controversial even among scientists, and some scientists have proven it wrong. | Virtually all scientists accept biological evolution and recognize its importance in explaining life on Earth. Scientists often argue and conduct experiments about the details of how biological evolution occurs — after all, this behavior is at the heart of scientific inquiry — but scientists don't question whether biological evolution is a fact. Darwin's central idea of biological evolution by natural selection is still accepted and has been supported by many lines of research. |

# How Humans Evolved

You're a member of the *Homo sapiens* species. Humans are the only living species of *hominids* (modern humans and their extinct relatives) on the planet, but scientists have found fossils of other hominid species that give clues to our evolutionary origins.

The closest living relatives to humans are other primates, such as apes and chimpanzees. *Primates* are the order of mammals that includes monkeys, apes, and humans. They have large brains, grasping hands, and three-dimensional vision. (See Chapter 10 for the full scoop on the various categories in the taxonomic hierarchy.)

In the sections that follow, we tell you all about the tools scientists use to fill in the blanks about how humans evolved as well as the discoveries and connections they've made over the years.

## Fossil finds

Perhaps the best clue to understanding why you have the physical structure you do is the fossil record of hominids. Scientists use fossils to piece together clues about where humans came from and what our relationship is to other primates. Hominid fossils are rare and often incomplete, but when new ones are found, they contribute new pieces of information to the story.

Scientific theories that are supported by lots of evidence from many lines of research, like the theory of biological evolution, don't usually change substantially in response to new evidence. Instead, new evidence helps refine the theories and point out important details of how processes work.

Scientists' ideas about the biological evolution of humans have developed over time. Following is a rundown of the different hominid fossil discoveries made since the late 17th century:

- In 1891, researcher Eugene Dubois discovered a few bones in Java, Indonesia, a large island off the southeastern coast of Asia. Calling his discovery Java Man, Dubois thought he'd found the link between ape and man. What he found certainly was an ancestor to modern *Homo sapiens,* but it wasn't apelike. Dubois had actually found a member of the species *Homo erectus,* one of the earliest walking hominids. Other *Homo erectus* bones have been found in China and Africa.
- During the 1930s, a researcher named Raymond Dart examined a small skull that was found in Taung, a town in South Africa. After studying the bone structure and realizing that the skull contained a petrified brain, Dart came to the conclusion that the skull belonged to a child who was about 6 years old and a member of a human ancestral species. The remnants were called Taung Child. Dart thought *he* had found a missing link between apes and men, but others disagreed. Dart was ridiculed for

suggesting that a human ancestor was "out of Africa," when the thinking at that time was that the first human species came from Asia (due to the hullabaloo surrounding Java Man). But Dart persevered, even though his belief was unpopular at the time. He classified his skull as *Homo habilis,* meaning handyman, because crude stone tools were found near the bones.

- In the 1930s, Louis and Mary Leakey began excavating the Olduvai Gorge in Tanzania, Africa. Three decades later, their son Richard noticed the jaw of a saber-toothed tiger sticking out of the archaeological site. Digging continued in the area, and eventually pieces of three skeletons were unearthed. The skeletons were also classified as those of *Homo habilis* and were dated at about 2 million years old. The Leakeys continued their work at Olduvai Gorge well into the 1980s, and in 1984, they unearthed a spectacular find: the first (and still the only) full skeleton of a *Homo erectus,* dated at 1.6 million years old.
- In 1994, Richard Leakey's wife, Meave, headed upstream from Olduvai Gorge to Kanapoi in northern Kenya and found a 4.2-million-year-old hominid ancestor. The lower jaw was complete, and the teeth were surprisingly like those of a modern human. However, the shape of the jaw was like that of a chimpanzee. Pieces of lower leg bone were also found, indicating that the creature walked on two legs. Meave named her find *Australopithecus anamensis.*
- In the years since Dart and the Leakeys, several more fossilized bones have been discovered along the southeastern coast of Africa, giving Africa the nickname "the cradle of civilization." In particular, a 3.2-million-year-old skeleton of the *Australopithecus afarensis* species was found in Ethiopia (part of northern Africa) in 1974 by Don Johanson; the skeleton was nicknamed Lucy.
- Bones from a 4.4-million-year-old skeleton found in Ethiopia are currently being studied. The skeleton is called *Ardipithecus ramidus,* and because it's the oldest known ancestral fossil, scientists are using it to try and determine whether this organism was in fact a direct ancestor to humans.

Scientists can use fossilized bones like those of early hominids such as Lucy to understand how our ancestors' more apelike features evolved into the features of modern humans. Table 12-2 gives you an overview of the physical changes that occurred as apes evolved into humans.

**Table 12-2 Changing from Apes into Humans**

| ***Anatomic Structure*** | ***Changes*** |
|---|---|
| Arms | Because apes walk on all four limbs, their front limbs don't straighten completely (if they did, the apes would suffer dislocations). Consequently, apes don't have elbows, which allow the arm to straighten, but humans do. |
| Brains | Modern humans have prominent foreheads. The size and shape of the skull has changed as the size and shape of the brain has changed. Human brains are now larger and more rounded than those of ancestral species. And the bony ridge above the eyebrows of humans has shrunk significantly in comparison with human predecessors. |
| Feet | Now that humans walk upright, the shape of the heel has changed to absorb the impact of the foot hitting the ground differently. |
| Hands | The hand of a human and a chimpanzee are amazingly similar. The anatomic structures are the same; the differences lie solely in the fingerprints. Humans and other primates have *prehensile* (grasping) thumbs, which allow the gripping of objects. Prehensile thumbs appeared in human predecessors about 18 million years ago. |
| Jaws | The human jaw and teeth have shrunk. Now that humans cook food (instead of eating it raw), their teeth don't have to tear and grind as much. Instead, humans have developed chins to help support the thinner jawbone. The changes in the jaw and flattening of the face have allowed humans to produce language. |
| Knees | Knees allow humans to walk upright. The ability to straighten the leg supports the weight of the body, and because the knee is positioned beneath the pelvic bones (rather than in front of them), humans don't waddle during movement. Waddling slows a human down, and humans occasionally need to run. |
| Tails | Apes don't have tails, and humans no longer have them either. This anatomic feature disappeared about 25 million years ago. However, the remnants of a tail are evident in the coccyx bone at the end of your spine. |

# Digging into DNA

The development of DNA technology has played a huge role in helping scientists read some of the human history encoded in DNA (we cover the complexities of DNA in Chapter 8). By simply comparing the DNA sequences of hominids, scientists can discover several pieces of information, such as

- **Which hominids are most closely related:** Species that are closely related have greater similarities in their DNA sequences than species that are more distantly related. Humans are most closely related to chimpanzees; our DNA sequences are about 97 percent identical. Today, the current line of thinking is that there was an apelike species alive 10 to 20 million years ago that branched into a line of gorillas around 7 million years ago. That species then branched off into two lines about 5 to 6 million years ago. One of these branches on the family tree led to chimpanzees, and the other evolved into humans.
- **How hominids migrated:** By comparing the genetic relationships between hominids, the age of certain fossils, and the geographic locations of these fossils, scientists can figure out where species originated and where they traveled.
- **When new species emerged:** In animals, species are defined based on their ability to interbreed successfully. Typically, two organisms that can successfully produce offspring are considered to be in the same species, whereas organisms that can't produce offspring together are considered to be unique species. With extinct species, scientists can't make direct observations of who could interbreed. However, scientists can look at the DNA sequences from fossils to see who was mingling DNA. Scientists have compared DNA sequences from humans and other living primates, as well as those from fossilized hominids, including fossilized Neanderthals.

# Check out the big brain on the Homo sapien

It's one thing to know how *Homo sapiens* evolved into today's modern humans. It's another thing entirely to understand *why* these changes happened. For example, why did the human brain become so much larger than that of other hominids? The clues about why things happened the way they did are pretty scarce, but they do exist in the form of tools found with a skeleton, evidence of burial of human remains, and evidence of the use of fire. From these types of clues, scientists can put together hypotheses to explain the evolutionary pathway of modern humans. Some of these hypotheses are as follows:

- As human ancestors began to walk upright, they soon began to hunt. Therefore, they went from being herbivores to carnivores. One factor that led to this development was climate change. As the Earth began to warm up, some of the forests disappeared and became open savannas. In an open savanna, it's much easier to see prey (especially if you're standing). So, human ancestors became successful hunters and ate plenty of meat.
- Eating plenty of meat, with all the fats and proteins meat contains, made hominid brains bigger, and bigger brains were selected for over time. (However, scientists still don't know why.)
- As the shape and size of brains changed and enlarged, ancestral females had to give birth earlier so the offspring's head could fit through the pelvic bones. Because babies were born earlier, they were much more dependent on their mothers for a longer period of time. This change meant that the mother couldn't contribute to hunting, but she still needed adequate nutrition to make milk to breastfeed her baby. The father and other members of the clan therefore had to help the mother by bringing her food. The fact that the mother had to rely on others for her survival and that of her baby led to the formation of close ties with other members of the clan.

Table 12-3 compares the brains of different hominid species.

**Table 12-3 The Evolution of Hominid Brains**

| ***Genus and Species Name*** | ***Brain Development*** |
|---|---|
| *Australopithecines anamensis, Australopithecines afarensis* | Brains were about 400 cubic centimeters (cc) in size, comparable to that of chimpanzees or gorillas. Both species could walk on two legs but were intellectually apelike. |
| *Homo habilis* | Brains were about 650 cc. This species was capable of using stone tools. |
| *Homo erectus* | Brains were 850 to 900 cc. This species began socializing with other members of the species. |
| *Homo neanderthalensis* | Brains were 1,300 cc. This species had larger bodies. |
| *Homo sapiens* | Brains are between 1,200 and 1,600 cc. This species has larger frontal lobes and broader foreheads because of increased brain capacity. |

# Part IV

# Systems Galore! Animal Structure and Function

## In this part . . .

Has anyone ever told you that you're an animal? Well you are — literally — because you're a member of the animal kingdom. The body plans of animals are very diverse, but they all have certain things in common, including the need for oxygen and food. Some animals, such as humans, are made up of many complex organ systems that coordinate the structure and function of the body.

In this part, we introduce the fundamentals of the many organ systems in the human body. We also take a peek at some of the different ways that other animals do things, including how they obtain the nutrients they need from food and how they send those nutrients throughout their bodies.

# Chapter 13

# Pondering the Principles of Physiology

### In This Chapter

- Connecting structure with function
- Understanding important physiological concepts such as evolution and homeostasis

*Physiology* is the study of the function of all living things in their normal state. The function of living things is closely tied to their structure and begins at the cellular level. In order to survive, living things must be able to regulate their functions and respond to changes in the environment. In this chapter, we help you see how the study of physiology is applicable at all levels of life and introduce you to some of the fundamental principles of physiology that apply to all the organ systems of the human body.

## Studying Function at All Levels of Life

To truly be able to understand a living being, you need to have a good mental picture of its structure and the function of its body and cells. Enter anatomy and physiology. *Anatomy* is the study of the structure of living things, and *physiology* is the study of how these structures function. These two branches of biology go hand in hand because the function of an organism is dependent upon its structure.

For example, the function of the heart is to pump blood around the body. The heart muscle contracts, putting pressure on the blood and squeezing it out of the heart and into the arteries. In order for the heart to function properly, flaps of tissue within the heart, called *valves,* must close off chambers within the heart so that the blood doesn't flow back into it. However, some people are born with a defect in their valves that prevents the valves from closing completely. In these people, the heart pumps the blood inefficiently because some blood flows back into the heart instead of going out into the arteries. A person with this type of heart defect may have poor circulation and tire easily. Heart valve defects are just one example of how differences in the structure of an organism can affect its function.

You can study anatomy and physiology at all levels of the organization of living things — from the smallest units of life (cells) to organisms. Some scientists even study *ecological physiology* by looking at how the physiology of organisms is interrelated with their environment. The complexity of anatomy and physiology grows as you move up the levels of the organization of living things, as you can see from the following:

- **Tissues are made of cells, which are made of molecules.** The foundation of physiology rests upon the function of cells, but to understand the function of cells, or the details of a physiological process, you need to be able to follow the interactions of molecules within the cell. (For more on cells, see Chapter 4; for more on molecules, see Chapter 3.)
- **Organisms are made of organ systems, which are made of organs, which are made of tissues.** In order to understand the function of an organism, you need to understand the functions of the organ systems and organs that make up the organism. And the functions of the organs depend upon the function of their tissues, which are groups of similar cells.
- **Organisms multiply to form populations, which interact with other populations to form communities, which interact with their environment to form ecosystems.** An organism's interaction with the living and nonliving things in its environment can influence its physiology. For example, environmental toxins such as polychlorinated biphenyls (PCBs) have estrogen-like properties and can affect the reproductive physiology of organisms. Likewise, interactions with pathogens can cause disease, which has a negative impact on an organism's physiology. (See Chapter 11 for the scoop on ecosystems and Chapter 17 for details on pathogens.)

To understand the function of a structure at any level in the organization of living things, you often need to know something about one of the lower levels. However, knowing everything about the lower levels doesn't necessarily tell you how the higher levels function. Sometimes the sum of all the parts is greater than what you expect. For instance, you likely wouldn't predict the intellectual and emotional properties of the human brain just from studying how individual neurons function.

The properties of an entire system that are greater than the functions of the individual parts have a special name — *emergent properties.*

# Wrapping Your Head around the Big Physiological Ideas

In the human body alone, ten different organ systems interact with each other to regulate physiology (we introduce you to these systems in the rest of Part IV). Each system has its own parts and processes and makes unique contributions to the whole.

In the following sections, we present a few physiological concepts that are central to the functioning of all of your organ systems. With these big ideas in mind, you can more easily see the similarities in the different systems and understand some of the fundamental processes that regulate their functions.

## Evolving the perfect form

*Biological evolution,* the study of how populations change over time, explains the relationship between structure and function that's at the core of physiology. Scientists can look at the structures and functions of different kinds of organisms and compare them to reveal how biological evolution creates variations on a theme to improve the functioning of a part of an animal (be it tissue, organ, or an entire organ system) so that the animal can better cope in its environment.

For example, today's scientists know that the function of the kidney is to reabsorb water into the body (we cover the kidney in more detail in Chapter 16). Within the kidney, a special tube called the loop of Henle helps set up conditions that allow mammals to reabsorb water from the fluid that enters the kidney, concentrating the urine and conserving water for the organism. Mammals that live in the desert are under strong selection pressure to conserve water (see Chapter 12 for more on biological evolution and selection pressures). Many desert mammals have an extralong loop of Henle in their kidneys that allows them to reabsorb most of their water. These mammals produce very concentrated urine, conserving water in a way that helps them survive in their desert environment. By comparing these desert mammals to their non-desert-dwelling relatives, scientists can discover evidence of the evolution of the loop of Henle's function.

## Balancing the body to maintain homeostasis

More than 100 years ago, the French physiologist Claude Bernard noted that two different environments are important to animals:

- **The external environment:** The external environment includes the Sun and the atmosphere that surrounds the animal. It experiences fairly large changes, such as temperature changes as the Sun rises and sets.
- **The internal environment:** The internal environment includes the fluids that surround the cells in an animal's tissues. It's mainly affected by the diet of the animal and the amount of water that the animal drinks.

If the internal environment of an animal changes too much, the conditions may kill the animal's cells. Animals therefore use control systems to respond to and counteract changes in their external environment in order to keep their internal environment within a certain range that allows them to survive. In other words, animals (including you) are constantly trying to maintain *homeostasis* (balance) within their bodies.

Many different homeostatic processes maintain the balance of variables such as pH level, glucose level, and body temperature in an animal's body. In order for homeostasis of a particular variable to be maintained, the animal must be able to

- Measure the change of the variable in the body
- Respond by changing the behavior of components in the body that regulate that variable

Most homeostasis control relies upon negative feedback. In *negative feedback,* a change triggers a response that reverses the change. For instance, after a meal, the amount of glucose in the blood increases. The body responds by releasing insulin into the bloodstream, which signals the body to transfer glucose from the blood into the cells and lowers the level of glucose in the bloodstream.

The range of values that an organism can tolerate for a particular variable act like a *set point* for maintaining homeostasis. If changes occur in the internal environment, the changes are measured and compared to the set point. The difference between the state of the variable and the desired set point is used to generate signals that trigger actions, like negative feedback, designed to return the body to the set point. For example, blood glucose must stay within a certain range or else a person will develop the disease diabetes. When blood glucose rises after a meal, insulin triggers negative feedback that lowers the levels back to the normal range.

To help you understand how homeostasis works, think of the body like a heating and cooling system. You determine the desired set point by setting the thermostat, and the thermostat measures the temperature of the room. If the temperature of the room is higher than the desired set point, the thermostat sends signals to turn on the cooling system. When the temperature of the room reaches the set point, the thermostat sends signals to turn off the cooling system. Just like your body, the heating system has a mechanism for measuring the change in the variable (in this case, temperature) and then responding to that change (by turning on the heating and cooling systems).

As you study the human body's organ systems, you can expect to encounter many examples of homeostatic control and negative feedback. Although the details of each control system are different, they all have the same three components:

- **A receptor:** The *receptor* measures changes in the variable, such as blood pressure, body temperature, or heart rate, and sends information to the control center.
- **A control center:** The *control center* can be a neuron or gland so long as it processes the information, initiates a response to keep the variable within its normal range, and sends the response to an effector.
- **An effector:** Often a muscle or a gland, an *effector* carries out the body's response.

Homeostasis doesn't keep conditions in the body exactly the same all the time. The set point for a variable can change depending on the situation the organism is in. Your body temperature, for example, changes throughout the day. It may drop low while you sleep, or it may be high when you exercise. So you see, homeostasis keeps your internal environment within a particular ideal range, but it doesn't keep it rigidly fixed at one point.

## Getting the message across plasma membranes

Cells communicate with other cells, with tissues, and with organs. This communication is vital to the integration of all the body systems and to the maintenance of homeostasis (covered in the preceding section). The plasma membranes of cells separate them from their environment, maintaining a delicate balance between the outside and the inside of the cell (for more on plasma membranes, see Chapter 4). In complex, multicellular organisms such as humans, each cell has a specialized function (head to Chapter 19 for details on how cells become specialized). The function of the entire organism depends upon the coordinated functioning of all the cells within the body.

Signals are received by cells at the plasma membrane and then passed inside the cell by a process called *signal transduction.* During signal transduction:

1. **A signal arrives at the plasma membrane of a cell and binds to a receptor in the plasma membrane.**

   Signaling molecules that bind to receptors are called *ligands.* Each ligand binds specifically to its unique receptor, so cellular responses to each signal are very specific.

2. **The receptor in the plasma membrane changes in response to the signal.**

   For example, the receptor may change shape.

3. **The receptor interacts with a messenger molecule inside the cell that receives the signal and changes in response.**

4. **The intracellular messenger interacts with a target protein that causes a change in the behavior of the cell.**

   The change in behavior is often the result of changes in gene expression (see Chapter 8 for more on this topic).

## Recognizing that what comes in, must go out

Organisms must take in matter and energy from their environment in order to survive, but they can't create (or destroy) either one. Instead, they must transform matter and energy from one form to another. The reactions that make this possible are the metabolism of an organism (see Chapter 5 for more on metabolism).

The *Law of Mass Balance* says that if the amount of a substance in the body is to remain constant, any input must be offset by an equal output. Simply stated, ins must equal outs.

Mass balance is the fundamental principle underlying the regulation of several systems in the human body, including

- The concentration of oxygen and carbon dioxide in the respiratory system
- The flow of blood through the heart
- The clearance of materials through the kidneys
- Water and electrolyte balance in the blood

In any of these systems, control mechanisms return the body to homeostasis when mass balance is disrupted.

# Chapter 14

# Moving and Shaking: Skeletal and Muscular Systems

### In This Chapter

- Seeing how animals move from one place to another
- Surveying the various skeletal systems
- Understanding what makes muscles so valuable

The coordinated efforts of muscles and skeletons are what make animal movements possible. Muscles pull or push, and skeletons give the muscles something to pull or push against. Prepare to find out all about how animals move from place to place as you discover the different types of skeletons and the fundamentals of muscle function in this chapter.

## Doing the Locomotion, Animal-Style

Fish swim, dogs run, frogs jump, worms crawl, and birds fly. Each of these types of *locomotion,* movement from one place to another, requires animals to use energy to overcome the forces of friction and gravity that would otherwise hold them to the Earth.

Each different animal is adapted for the environment it lives in and the type of locomotion it performs.

- Swimming animals, such as fish and whales, have bodies that are shaped to minimize resistance as they move through the water. They're often coated with water-resistant mucus and may have structures that make them more buoyant.
- Birds have hollow bones and wings that are shaped like those of an airplane to create added lift during flight.
- Animals that walk or run on land, such as lions and elephants, have a strong skeleton and muscles to support them against the force of gravity.

- Rabbits, kangaroos, and other animals that jump or hop have extralarge leg muscles and strong tendons to help put some spring in their hop.
- Animals that slither or crawl, think worms and snakes, have smooth, tubular bodies to lessen resistance due to friction as they move over or through soil.

# The Types of Skeletal Systems

Skeletons support animals, give their bodies shape, and protect their internal organs, but not all animals have the same type of skeleton. Following are the three different kinds of skeletons you may see in your study of biology:

- **Hydrostatic skeletons:** Found in creatures such as worms and jellies, *hydrostatic skeletons* are basically chambers filled with water. Animals with this skeleton type move and change their shape by squeezing their water-filled chambers — just like what happens when you squeeze a water balloon.
- **Exoskeletons:** These are exactly what they sound like — skeletons on the outside of the body. You're probably quite familiar with these hard exterior coverings because they're found on crabs, lobsters, and many insects. Exoskeletons are rigid and can't expand as animals grow, so animals must *molt,* or shed, their exoskeletons periodically. After an animal molts, its new exoskeleton is soft — as in a soft-shelled crab.
- **Endoskeletons:** The most familiar of all skeleton types is the endoskeleton. After all, it's the kind of skeleton you have. An *endoskeleton* exists within an animal's body. The human endoskeleton is hard because it's partially constructed of the mineral calcium. The endoskeletons of other animals may be more flexible — for example, the endoskeleton of a shark is made of cartilage, the same material that makes up the soft parts of your nose.

Animals with hydrostatic skeletons and exoskeletons are considered *invertebrates,* meaning they don't have a backbone. Animals with endoskeletons, like you, are considered *vertebrates* because they have a backbone.

The following sections not only break down the parts of a vertebrate animal's skeleton but they also get you more familiar with the important components of your skeleton — bones and joints.

## Splitting apart vertebrate skeletons

All vertebrate skeletons— whether they belong to humans, snakes, bats, or whales — developed from the same ancestral skeleton (which explains why you may notice similarities between your skeleton and that of your pet dog or cat). Today, these animals show their relationship to each other in part due to *homologous structures* — structures that are equivalent to each other in their origin (see Chapter 12 for more on homologous structures and their importance to the study of evolution).

All vertebrates' skeletons, whether yours (see Figure 14-1), a whale's, or a cat's, have two main parts:

- **The axial skeleton:** This part supports the central column, or *axis,* of the animal. The axial skeleton includes the skull, the backbone (also called the *vertebral column*), and the rib cage. The skull protects the brain, the backbone protects the spine, and the rib cage protects the lungs and heart.
- **The appendicular skeleton:** This part extends from the axial skeleton out into the arms and legs (which are also known as *appendages*). It includes the shoulders, pelvis, and bones of the arms and legs.

In some vertebrates, such as snakes, the appendicular skeleton has become extremely reduced or nonexistent.

# Boning up on bones

If you've ever watched an old Western movie, you've probably seen images of bones bleached white by the Sun and scattered alongside a pioneer trail. The dry white bones of these images are very different from the living bones that are in your body right now. Bone is actually a moist, living tissue that contains different layers and cell types.

- Fibrous connective tissue covers the exterior of bones and helps heal breaks in an injured bone by forming new bone.
- Bone cells, which are embedded in a bone matrix, give cells their hard nature. The cells actually make the matrix, which consists of collagen that has been hardened by the attachment of calcium and phosphate crystals.
- Cartilage covers the ends of bones and protects them from damage as they rub against each other.

The tissues found within living bone fall into two categories:

- **Spongy bone tissues** are filled with little holes, similar to those you see in volcanic rocks. These holes are filled with *red bone marrow,* which is the tissue that produces your blood cells.

- **Compact bone tissues** are hard and dense. A cavity within compact bone is filled with *yellow bone marrow,* which is mostly stored fat. If the body suddenly loses a large amount of blood, it converts the yellow bone marrow to red bone marrow so that blood cell production can be increased.

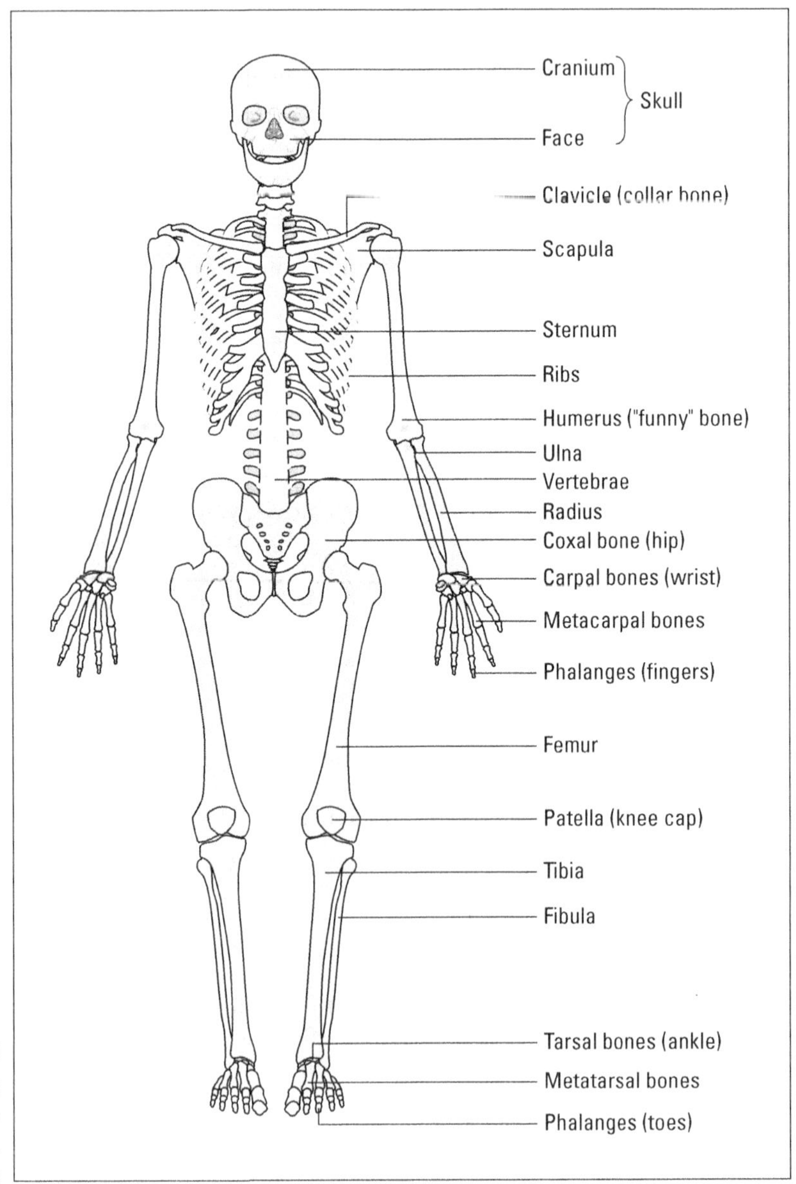


**Figure 14-1:** The human skeleton.

*From LifeART®, Super Anatomy 1, © 2002, Lippincott Williams & Wilkins*

### Got broken bones?

The number of broken bones in American children is on the rise, and doctors think this increase may be due to kids exercising less and drinking less milk. Exercise is important because it puts weight on your bones, which encourages bone growth and helps keep bones dense. Milk is important because the calcium and vitamin D found within it are vital for the development of healthy, strong bones. Doctors recommend milk as an excellent source of concentrated calcium that's easily absorbed by the body. To get your daily dose, drink four 8-ounce glasses per day, or take a calcium supplement (with your doctor's permission, of course). (***Note:*** Bone mass stops increasing after age 20, so get your calcium in before then!)

## Joining the movement fun

*Joints* are structures where two bones are attached so that bones can move relative to each other. Bones are held together at joints by *ligaments,* which are strong, fibrous, connective tissues.

Three different types of joints enable the many movements of animals:

- **Ball and socket joints** consist of one bone, with a rounded, ball-like end, that fits into another bone, which has a smooth, dishlike surface. Your arms and legs fit into your skeleton with ball and socket joints, which is why you're able to rotate your arms and legs in all directions.
- **Pivot joints** allow you to swivel a bone. When you rotate your arm so that your palm faces up, then down, then up again, you're using a pivot joint.
- **Hinge joints** allow you to bring two bones close together or move them farther apart, much like you open and close a book. Your elbows and knees have pivot joints that allow you to extend and contract your arms and legs.

# Why Muscles Are So Essential

Muscles are extremely important to your body — and not just because toned ones make your body look better and grow stronger. Without muscles, you couldn't walk, run, or play sports. You couldn't even obtain nutrients from your food or send blood to all the organs and tissues throughout your body.

Following is a rundown of all the things muscles do for you:

- **Muscles allow you to stand upright.** The force of gravity is strong; without your muscles, it'd keep you pinned to the ground. Your muscles contract so you can push against the surface of the Earth and stand upright and assume different positions.
- **Muscles make it possible for you to move.** Every little movement that your body performs, including blinking and smiling (or frowning), is controlled by your muscles.
- **Muscles allow you to digest.** Muscles all along your digestive tract keep food moving downward and outward. *Peristalsis,* the squeezing of food down through the esophagus, stomach, and intestines, is due to the contraction of these muscles. If they didn't contract to squeeze food throughout your digestive system, you'd never be able to obtain the nutrients you need to survive.
- **Muscles affect the rate of blood flow.** Blood vessels contain muscle tissue that allows them to dilate so blood can flow faster or to contract so blood flow slows down. Muscle contraction is also responsible for the movement of blood through your veins. (Not to mention that your heart is actually a muscle; without it, your blood wouldn't be flowing anywhere, period!)
- **Muscles help you maintain a normal body temperature.** Muscles give off heat when they contract, and your body uses that heat to maintain your body temperature because you continually lose some heat through your skin. This fact explains why you shiver when you're cold; shivering is your body's way of trying to generate heat.
- **Muscles hold your skeleton together.** The ligaments and tendons at the ends of your muscles wrap around joints, holding together the joints — and therefore the bones of your skeleton.

Figure 14-2 gives you an up-close view of the human muscular system. To find out about the specific types of muscle tissue in your body and how your muscles contract, check out the next sections.

## Muscle tissue and physiology

Muscle tissues are made up of muscle fibers, and muscle fibers contain many, many *myofibrils* — the parts of the muscle fiber that contract. Myofibrils are perfectly aligned, which makes muscles look striped, or *striated.* The repeating unit of these striations — containing light and dark bands — is called a *sarcomere.*

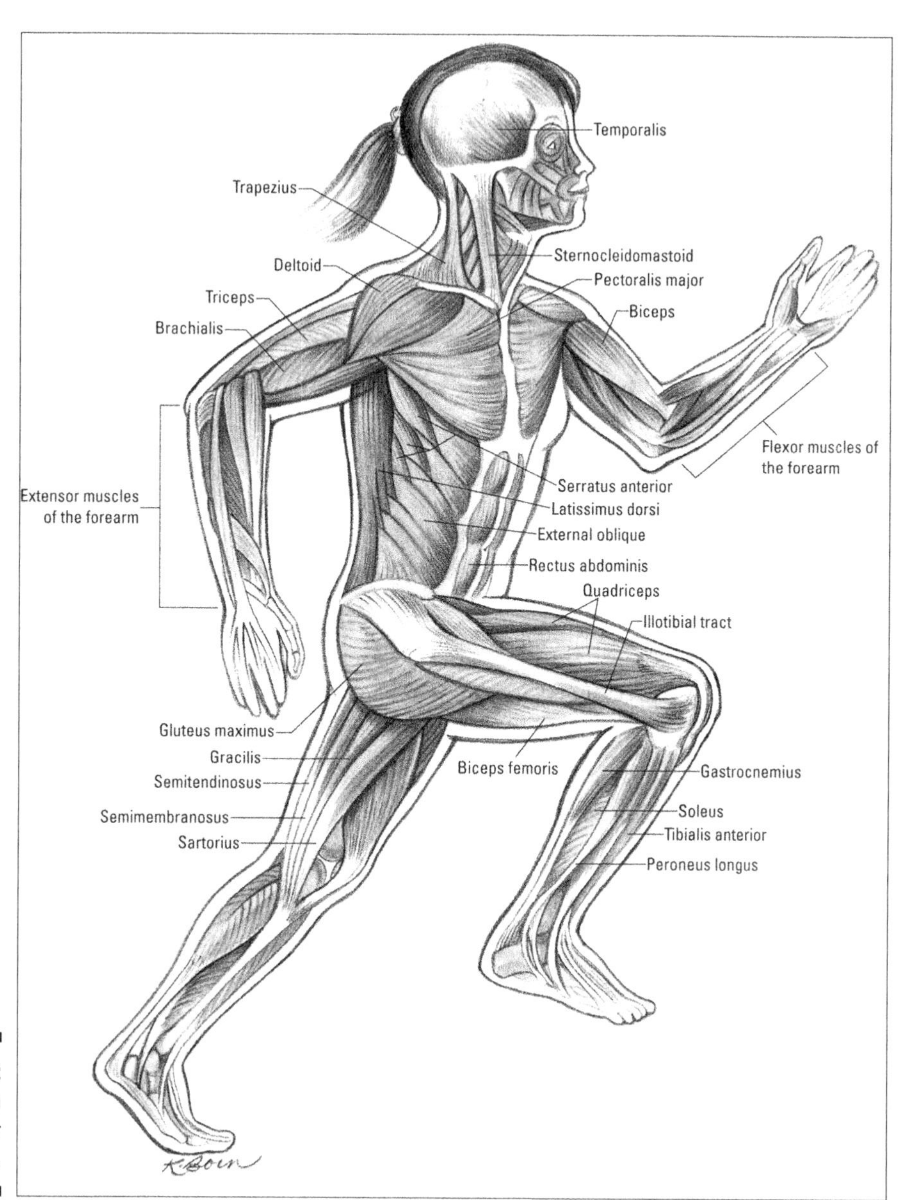


**Figure 14-2:** The human muscular system.

Three types of muscle tissue exist within your body:

- **Cardiac muscle** is found in the heart. The fibers of cardiac muscle have one nucleus (so they're *uninucleated*), are striated (so they have light and dark bands), and are cylindrical in shape and branched. The fibers interlock so contractions can spread quickly through the heart. Between contractions, cardiac fibers relax completely so the muscle doesn't get fatigued. Cardiac muscle contraction is totally *involuntary,* meaning it occurs without nervous stimulation and doesn't require conscious control.
- **Smooth muscle** is found in the walls of internal organs that are hollow, such as the stomach, bladder, intestines, or lungs. The fibers of smooth muscle tissue are uninucleated, shaped like spindles, and arranged in parallel lines; they form sheets of muscle tissue. Smooth muscle contraction occurs involuntarily and more slowly than skeletal muscle contraction, which means smooth muscle can stay contracted longer than skeletal muscle and not fatigue as easily.
- **Skeletal muscle** is probably what you think of when you picture a muscle. The fibers of skeletal muscle have many nuclei (so they're *multinucleated*) and are both striated and cylindrical; they run the length of the muscle. Skeletal muscle is controlled by the nervous system (which we describe in Chapter 18). The movement and contraction of skeletal muscle can be stimulated consciously, which means you consciously decide that you're going to stand up and walk across the room, an action that requires the use of muscle. Therefore, skeletal muscle contraction is said to be *voluntary.*

## Muscle contraction

Muscle contractions rely on the movements of the filaments that make up myofibrils. Basically, the different filament types slide over each other to cause a muscle contraction; this general theory is referred to as the *sliding-filament theory.*

The two filament types found in myofibrils are

- **Actin (thin) filaments:** An *actin,* or *thin, filament* is made up of two strands of *actin,* which is a protein that's wound in a double helix (just like DNA). It has molecules of troponin and tropomyosin at binding sites along its actin double helix.
- **Myosin (thick) filaments:** A *myosin,* or *thick, filament* contains groups of *myosin,* a type of protein with a bulbous end. Multiple strands of myosin are mixed together in opposite directions in muscle tissue, so it appears that both ends of myosin filaments are bulbous.

The actin filaments are attached to something called a Z-line, and the myosin filaments lie between the actin filaments, unattached to Z-lines. From Z-line to Z-line is one *sarcomere,* which is a unit of contraction. You can see all of these parts and more in Figure 14-3, which shows you how skeletal muscle is connected to the nervous system and how it contracts.

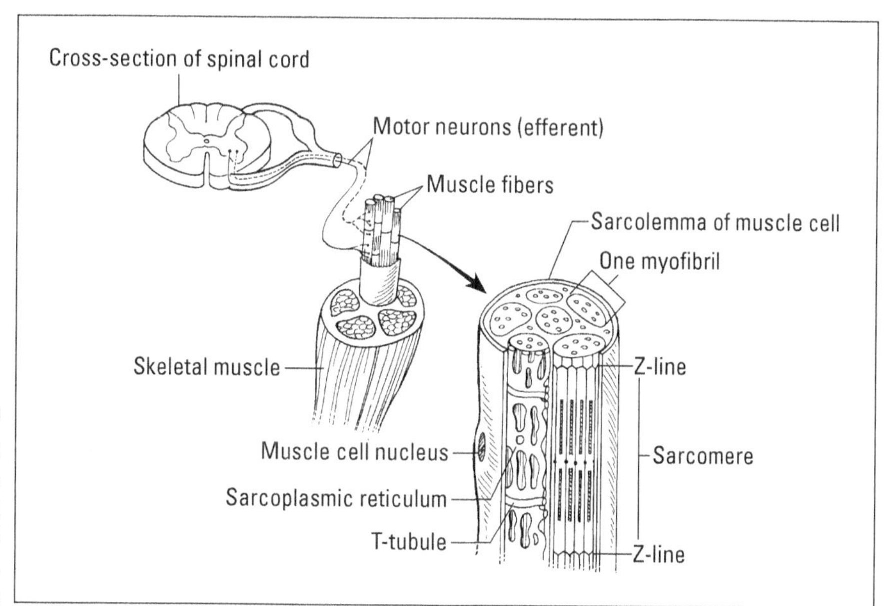


**Figure 14-3:** Structure and function of a skeletal muscle.

The other element necessary for muscle contractions is adenosine triphosphate, or ATP for short (we cover this energy-storing molecule in greater detail in Chapter 5). A muscle fiber contains only enough ATP to sustain a contraction for about one second. After muscle cells use up their available ATP, they get more by

- **Using energy from stored phosphocreatine molecules:** Phosphocreatine, which is made up of ATP and creatine, forms during periods of no contraction. It's quickly broken down to release more ATP as the low amounts of ATP in a muscle cell are used up.
- **Increasing the rate of cellular respiration:** Muscle cells are loaded with *mitochondria,* organelles that perform cellular respiration. During cellular respiration, the mitochondria use oxygen to break down food molecules and transfer their energy to ATP (described in Chapter 5). As your muscle cells use up their ATP, you breathe harder to supply them with more oxygen so they can do more cellular respiration.

- **Recycling ADP molecules into ATP:** Every time a muscle cell uses an ATP molecule for energy, a phosphate is removed, producing adenosine diphosphate, or ADP. Human muscle cells take every two molecules of ADP produced during contraction and recombine them to make a new ATP molecule plus a molecule of adenosine monophosphate, or AMP.
- **Resorting to lactic acid fermentation:** *Lactic acid fermentation,* which produces a small amount of ATP through the partial breakdown of glucose, is a worst-case scenario for muscle cells. Cells make a lot more ATP for each glucose molecule they break down by cellular respiration than they do by lactic acid fermentation. So, cells resort to lactic acid fermentation only when ATP can't be obtained through cellular respiration, a situation that can occur when the body's oxygen stores are depleted.

According to the sliding-filament theory, muscle contraction occurs via the following process:

1. **ATP binds to the bulbous end of a myosin filament and splits into a molecule of ADP plus a molecule of inorganic phosphate ($P_i$).**

   The ADP and $P_i$ stay attached to the myosin.
2. **Calcium binds to the troponin in the actin filament, which causes the tropomyosin in the actin filament to move out of the way so the binding sites on the actin filament can open.**
3. **After the actin filament's binding sites are exposed, myosin binds to the actin, which causes myosin to release the ADP and $P_i$.**
4. **When myosin releases the ADP and $P_i$ so it can link to actin, the shape of the bulbous end of the myosin filament changes and the actin filament slides toward the middle of the sarcomere, pulling the Z-lines at the end of the sarcomere closer together.**

   The result? A shortening, or contraction, of the muscle fiber.
5. **The connection between the actin and myosin filaments breaks when another ATP molecule attaches to the bulbous end of the myosin filament.**

## Chapter 15

# Going with the Flow: Respiratory and Circulatory Systems

### *In This Chapter*

- Discovering the four ways animals exchange gases
- Understanding how circulatory systems feed cells and take out their trash
- Examining the heart and circulatory systems of not-so-complex animals
- Tracing the path of blood through your heart and blood vessels
- Surveying the details of your body's most important fluid — blood

Every living thing needs to be able to exchange materials (think food, oxygen, and waste products) with the environment and circulate these materials around their bodies. Complex animals use respiratory systems to exchange gases such as oxygen and carbon dioxide with their environment and circulatory systems (often called *cardiovascular systems* when they're in animals with backbones) to move nutrients and gases around their bodies.

In this chapter, we present some of the different processes used by animals to exchange and move important materials. In particular, we focus on the details of the human respiratory and circulatory systems.

## Passing Gas: How Animals "Breathe"

All animals, from worms and fish to insects and humans, must exchange gases between themselves and their environment on a continual basis throughout every moment of their lives. Simple animals exchange gases throughout their entire body surfaces, whereas more complex animals have respiratory systems that process the air from the environment. This latter group relies on breathing to simultaneously move oxygen gas from the air into the body and remove carbon dioxide from the body and return it to the air. Ultimately, whichever way an animal does it, the actual exchange of oxygen and carbon dioxide gases between the animal and its environment occurs across a moist surface.

*Respiration* is the entire process of taking air in, exchanging needed gases for unnecessary gases, using the needed gases, and releasing the waste form of gases.

Animals use four types of gas-exchange systems, which we cover in detail in the next sections:

- **Integumentary exchange** occurs through the outer surface of an animal. Worms and amphibians employ this system.
- **Gills** are external structures that exchange gases in watery environments, which is why you find gills on many forms of marine life.
- **Tracheal exchange systems** rely on a network of tubes that end in holes to move oxygen and carbon dioxide throughout the bodies of certain types of insects.
- **Lungs** are internal structures that use diffusion to transport gases into and out of the body. Land animals, including humans, and marine mammals, such as dolphins and whales, have lungs.

## Integumentary exchange

To understand integumentary exchange, you first need to know what the heck an integument is. The *integument* is the outer covering of an animal. In worms, frogs, and salamanders — all of which respire through integumentary exchange — the integument is more of an outer membrane; in you, the integument is your skin.

Small animals that constantly stay moist may "breathe" right through their skin. Oxygen from the air diffuses through such an animal's moist surface and into the fluids in its body; at the same time, carbon dioxide diffuses out. (Refer to Chapter 4 for a full explanation of diffusion.) Because oxygen and carbon dioxide are swapped across the integument, this process is called *integumentary exchange.*

Earthworms are a perfect example of integumentary exchange in action. They have small blood vessels called capillaries right under their "skin." As an earthworm moves through the soil, it loosens the soil, creating air pockets. The worm takes in oxygen from the air pockets and releases carbon dioxide right through its outer surface.

You know how worms get flooded out of the ground when it rains and end up all over your driveway and sidewalk? Well, they head right back into the soil as soon as they can (and not just because they're potential bird food). If they lingered on your driveway, their outer surfaces would dry out, preventing them from taking in oxygen and getting rid of carbon dioxide. When this happens, they die. (This is also the reason why pouring salt on a slug stops it in its slimy tracks. The salt dehydrates its outer surface, which prevents it from exchanging gases.)

## Gills

Animals that live in water, including lobsters and starfish, have *gills,* which are extensions of their outer membranes. The membranes in gills are very thin (usually just one cell thick), which allows for easy gas exchange. Capillaries connect to the cells in the gills so that gases can be taken in from the water and passed into the bloodstream of the aquatic animal. Also, gaseous waste can diffuse from the capillaries into the cells of the gills and pass out into the watery environment.

The type of gill you're probably most familiar with is that of a fish. In fish, the gills are membranous filaments covered by a flap called an *operculum,* which is the flap you can see opening and closing on the head of a fish. A fish opens and closes the flap by opening and closing its mouth. After water enters the mouth, it's forced over the gills and then out the back of the operculum. As the water passes over the gills in one direction, the blood inside the gills moves in the opposite direction. Oxygen from the water diffuses into the capillaries in the gills, and carbon dioxide diffuses out of the capillaries in the gills. After the oxygen is in the capillaries, it can be transported throughout the fish's body so all the cells can get some of the needed gas.

Because the water outside the gill and blood inside the gill are moving in opposite directions, the exchange of gases in a fish gill is referred to as *countercurrent exchange.* Countercurrent exchange improves the efficiency of gas exchange.

## Tracheal exchange systems

Tracheal exchange systems are made possible by a network of tubes called a *trachea.* The holes at the ends of the tubes, which open to the outside surface, are called *spiracles.* You can find this system in insects.

The trachea found in insects is different from the trachea found in your body. In insects, the trachea is the network of tubes that runs through the entire body and opens to the air; in humans, the trachea is a tube that carries air down into the lungs.

In a tracheal exchange system, oxygen diffuses directly into the trachea, and carbon dioxide exits through the spiracles. The cells of the body exchange air directly with the tracheal system, and the oxygen and carbon dioxide don't need to be carried through a circulatory system because the tracheal system runs through all parts of the insect's body.

Some insects, such as bees and grasshoppers, combine a breathing process with a tracheal exchange system. They contract muscles to pump air in and out of their tracheal systems. A grasshopper even has air sacs on some of the air tubes in its tracheal system. The bags "pump" like fireplace bellows after pressure from muscles is applied.

## Lungs

Lungs are the opposite of gills (which we describe earlier in this chapter). Gills extend out off of an organism, and lungs are internal growths of the surface of the body. Animals' lungs are housed inside their bodies in order to keep them moist. They basically work by providing lots of moist surface area for the diffusion of oxygen and carbon dioxide. Lungs may be different shapes and sizes in various land animals, but they function essentially the same as they do in humans. We use humans as the model for the mechanics of the lungs in order to give you a better understanding of how your body works.

Humans have a pair of lungs that lie in the chest cavity (as shown in Figure 15-1); one lung is on the left side of the *trachea* (the tube that connects the nose and mouth to the lungs), and the other is on the right side of it. Inside the lungs, the trachea branches off into *bronchi* (small passageways that move air into the lungs), which then branch and rebranch off into smaller *bronchioles* (smaller versions of bronchi). The bronchioles end in little clusters of sacs called *alveoli* (tiny, moist sacs where gas exchange occurs in the lungs) that look a little bit like raspberries. Each *alveolus* (that's the singular word for alveoli) is wrapped with capillaries so that gas exchange can occur between the lungs and the blood. The muscle found underneath the lungs is called the *diaphragm,* and it's responsible for contracting and creating negative pressure to draw air into the lungs. A pair of ribs surrounds the chest cavity to protect the lungs (and heart) and to assist in the motions of breathing.

Because lungs are the most complex gas-exchange system employed by animals, we break down the details for you in the following sections.

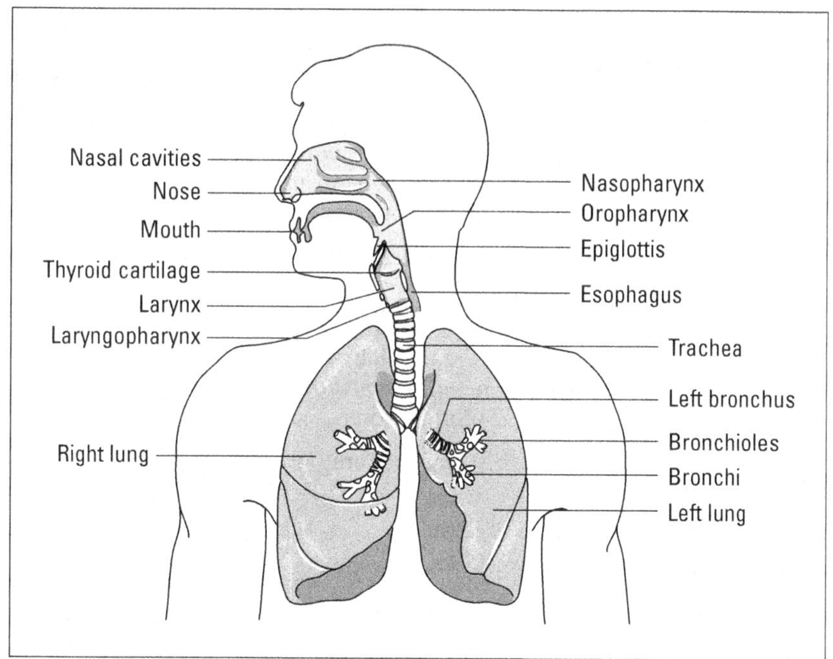


**Figure 15-1:** Anatomic structures of the human respiratory system.

*From LifeART®, Super Anatomy 1, © 2002, Lippincott Williams & Wilkins*

### Taking a peek at what happens when you breathe

When you breathe in, called *inhalation* or *inspiration,* the diaphragm muscle contracts (meaning it becomes smaller and moves down), allowing your rib cage to move upward and outward. Because the lungs have more room when your chest is expanded, they open up, similar to how a balloon blows up when it's filled with air. The opening up of the lungs means there's more room in the lungs, so air rushes in to fill the space. When your diaphragm relaxes, your rib cage moves back downward and inward, increasing air pressure inside your lungs and forcing air out. The process of breathing out is called *exhalation* or *expiration.*

Following is a breakdown of how oxygen passes through all the branches of your respiratory system when your lungs fill up:

1. **Oxygen enters through your nostrils and flows through the top part of your throat.**

   Inside your nasal cavity, hair, cilia, and mucus trap dust and dirt particles, purifying the air that enters your lungs. Occasionally, you must cough and either spit or swallow to move the trapped particles out of your throat. (Don't worry about swallowing dirt; it enters your stomach where it's digested and excreted.)

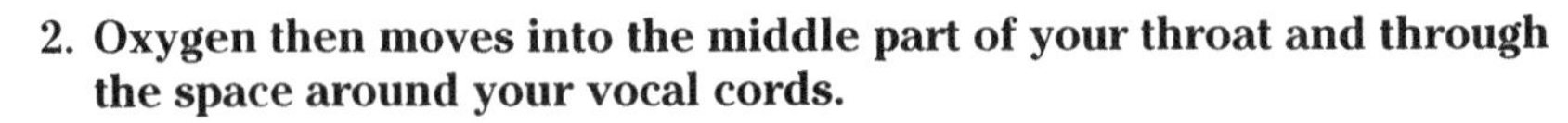

2. **Oxygen then moves into the middle part of your throat and through the space around your vocal cords.**

When you eat, food passes through your throat (or *pharynx*) on its way to your stomach. When you breathe, air passes through your pharynx on its way to the lungs. So, both your mouth and your nose connect to the pharynx. The place where your mouth connects is called the *oro-pharynx;* the place where your nose connects is called the *nasopharynx.*

3. **Next, oxygen enters your trachea, flows through the bronchi and bronchioles, and then flows into the alveoli.**

Capillaries surround the alveoli, ready to accept oxygen and give up carbon dioxide (see Figure 15-2). Oxygen and carbon dioxide diffuse across the membranes of the alveoli, and the capillaries move the freshly oxygenated blood into your circulatory system.

Gas exchange in the lungs occurs only in the alveoli.

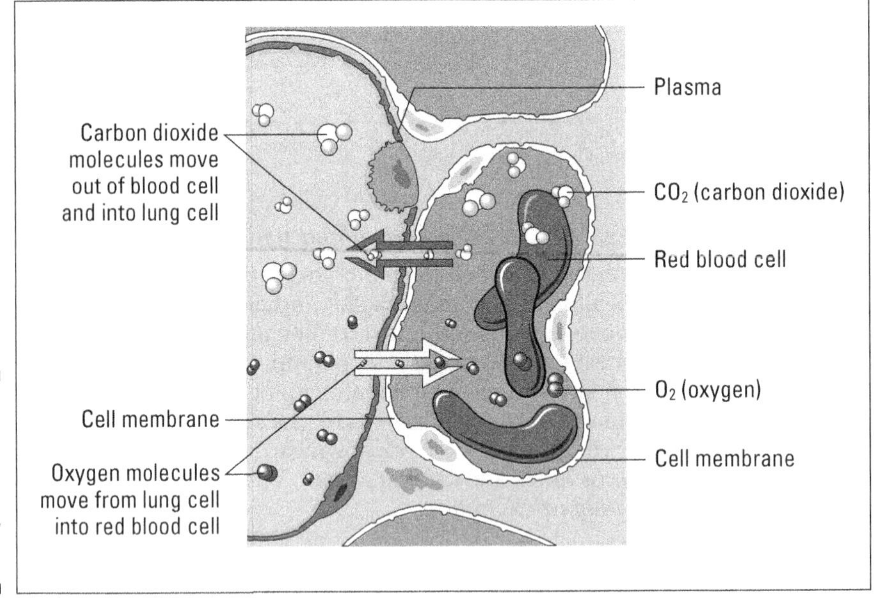


**Figure 15-2:** Oxygenation of blood at the respiratory membrane.

*From LifeART®, Super Anatomy 2, © 2002, Lippincott Williams & Wilkins*

### *Understanding the concept of diffusion*

As explained in Chapter 4, *diffusion* is when a substance moves from an area of higher concentration to an area of lower concentration. Gas exchange in the lungs is an example of diffusion.

Oxygen is more concentrated within each of the 300 million alveoli in your lungs than it is in the capillaries surrounding the alveoli. Your capillaries in turn have a higher concentration of carbon dioxide than oxygen. The oxygen in the alveoli diffuses across the alveoli's membranes and the capillaries' membranes and into the capillaries. Meanwhile, the carbon dioxide diffuses across the capillaries' membranes and the alveoli's membranes into the lungs for expulsion from your body. After oxygen is in the capillaries, your red blood cells snatch it up and transport it throughout your body. While this happens, your diaphragm relaxes, and the carbon dioxide waste leaves your body as you exhale.

# Circulation: Nutrients In, Garbage Out

Every animal alive possesses a circulatory system. This system makes sure that all the animal's cells obtain the nutrients they need to function and that all the cells dispose of waste so it doesn't build up within the animal's body and cause illness (or, yikes, death). Some of the additional responsibilities of a circulatory system include

- Delivering oxygen to cells and picking up carbon dioxide
- Distributing hormones to cells
- Maintaining body temperature by transporting heat
- Transporting cells to fight infection (more on this in Chapter 17)

Two kinds of circulatory systems exist: open and closed. We fill you in on both types in the next sections.

## Open circulatory systems

In an open circulatory system, the animal's heart pumps a bloodlike fluid called *hemolymph* into an open cavity (called a *hemocoel*) through openings in the heart called *ostia*. When the hemolymph flows into the hemocoel, it directly bathes the tissues of the organism with nutrients — no blood vessels are involved. Muscle contractions push the hemolymph back toward the heart so it can be circulated throughout the animal again and again. Insects and some mollusks (specifically snails and clams) possess an open circulatory system.

In insects, hemolymph carries nutrients but not oxygen to the cells. Oxygen is circulated via the tracheal exchange system (which we describe earlier in this chapter).

## Closed circulatory systems

A closed circulatory system is the type of circulatory system you're most familiar with. It has a network of vessels (think of them as highways that connect one organ to another) that perform the transportation and keep your blood from seeping out. In animals, each blood vessel in the network is responsible for transporting nutrients and oxygen to the cells and removing wastes and carbon dioxide from them.

The three types of blood vessels are

- Arteries
- Veins
- Capillaries

Animals with backbones, called *vertebrates,* possess a closed circulatory system. Some animals without backbones, called *invertebrates,* also have one; these include some worms, octopuses, and squids.

Closed circulatory systems, which are said to be closed because they have vessels that contain the fluid, are more efficient than open circulatory systems because they need to meet the dual demand of delivering both oxygen and nutrients to cells.

# Getting to the Heart of Simpler Animals

Hearts come in different sizes and shapes, but a heart's function remains the same in all organisms that have one: It pumps fluid throughout the circulatory system. That fluid is either hemolymph or blood, depending on the type of circulatory system an animal has, and it transports either nutrients or a combination of nutrients and oxygen to the animal's cells.

In animals with an open circulatory system, this process occurs the same way. However, in animals with a closed circulatory system, the process happens differently depending on how the system is set up. We give you a peek at how the process occurs in two of the simpler animals — worms and fish — in the sections that follow.

## A worm's heart and circulatory system

Although you may think earthworms are insects that possess an open circulatory system, they're not (technically they're *annelids*). These little guys and gals have a closed circulatory system that's a bit simpler in design than that of a human.

Earthworms have just one *dorsal* (top side) blood vessel and one *ventral* (bottom side) blood vessel, plus a network of capillaries. The heart of an earthworm is a series of muscular rings near the thicker tip of the worm. It pumps blood away from the heart through the ventral blood vessel. From there, the blood oozes into all the capillaries to reach all the cells of the worm before traveling back to the heart through the dorsal blood vessel.

## A fish's heart and circulatory system

A fish's heart has two separate chambers, one that receives blood from the body and another that pumps the blood out over the gills. Its closed circulatory system makes a single loop through the body of the fish. Overall, the process of how blood passes through a fish is rather simple.

1. **When a fish's heart pumps, blood leaves the heart through the ventral aorta that runs along the underside of the fish.**
2. **The ventral aorta carries the blood to the gills, and the blood then passes through the capillaries along the gills to pick up oxygen.**

   This part of the circulatory loop is called *gill circulation*.
3. **The oxygenated blood immediately flows from the gills into the dorsal aorta, which runs along the upper side of the fish.**
4. **The dorsal aorta carries the oxygenated blood to the rest of the fish's capillaries.**

   This part of the loop is referred to as *systemic circulation*.
5. **After the blood has reached all the cells within the fish, it returns to the heart.**

The circulatory system in a fish is simple and effective, but because the blood passes through the heart only once during its travels, a fish's blood pressure is quite low. (*Blood pressure* is the force that sends blood through an animal's circulatory system.)

# Exploring the Human Heart and Circulatory System

The heart and circulatory system of a human, as well as some other mammals, are complex. Because these animals are larger, they need to have a higher blood pressure to push the blood throughout their entire bodies. This need results in a *two-circuit circulatory system,* a system that has two distinct pathways (as you can see in Figure 15-3):

- One pathway is for *pulmonary circulation,* which first delivers deoxygenated blood to the lungs so it can become oxygenated and then delivers oxygenated blood back to the heart.
- The other circuit is for *systemic circulation,* which carries oxygenated blood from the heart to the rest of the body.

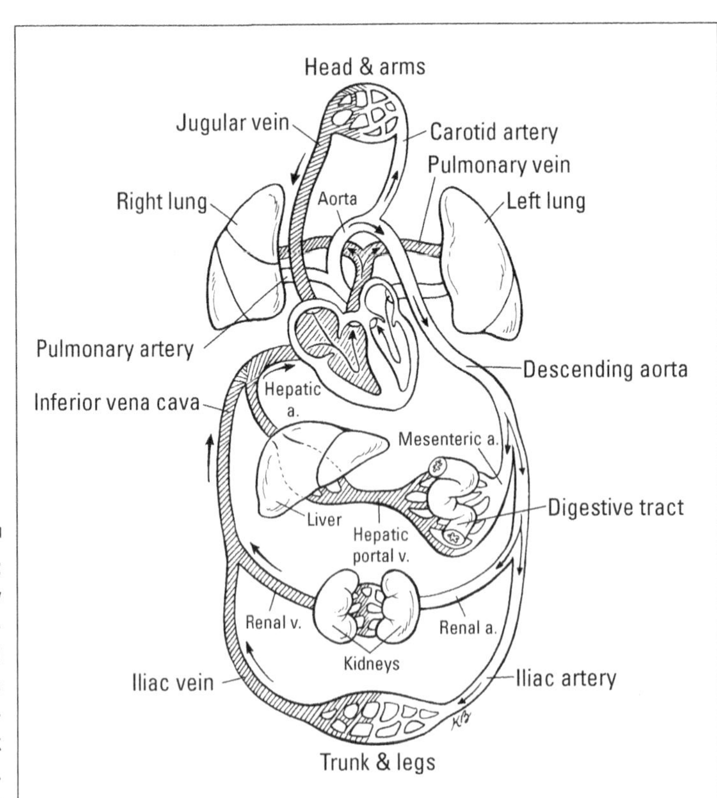


**Figure 15-3:** Pulmonary circulation and systemic circulation work together.

The human heart (depicted in Figure 15-4) has four chambers: two *ventricles,* muscular chambers that squeeze blood out of the heart and into the blood vessels, and two *atria,* muscular chambers that drain and then squeeze blood into the ventricles. The heart is divided into left and right halves, so there's a left atrium and a left ventricle, as well as a right atrium and a right ventricle. The two atria reside at the top of the heart; the two ventricles are at the bottom.

***Note:*** If Figure 15-4 seems confusing to you because it looks as if the right atrium is on the left side of the heart, pretend you're looking at someone's heart through her chest. Her right atrium is on the right side of her body, which is how we've depicted it in the figure.

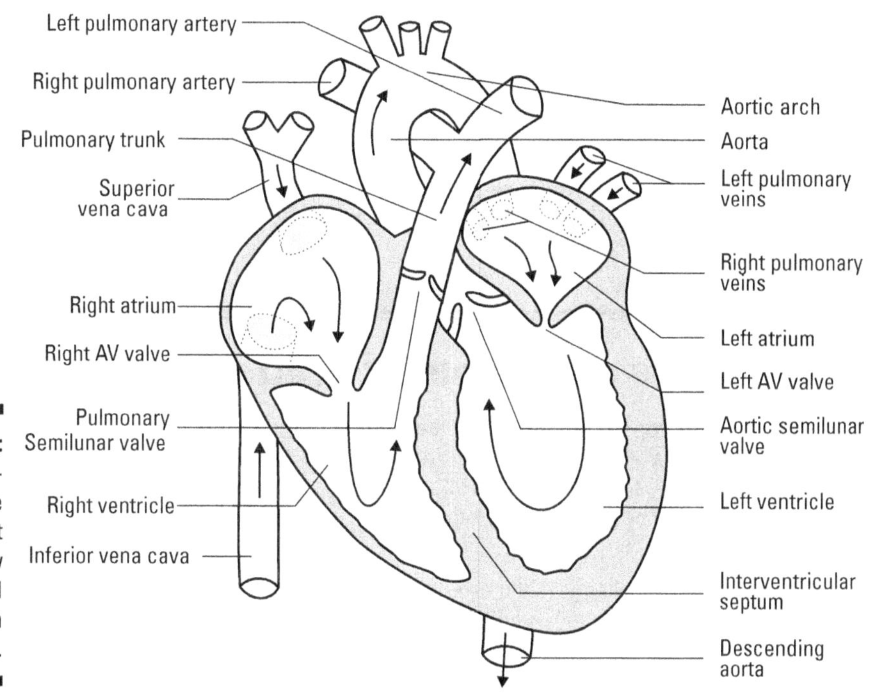


**Figure 15-4:** The structures of the human heart and the flow of blood through them.

Your heart is divided into halves because of your two-circuit circulatory system. The right side of your heart pumps blood to your lungs, while the left side of your heart pumps blood to the rest of your body. Blood goes into both pathways with each and every pump.

*Valves* separate one chamber of the heart from another. Each valve consists of a pair of strong flaps of muscle tissue, called *cusps* or *leaflets.* When your heart is working properly, the valves open and close fully so blood can flow through it in only one direction.

Four valves separate the four chambers of your heart from one another and from the major blood vessels that are connected to it (refer to Figure 15-4 for the visual):

- The *right atrioventricular (AV) valve* is located between the right atrium and right ventricle. This valve is also referred to as the *tricuspid valve* because it has three flaps in its structure.
- The *pulmonary semilunar valve* separates the right ventricle from the pulmomary artery. *Semilunar* means "half-moon" and refers to the shape of the valve.
- The *left atrioventricular (AV) valve* is between the left atrium and left ventricle. This valve is also called the *bicuspid valve* because it has only two flaps in its structure.
- The *aortic semilunar valve* separates the left ventricle from the aorta. Like the pulmonary semilunar valve, this valve has a half-moon shape.

The sections that follow take you on a journey through the human circulatory system. You find out just how blood moves through your heart and the rest of your body, and you discover what gets your heart beating in the first place. Bon voyage!

## The many faces of heart disease

Heart disease is the top cause of death in the United States. Deaths from heart disease are usually attributed to heart attacks, but heart attacks can be caused by many factors, including the following:

- **Atherosclerosis:** Blockages in the arteries occur when fats, especially cholesterol, accumulate in the lining of the arteries. Cholesterol serves a necessary function within the body, but when too much cholesterol is present — which can be caused by diet or by genetic factors — it starts to stick to the vessels instead of passing through them. These fatty deposits are called *plaques.* As the plaques increase in size, they fill more and more of the artery, eventually affecting blood flow.
- **Hypertension:** More commonly referred to as "high blood pressure," *hypertension* puts added stress on the arteries of the heart, increasing the risk of damage and fatty deposits along artery walls.
- **Ischemic heart disease:** *Ischemia* is a lack of oxygen, and this type of heart disease can occur when the arteries are partially blocked. People with this condition have difficulty breathing during exercise or times of stress because the blocked arteries slow the flow of blood, preventing enough oxygen from being delivered to the heart muscle's tissues. This lack of oxygen can cause a pain in the chest that radiates to the left arm; this pain is called *angina pectoris.*
- **Thromboembolism:** If an artery is blocked by a plaque, blood cells can stick to the plaque, eventually forming a blood clot that can travel through the bloodstream and block blood vessels. A blood clot stuck in a blood vessel is called a *thrombus.* If a thrombus breaks free and moves around the bloodstream, it's called an *embolism.*

## Entering the cardiac cycle

Your heart is an impressive little organ. Even though it's only as big as a clenched adult fist, it pumps 5 liters of blood (the equivalent to 2½ big bottles of soda) throughout your body 70 times a minute. Did we mention it's a hard worker too? Your heart never stops working from the time it starts beating (when you were nothing but a wee little embryo in your mother's womb) until the moment you die. It doesn't even get an entire second to rest. It beats continuously every 0.8 seconds of your life; this is known as the *cardiac cycle.*

During the cardiac cycle, your heart forces blood into your blood vessels and takes a quick nap. Here's exactly what happens:

- The left and right atria contract.
- The left and right ventricles contract.
- The atria and ventricles rest (for just 0.4 seconds).

When the atria and ventricles are resting, the muscle fibers within them aren't contracting. Therefore, the relaxed atria allow the blood within them to drain into the ventricles beneath them. With most of the blood from the atria now in the ventricles, the atria contract to squeeze any remaining blood down into the ventricles. Then the ventricles immediately contract to force blood into the blood vessels.

The period of relaxation in the heart muscle is referred to as *diastole,* and the period of contraction in the heart muscle is called *systole.* If these terms sound familiar, it's probably because you've heard them used in terms of blood pressure.

In a blood pressure reading, such as the normal value of 120/80 mmHg, 120 is the *systolic blood pressure,* the pressure at which blood is forced from the ventricles into the arteries when the ventricles contract, and 80 is the *diastolic blood pressure,* the pressure in the blood vessels when the muscle fibers are relaxed. The abbreviation *mmHg* stands for millimeters of mercury (Hg is the chemical symbol for mercury).

If your blood pressure is 140/90 mmHg, which is the borderline value between normal and high blood pressure, that means your heart is working harder to pump blood through your body and not relaxing as well between pumps. This reading indicates that something is causing your heart to have to work at a much higher level all the time to keep blood flowing through your body, which stresses it out. The culprit could be a hormonal imbalance, a dietary problem (too much sodium or caffeine), a mechanical problem in the heart, a side effect of medication, or blockages in your blood vessels.

## Navigating the path of blood through the body

The *cardiac cycle,* which describes the rhythmic contraction and relaxation of the heart muscle (see the preceding section), coincides with the path of blood through your body. As each atrium and ventricle contracts, blood is pumped into certain major blood vessels that connect to your heart and then continues flowing throughout your circulatory system. In other words, here's where your two-circuit circulatory system really comes into play.

The following sections describe the process of pulmonary and systemic circulation, as well as the process of capillary exchange (which gets nutrients into and wastes out of your cells).

### Oxygenating the blood: Pulmonary circulation

*Pulmonary circulation,* the first pathway of your two-circuit circulatory system, brings blood to your lungs for oxygenation. Following is a rundown of how blood moves during pulmonary circulation (trace the path in Figure 15-4 as you read):

1. **Deoxygenated blood from your body enters the right atrium of your heart through the superior vena cava and the inferior vena cava.**

   *Superior* means "higher," and *inferior* means "lower," so the superior vena cava is at the top of the right atrium, and the inferior vena cava is at the bottom of the right atrium.

2. **From the right atrium, the deoxygenated blood drains into the right ventricle through the right AV valve.**

   When the ventricles contract, the right AV valve closes off the opening between the ventricle and the atrium so blood doesn't flow back up into the atrium.

3. **The right ventricle then contracts, forcing the deoxygenated blood through the pulmonary semilunar valve and into the pulmonary artery.**

   The pulmonary semilunar valve keeps blood from flowing back into the right ventricle after it's in the pulmonary artery.

4. **The pulmonary artery carries the blood that's very low in oxygen to the lungs, where it becomes oxygenated.**

### Spreading oxygenated blood around: Systemic circulation

*Systemic circulation* brings oxygenated blood to the cells of your body. Here's how blood moves through this pathway (you can follow along by tracing the path in Figure 15-4):

1. **Freshly oxygenated blood returns from the lungs to the heart via the pulmonary veins.**

   Note that your pulmonary veins are the only veins in your body that contain oxygenated blood; all of your other veins contain deoxygenated blood.

2. **The pulmonary veins push the oxygenated blood into the left atrium, which then relaxes, allowing the blood to drain into the left ventricle through the left AV valve.**
3. **As the left ventricle contracts, the oxygenated blood is pumped into the main artery of the body — the aorta.**

   To get to the aorta, blood passes through the aortic semilunar valve, which serves to keep blood in the aorta from flowing back into the left ventricle.

4. **The aorta branches into other arteries, which then branch into smaller arterioles, carrying oxygenated blood all around your body.**

   Throughout your body, arterioles meet up with capillaries where oxygen is exchanged for carbon dioxide.

The blood vessels in order of decreasing oxygen content are as follows:

- Arteries
- Arterioles
- Capillaries
- Venules
- Veins

### Exchanging the good and the bad: Capillary exchange

For being the teeniest of blood vessels, *capillaries,* which bridge the smallest of the arteries and the smallest of the veins, have a rather important role: to facilitate the exchange of materials between capillaries and cells via diffusion — in other words, to make *capillary exchange* possible.

Your capillaries are only as thick as one cell, so the contents within them can easily exit by diffusing through the capillaries' membranes (see Chapter 4 for more on diffusion). And, because the capillaries' membranes touch the membranes of other cells all over the body, the capillaries' contents can easily continue moving through adjacent cells' membranes.

Through capillary exchange, oxygen leaves red blood cells in the bloodstream and enters all the other cells of the body. Capillary exchange also allows nutrients to diffuse out of the bloodstream and into other cells. At the same time, the other cells expel waste products, including carbon dioxide, that then enter the capillaries.

After the capillaries "pick up" the garbage from other cells, they carry the wastes and carbon dioxide through the deoxygenated blood to the smallest of the veins, which are called *venules.* The venules branch into bigger vessels called *veins,* which then carry the deoxygenated blood toward the main vein — the *vena cava.* The two branches of the vena cava enter the right atrium, which is where pulmonary circulation begins.

The pressure created when the ventricles contract is what forces blood through the arteries. However, this pressure declines as the blood gets farther away from the heart and into the capillaries. Blood pressure doesn't force blood through the veins as it does through the arteries. What makes blood travel through the veins are contractions of your skeletal muscles. As your limbs and trunk move, deoxygenated blood is pushed farther along the venules and veins, eventually returning to the heart. Without movement, blood pools in the veins, creating poor circulation.

## Seeing what makes your ticker tick

Electrical impulses from your heart muscle cause your heart to beat. These impulses begin in special areas of tissue within the heart, called *nodes,* that are infused with nerves. The nodes send out signals that stimulate contraction of the heart muscle cells, causing your heart to beat. For particulars on the process, check out the following:

1. **Each beat of your heart is started by an electrical signal from the sinoatrial (SA) node in your right atrium.**

   The SA node is also called your *natural pacemaker* because it sets the pace of your heartbeats. (Yes, your heart already contains a pacemaker. People who have pacemakers "installed" through surgery have it done because their natural pacemaker has stopped working correctly.)

2. **The signal from the SA node spreads across the left and right atria, causing them to contract and push the blood into the ventricles.**

3. **As the electrical impulse passes through the atria, it signals the atrioventricular (AV) node to take action.**

   The AV node is located in the lower part of the right atrium. The signal is temporarily slowed in this node so that your ventricles have time to fill with blood.

4. **The signal in the AV node stimulates an area of tissue called the bundle of His.**

The bundle of His lies between the right and left ventricles and connects with specialized fibers called Purkinje fibers.

5. **When the impulse reaches the Purkinje fibers, it causes the ventricles to contract, completing the heartbeat.**

The sound your heart makes — described as lub-dub, lub-dub — is attributed to the closing of the heart's valves. The first heart sound, the lub, is caused by AV valves closing to prevent backflow from the ventricles into the atria. The second heart sound, the dub, occurs when the semilunar valves close to keep blood in the aorta from flowing back into the left ventricle.

# A Bloody-Important Fluid

Blood is the fluid that sustains life in animals with a closed circulatory system — including you. Some blood cells carry oxygen, which is necessary for metabolic reactions; some blood cells fight off invading substances that could destroy your cells; and other blood cells help form clots, which keep your body from losing too much of this precious fluid and assist with wound healing. The following sections introduce you to the elements that make up your blood and the special process that keeps you from losing too much blood when you cut yourself.

## The solids found in your essential fluid

Believe it or not, your blood, although it's a fluid, contains solid parts called *formed elements.* These solid parts are your red blood cells, white blood cells, and platelets. You can find out more about all three in the sections that follow.

***Note:*** We use the word *solid* simply to differentiate platelets and blood cells from the liquid portion of blood. Platelets and blood cells definitely aren't hard and solid. If they were, they wouldn't be able to squeeze through your capillaries.

### Red blood cells

Your red blood cells, which are also called *erythrocytes,* have the important responsibility of carrying oxygen throughout your body. *Hemoglobin,* the iron-containing molecule that harnesses oxygen, exists in the red blood cells. It not only binds oxygen and transports it to capillaries but it also helps transport carbon dioxide from the capillaries back to the lungs to be exhaled. By transporting oxygen and hemoglobin, your red blood cells are an extremely important part of *homeostasis* — how your body tries to constantly achieve and maintain balance.

If a person has too few red blood cells, as determined by a lab test that measures red blood cell count, or if her red blood cells don't have enough hemoglobin, she has *anemia,* a disease characterized by too few red blood cells or a hemoglobin deficiency that often leads to feelings of fatigue. Anemia can be caused by dietary deficiencies, metabolic disorders, hereditary conditions, or damaged bone marrow.

Red blood cells are created in the red bone marrow. They live about 120 days shuttling oxygen and carbon dioxide, and then certain white blood cells destroy them in the liver and spleen. As the red blood cells are destroyed, the iron they contain is recycled back to the red bone marrow to be used in new cells. The rest of the material in the old red blood cells is degraded and transported to the digestive system, where much of it ends up in fecal matter.

### White blood cells

Your white blood cells, which are also called *leukocytes,* are involved in functions controlled by your immune system, which is responsible for fighting infections (head to Chapter 17 for details on the immune system). If a person has a low white blood cell count, her immune system isn't functioning properly. If her white blood cell count is too high, that indicates she has some type of infection.

Following are the five important types of white blood cells you should know:

- **Basophils** release *histamines,* those annoying little chemical molecules that cause you to swell up with hives, itch like crazy, sneeze, wheeze, and get teary-eyed when you're around something you're allergic to. All of these reactions are side effects of *inflammation,* a very important defensive process that helps you clear damaging agents out of your body.
- **Eosinophils** help defend the body against invading organisms, particularly parasitic worms.
- **Lymphocytes** are key players in your *adaptive immune response,* the response your body makes to defend you against invading microbes. Two of their important functions are to destroy virally infected cells and to make defensive proteins called *antibodies.*
- **Monocytes** are precursors to macrophages. *Macrophages* digest bacteria and viruses (*macro-* means "big," and *phago* means "to eat," so a *macrophage* is literally a big eater).
- **Neutrophils** are the most abundant white blood cells in the body. These cells eat bacteria; in doing so, they keep your system from being overrun by every germ with which it comes in contact.

### Platelets

Platelets, which are also called *thrombocytes,* are pieces of cells that work to form blood clots (we detail the clotting process later in this chapter). Platelets form when pieces are torn off of cells called megakaryocytes. Because they're just cell fragments, platelets are smaller than red or white blood cells. They survive in the blood for about ten days.

The number of platelets in the blood is often determined as part of a complete blood count. Low numbers of platelets can indicate certain cancers and chronic bleeding disorders. Increased numbers of platelets may be a sign of chronic infection or certain blood diseases.

## The plasma "stream" in your bloodstream

The liquid portion of your blood is *plasma.* Your blood cells and platelets flow in your plasma much like leaves float in a stream. In fact, when you think about it, plasma literally puts the "stream" in bloodstream.

Plasma contains many important proteins, without which you'd die. Two major proteins found in plasma are

- **Gamma globulin:** Also called *immunoglobulin, gamma globulin* is a broad term for a class of defensive proteins that make up the different types of antibodies. The production of antibodies, which help to fight infections, is controlled by your immune system (as explained in Chapter 17).
- **Fibrinogen:** This protein is involved in blood clotting.

## How blood clots form

When you cut your finger chopping an onion or picking up a piece of broken glass, your body embarks on a mission to form a *blood clot* (a semisolid plug made of blood cells trapped in a protein mesh) to prevent you from bleeding to death. First, the injured blood vessel constricts, reducing blood flow to the injured blood vessel, which helps limit blood loss. (Tourniquets help squeeze off blood flow in much the same way when major blood vessels are damaged.) With the injured blood vessel constricted, the platelets present in the blood that's passing through that vessel start to stick to the collagen fibers that are part of the blood vessel wall. Eventually, a platelet plug forms, and it fills small tears in the blood vessel.

After the platelet plug is formed, enzymes called *clotting factors* (your body has 12 of them) initiate a chain of reactions to create a clot. The process is rather complex, but you don't need to know all the nitty-gritty details; just focus on these highlights:

- After a platelet plug forms, the coagulation phase begins, which involves a cascade of enzyme activations that lead to the conversion of inactive prothrombin to active thrombin. (Calcium is required for this reaction to occur.)
- Thrombin itself acts as an enzyme and causes fibrinogen — one of the two major plasma proteins — to form long fibrin threads.
- Fibrin threads entwine the platelet plug, forming a meshlike framework.
- The fibrin framework traps red blood cells that flow toward it, forming a clot. (***Note:*** Because red blood cells are tangled in the meshwork, clots appear to be red. As the red blood cells trapped on the outside dry out, the color turns a brownish red, and a scab forms.)

## Chapter 16

# Checking Out the Plumbing: Animal Digestive and Excretory Systems

### In This Chapter

- Looking at how different types of animals consume and remove food
- Taking an in-depth look at the human digestive system
- Playing with the building blocks of healthy food choices
- Discovering how waste products leave the human body

After an animal ingests or absorbs food, its digestive system immediately starts breaking down the food to release the nutrients within it. After the useful nutrients are absorbed into the bloodstream, the animal eliminates solid wastes through its large intestine and nitrogenous wastes through its urinary system.

Want to know more about the ins and outs of digestive and excretory systems? Then you're in the right place. In this chapter, we give you the lowdown on the various ways animals obtain food and eliminate wastes. We then present the workings of the human digestive system and the fate of food and wastes in your body. After all, everyone should know the scoop on their poop.

## Obtaining Food and Breaking It Down

All animals need food as a source of energy and materials for growth (as explained in Chapter 5), but the various types of animals have different strategies for obtaining the food they need.

- **Bulk feeders take big bites out of food.** They use teeth, claws, tentacles, or pinchers to tear off pieces of food and ingest them. Eagles, people, frogs, and snakes are all examples of bulk feeders.

- **Filter feeders strain liquids to capture tiny particles suspended in the fluid.** They have structures such as gills or very fine tentacles that trap small particles out of the water. Clams, for example, pull seawater through their bodies and pass it over their mucus-covered gills, trapping the small organisms and organic matter that were suspended in the water. Other examples of filter feeders include sponges, whale sharks, and gray whales.
- **Fluid feeders suck nutrient-rich fluids from other organisms.** They're often parasites that live inside a host and use the host's fluids as a source of food. Aphids and mosquitoes, for example, puncture other organisms and draw fluid from them — aphids draw the sugary sap out of plants, and mosquitoes draw blood out of animals. Ticks, leeches, and lampreys are some additional fluid feeders you may be familiar with.
- **Substrate feeders live right in or on their food and eat as they move through it.** Despite its complex name, this type of animal obtains its food pretty simply. Just think of earthworms and maggots. As earthworms burrow through the soil, they ingest the soil, digest the small organisms and organic matter within it as it passes through them, and then release the undigested matter as worm casings out their back ends. Similarly, when flies lay their eggs in bodies, the maggots that hatch out of the eggs eat their way through the food (the decomposing organism) until they reach the outside.

Four main events occur from the moment food enters an animal's body until the time its wastes are released:

- **Ingestion** occurs when an animal takes food into its digestive tract.
- **Digestion** occurs when the animal's body gets busy breaking down the food. Two types of digestion exist:
  - **Mechanical digestion** physically breaks down food into smaller and smaller pieces. It begins when an animal consumes the food and continues until the food enters its stomach.
  - **Chemical digestion** uses enzymes and acids to break down chewed or ground-up food into even smaller pieces. It also begins as soon as food is consumed and the enzymes in the mouth go to work. Chemical digestion continues as the food moves through the stomach and small intestine and encounters enzymes and acids in the stomach and enzymes in the small intestine.
- **Absorption** occurs when cells within the animal move small food molecules from the digestive system to the insides of the cells.
- **Elimination** occurs when material that can't be digested passes out of an animal's digestive tract.

# The Ins and Outs of Digestive Systems

The basic way an animal's digestive system works has a great deal to do with whether it can spend a few hours between meals or whether it has to keep consuming food constantly just to stay alive. The following sections take a look at the two types of digestive systems and help you distinguish between animals that need to take in food all the time and those that don't.

## Incomplete versus complete digestive tracts

Of the animals that you can see without a microscope, the ones with the most primitive digestive system are animals with *incomplete digestive tracts,* meaning they have a gut with just one opening that serves as both mouth and anus. (Gross, we know.) Jellyfish are a classic example of an animal with an incomplete digestive tract.

Increasing in complexity are animals that have gut tubes (where foods are digested and nutrients are absorbed) with a mouth at one end and an anus at the other. Although simple, this type of system is considered a *complete digestive tract.* You're probably quite familiar with one particular animal that possesses this digestive system — you.

The benefit of a complete digestive tract is that it allows thorough digestion before excretion occurs. Organisms with incomplete digestive tracts release undigested food along with their wastes. An organism with a complete digestive tract doesn't have to take in food constantly to replace food that's excreted before the nutrients could be acquired from it.

## Continuous versus discontinuous feeders

Animals that must "eat" constantly because they take food in and then push it out soon afterward are called *continuous feeders.* Most of these animals are either permanently attached to something (think clams or mussels) or incredibly slow movers. Animals that are *discontinuous feeders* consume larger meals and store the ingested food for later digestion. These animals are generally more active and somewhat nomadic.

The ability to "eat and run" serves a preying carnivore well. If an animal such as a lion were a continuous feeder that had to hunt and eat constantly, it'd be exhausted and spend much more time out in the open savanna, increasing the chance that it could become prey for another predator.

Although you may find yourself snacking and grazing constantly throughout a day, you're really a discontinuous feeder. You can consume food rapidly, but you digest it gradually so you don't have to eat again for several hours.

You and all the other animals that are discontinuous feeders must have a place in the body to store food as it slowly digests. In humans, this organ is the *stomach.*

# Traveling through the Human Digestive System

You know that the mouth is where you put your food, but did you realize it's also part of your digestive system (see Figure 16-1)? Well, it is. The act of chewing (the technical term for it is *mastication*) is the first step in food digestion for humans. Chewed-up food travels from your mouth, down your esophagus, and into your stomach. From your stomach, it passes into the small intestine. Materials that can't be digested and absorbed move into the large intestine and then pass out of your body through your rectum and anus. In the sections that follow, we fill you in on the role that each of these parts of the digestive system plays in digestion.

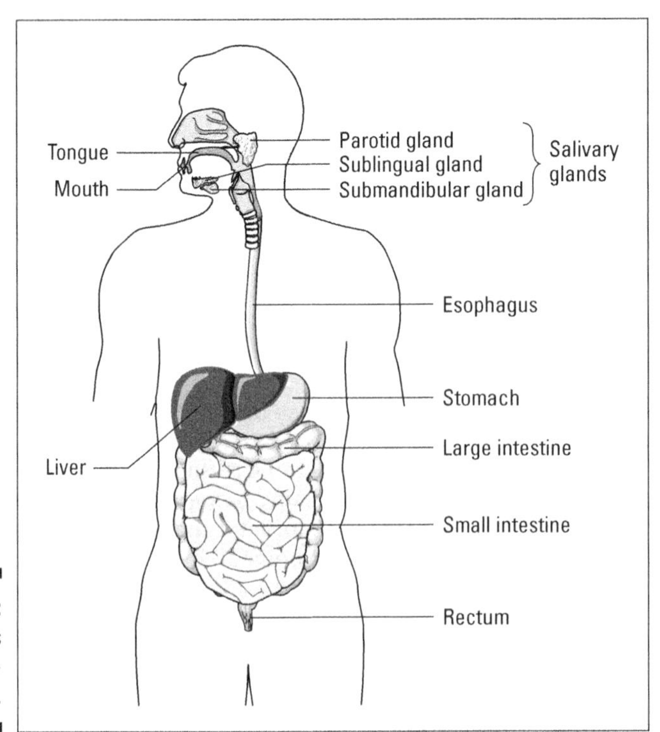


**Figure 16-1:** The organs of the digestive system.

## The busiest stop of all — your mouth

Your mouth does more than let you shout at the television during sporting events or talk to that cute guy or gal sitting next to you in biology class. It also kicks off the entire process of digestion. And don't think your teeth are the only participants in this process just because they mash your food into smaller and smaller bits. Your whole mouth gets a piece of the pie, so to speak. In addition to what your teeth do, your

- Taste buds detect the nutrients that make up the food you're eating — such as carbohydrates, proteins, and fats — so the cells of your digestive system can release the right enzymes to chemically digest your food.
- Saliva starts chemically digesting your food thanks to *salivary amylase,* an enzyme found within saliva that splits apart the bonds between glucose molecules in a long chain of starch.

You know how you salivate just before you're about to eat something? That's the effect of your eyes or nose sensing something delicious and sending a message to your brain that you're about to open your mouth and take a bite. It's also your mouth's way of preparing for digestion by producing saliva containing salivary amylase.

After your teeth have chewed, your taste buds have sent along the information about what you're eating, and the enzymes in your saliva have started breaking apart starches, you're ready to swallow. This action involves your tongue pushing the chewed food to the back of your throat and you squeezing food down your esophagus and into your stomach. (The *esophagus* is the tube that connects your mouth to your stomach.)

The squeezing of food through a hollow muscular structure is called *peristalsis,* and it occurs throughout the digestive system to move food down the esophagus and into the stomach and then again to move food through the intestines.

When the swallowed food drops into your stomach, it's referred to as a *bolus.* At this point, salivary amylase stops breaking apart the starch molecules, and the gastric juice in your stomach, which contains enzymes as well as hydrochloric acid (HCl), takes over chemical digestion.

If you eat too much, your stomach produces more acid, and the contents of your overly full stomach can be forced back up into your esophagus, which runs in front of the heart. The unpleasant result? Heartburn.

## The inner workings of your stomach

When food particles reach your stomach, the organ churns them up, and the enzyme pepsin starts breaking down the food's proteins into smaller chains of amino acids (see Chapter 4 for more on enzymes). Then the whole goopy substance gets squirted into the top of your intestines via the *pyloric valve*, otherwise known as the gate between your stomach and small intestine. Your pyloric sphincter muscle occasionally opens the valve, allowing your stomach's contents into your small intestine a little bit at a time.

Perhaps you're wondering why pepsin doesn't destroy the proteins that make up the tissues in your digestive tract. Well, when pepsin is secreted, it's in its inactive form, called *pepsinogen*. Because pepsinogen is inactive, it doesn't damage the cells that make it. When it's in the cavity of the stomach, pepsinogen gets converted to its active form, pepsin, by losing a few dozen of its amino acids. The lining of the stomach isn't affected by pepsin because pepsin acts only on proteins, and the stomach is covered with mucus (a substance made from fats) that protects the protein-containing tissues — that is, unless you have a stomach ulcer (see the related sidebar nearby for more on stomach ulcers).

## The long and winding road of your small intestine

When food molecules hit your small intestine, they get broken down into even smaller units (so your cells can absorb them) with a little help from your liver and pancreas.

### The true cause of stomach ulcers

When the lining of the stomach is eroded by digestive enzymes and stomach acid, the resulting condition is referred to as a *stomach ulcer*. For years the medical community thought stomach ulcers were largely caused by stress, worry, frustration, and other negative emotions. That belief changed in 1982 when two Australian doctors, Barry Marshall and Robin Warren, detected a bacterium called *Helicobacter pylori* (or *H. pylori*) in the stomachs of people with ulcers. They proposed that *H. pylori* in fact caused the ulcers, but no one believed them. To prove their point, Dr. Marshall, who didn't have any stomach ulcers, did the unthinkable — he drank a culture of live *H. pylori!* After drinking the culture, Dr. Marshall developed ulcers, proving that infection with the bacterium could indeed cause stomach ulcers.

So just how does a stomach ulcer form? Well, *H. pylori* inflames the lining of the stomach, causing the organ's protective mucus to disappear and allowing the enzymes and stomach acids secreted during digestion to eat away at the proteins in the tissue of the stomach wall. Stomach ulcers can be painful, and if they bleed or cause a hole (or *perforation*) through the wall of the stomach, they can become a medical emergency. Fortunately, antibiotic therapy can often successfully treat the condition by removing the *H. pylori* bacteria from the stomach.

The liver secretes *bile* (a yellow-brown or greenish fluid) into the small intestine. Bile salts *emulsify* (suspend in water) fats, helping them mix into the liquid in the intestine so they can be digested more easily. Meanwhile, the pancreas releases pancreatic juice into the mixture that contains the following enzymes to help chemically digest fats and carbohydrates:

- **Lipase** breaks apart fat molecules into fatty acids and glycerol.
- **Pancreatic amylase** breaks long carbohydrates into *disaccharides,* which are short chains of two sugars. The disaccharidases then break apart into monosaccharides that can be absorbed by the cells lining your small intestine.
- **Trypsin and chymotrypsin** are enzymes that break apart peptide fragments. After they break the peptides down into small chains, *aminopeptidases* finish them off by breaking apart the peptides into individual, absorbable amino acids.

Just like pepsin in the stomach, trypsin and chymotrypsin are originally secreted in their inactive forms, *trypsinogen* and *chymotrypsinogen.* To protect the cells of your digestive system, the enzymes aren't activated until they reach the cavity of your small intestine.

Don't let the word *small* fool you. The small intestine is much longer than the large intestine (10 feet long versus 5 feet long). The term *small intestine* refers to the fact that this part of the intestine is narrower in diameter than the large intestine; the large intestine is wider in diameter but shorter in length.

After several hours in your digestive system, the carbohydrates, fats, and proteins from your food are all in their smallest components: monosaccharides (such as glucose), fatty acids and glycerol, and amino acids (see Chapter 3 for details on these molecules). Now they can leave your digestive system in order to be used by all the cells in your body (as explained in the next section).

# Absorbing the Stuff Your Body Needs

During the digestion process, the nutrients your body can use are absorbed into the cells lining your small intestine. The rest of the material that can't be further digested or used passes on to the large intestine. The next sections outline how nutrients move throughout your system and how your liver constantly works to make sure you have the right amount of *blood glucose* (blood sugar) necessary to keep you going.

## How nutrients travel through your body

If digestion is all about obtaining the nutrients your cells need to function, then how exactly do those nutrients get out of your digestive system and into the rest of your body?

First, they pass directly into the cells of your small intestine via active transport. In other words, energy obtained from *adenosine triphosphate* (ATP) molecules is expended to move sugars (which came from consumed carbohydrates) and amino acids (which came from consumed proteins) into the intestinal cells (see Chapter 4 for the full scoop on active transport). From there, the nutrients are able to participate in *capillary exchange,* a type of trading system that allows cells to exchange nutrients and waste.

Capillary exchange relies on two things: *capillaries,* teeny-tiny blood vessels with extremely thin walls, and *interstitial fluid,* the fluid that fills every space between every cell in your body, cushioning and hydrating the cells.

Did you know that 60 percent of a human's body weight is from fluid? Of that 60 percent, 20 percent is *extracellular fluid,* fluid that exists outside the cells. This extracellular fluid is made up mostly of interstitial fluid (16 percent) and plasma (4 percent; we fill you in on plasma in Chapter 15). The remaining 40 percent of fluid in your body is *intracellular fluid,* fluid that exists inside the cells of your body; it's called cytoplasm.

The nutrients gained from digested food diffuse through the walls of your small intestine, through the capillaries' walls, across the interstitial fluid, and into your cells. At the same time, waste produced by your cells' metabolic processes diffuses out of the cells, across the interstitial fluid, and into the capillaries, where it can be carried to your kidneys for excretion. (To discover how your kidneys remove wastes from your body, head to the later related section in this chapter.)

Although the sugars and amino acids inside the capillaries get shuttled through the bloodstream to the liver, the products of fat digestion have a different fate. They get coated with proteins and acquire a new name: *chylomicrons.* Instead of being carried through the bloodstream, the chylomicrons get transported through the lymph system, which deposits lymph fluid into veins near the heart.

## Glucose regulation

Perhaps the most important stop digested sugars make as they travel through your bloodstream is your liver. The liver can detect abnormalities in the levels of various substances in your blood, such as glucose, and correct them.

- If the level of glucose is too high (a condition called *hyperglycemia*), the liver removes some of the glucose from the blood and turns it into the storage polysaccharide *glycogen*. If excess glucose is still in the blood after the liver has made enough glycogen, the liver switches its metabolic process to storing the extra glucose as fat. The fat molecules are then carried away by your bloodstream and deposited around your body in your *adipose tissue* (fat).
- If the level of glucose in the blood is too low (a condition called *hypoglycemia*), the liver converts some of its stored glycogen back into glucose and puts the glucose into the blood. If all the glycogen stores are used up, the liver starts to break down stored fats to obtain glucose for your cells.

Having just the right amount of glucose in your bloodstream is essential because glucose is your brain's main source of fuel. Glucose is so important, in fact, that your body will literally digest itself in order to get glucose to your brain. In extreme cases, such as starvation, when the breakdown of glycogen and fat isn't enough to restore blood glucose to normal levels, the body starts breaking down proteins to get the energy molecules it so desperately needs. Proteins in the muscles are broken down into amino acids, which can be converted into glucose. This may not sound so bad until you realize that the heart is a muscle. When your body starts breaking down proteins in your heart, death is a serious possibility.

## What's for Dinner? Making Wise, Nutritious Food Choices

It's a shame humans have such sensitive taste buds. If you didn't have such an evolved sense of taste, maybe you'd be like other animals and just eat what's part of your natural diet only when you're truly hungry. But, alas, food tastes good, and humans are often tempted to put really cheap fuel in their systems. Would you do that to your car on a regular basis? Or would you rather use the premium stuff to make sure your car's engine doesn't knock and ping?

If you want to keep your bodily systems from knocking and pinging, we strongly encourage you to follow the nutrition recommendations made by the United States Department of Agriculture (USDA). Its *MyPyramid,* an updated version of the classic Food Guide Pyramid, serves as a way of visualizing the proportion of items from different food groups that should make up your diet. More detailed information, such as the difference between whole grains and refined carbohydrates in the bread, cereal, and pasta group, are available on the USDA's MyPyramid Web site at `www.mypyramid.gov`.

For now, why not check out the following sections? They provide the basic facts about the essential nutrients your body needs: carbohydrates, proteins, fats, and minerals and vitamins.

## Carbohydrates: The culprits of your food cravings

*Carbohydrates* are compounds of carbon, hydrogen, and oxygen that supply your body with short-term energy (we cover the chemical structure of carbs in greater detail in Chapter 3). Carbohydrate molecules, such as sugars, break down quickly. You've witnessed this yourself if you've ever held a marshmallow over a campfire too long.

*Glucose* is the most important carbohydrate molecule. You can acquire it directly from foods containing carbohydrates (such as breads, pastas, sweets, and fruits). However, your body also creates glucose when it breaks down proteins and fats.

When it comes to your diet, all carbohydrates are not created equal.

- Foods made with *whole grains* are high in fiber and vitamins. They help prevent heart disease and constipation. Eating breads and cereals made with whole grains, as well as the grains themselves (think brown rice, quinoa, and bulgur wheat), is a way to get plenty of good carbs in your diet.

  Lots of products advertise that they're "made with whole grains," but that assertion doesn't tell you what the proportion of whole grains in the food actually is. When in doubt, check the fiber content of the food. Foods that are high in fiber are made with lots of whole grains.

- Foods made with *refined grains* are low in fiber and vitamins. They break down very quickly and cause a rapid rise in blood glucose. White bread, flour, and pasta — as well as cookies, cakes, and scones — are made with refined grains. It's best to avoid these types of carbs as much as you can.

## Proteins: You break down their chains; they build yours

Every muscle, cell membrane, and enzyme in your body is made from proteins. So, to create more muscle fibers, new cells, and other elements that help your body run, you need to take in protein (to get an idea of a protein's chemical structure, see Chapter 3).

Proteins are made from amino acids. Your body requires nine amino acids in particular to construct proteins, which is why those nine amino acids are referred to as *essential amino acids* (we list them for you in Table 16-1). Humans can synthesize 11 amino acids from a variety of starting compounds that aren't necessarily derived from amino acids themselves. Because these amino acids are made in the body, they're considered *nonessential amino acids;* it's not essential that you consume foods containing them because you can get them another way. We list nonessential amino acids in Table 16-1 too.

**Table 16-1 The Amino Acids Humans Can Consume**

| *Essential Amino Acids* | *Nonessential Amino Acids* |
|---|---|
| Histidine | Alanine |
| Isoleucine | Arginine |
| Leucine | Asparagine |
| Lysine | Aspartate |
| Methionine | Cysteine |
| Phenylalanine | Glutamate |
| Threonine | Glutamine |
| Tryptophan | Glycine |
| Valine | Proline |
| | Serine |
| | Tyrosine |

When you think of protein sources, you probably think of meat. There's a good reason for that. Both your muscles and the muscles of other animals are made from protein. When you eat meat — whether it's beef, chicken, turkey, pork, or fish — you're consuming the protein-containing muscle tissue of another animal.

Beans, nuts, and soy are additional sources of protein, but the protein they contain is plant protein, not animal protein. Plant proteins are considered incomplete because they don't contain enough of some of the amino acids humans need. Because animals acquire essential amino acids plus make their own nonessential amino acids, animal protein is considered complete — it has all the amino acids humans need.

Vegetarians must combine certain foods to make sure they're getting all the necessary amino acids.

## Fats: You need some, but don't overdo it

Your body needs fats to make tissues and hormones and insulate your nerves (just like a rubber coating often insulates wires). Fat is also a source of stored energy. It gives your body shape, reduces heat loss by insulating your organs and muscles, and cushions your body and organs (much like shock absorbers).

Fat supplies long-term energy, which is why you don't start to "burn fat" until 20 minutes or more into an aerobic workout. First, your body quickly burns off the glucose that's readily available in your cells. Then it starts breaking down fat molecules and converting them into glucose for fuel.

### Lipoproteins and your risk for heart disease

*Lipoproteins* are compounds made from a fat and a protein. Their job is to carry cholesterol around your body through the bloodstream. You're capable of producing four types of lipoproteins:

- High-density lipoproteins (HDLs)
- Low-density lipoproteins (LDLs)
- Very low-density lipoproteins (VLDLs)
- Chylomicrons

Sometimes in the news you read or hear about HDL being "good" cholesterol and LDL being "bad" cholesterol. However, HDL and LDL are lipoproteins, not cholesterol molecules. They just attach to and transport cholesterol. Here's what's "good" and "bad" about the lipoproteins.

Chylomicrons are very small, newly created lipoproteins that fall into the VLDL category. VLDLs have very little protein and a lot of fat. As VLDLs travel through your bloodstream, they lose some lipids, pick up cholesterol, and become LDLs. The LDLs deliver the cholesterol to cells in your body that need it, but along the way, VLDLs and LDLs can squeeze through blood vessel walls. While doing that, the cholesterol can get stuck to the wall of the blood vessel, causing deposits called plaque to form. If enough cholesterol gets stuck, an artery may get clogged, which means blood can't flow through. If that happens, a heart attack or stroke may occur. So, although LDLs help the body by transporting cholesterol, if you have too many of them, the cholesterol may start to block your blood vessels, increasing your risk of heart disease, heart attack, and stroke.

HDLs, on the other hand, are the lipoproteins that contain more protein than lipid, which makes them denser and gives them their name. Because they're so dense, HDLs can't squeeze through the blood vessel walls. Instead, they shuttle cholesterol right out of the body. They aren't able to deposit cholesterol in blood vessels because they can't get into them, so they don't increase your risk of heart disease, heart attack, or stroke. ***Remember:*** You always want to have more of these dense little guys floating in your blood than LDLs or VLDLs.

Although fats are important nutrients for your body, good and bad kinds exist.

- **Unsaturated fats** are good for you. Plant oils, such as olive and flaxseed, are excellent sources of unsaturated fats, as are fish oils.
- **Saturated fats** are unhealthy. Animal fat, like the fat on meat, and the fat found in butter are saturated fats.

Contrary to popular belief, fat doesn't make people fat. Consuming more fuel than you burn leads to the production and deposit of fatty tissue, whether that fuel comes from fats, proteins, or carbohydrates.

## Minerals and vitamins: The fuel for your enzymes

In addition to carbohydrates, proteins, and fats, your body needs certain minerals and vitamins to help your enzymes function (see Chapter 4 for more on enzymes).

*Minerals* are inorganic molecules that are part of the Earth (think iron, zinc, and calcium). Your body doesn't need a ton of minerals to stay healthy, but some are essential for its proper functioning. The essential minerals are called *major minerals,* and the minerals that you need only in very small amounts are called *trace elements.* Table 16-2 lists all the major minerals and trace elements your body needs.

**Table 16-2 The Minerals Humans Need**

| *Major Minerals* | *Trace Elements* |
|---|---|
| Calcium | Chromium |
| Chloride | Copper |
| Magnesium | Fluoride |
| Phosphorus | Iodine |
| Potassium | Iron |
| Sodium | Manganese |
| Sulfur | Molybdenum |
| | Selenium |
| | Zinc |

*Vitamins* are organic molecules that exist naturally in all living things. They're made up of the same carbon, hydrogen, oxygen, and nitrogen atoms as carbohydrates, proteins, and fats are. Any given vitamin falls into one of two categories:

- **Fat-soluble:** These vitamins (which include vitamins A, D, E, and K) need to be "dissolved" in fat molecules (or *phospholipids*) so that cells can use them. The phospholipids carry the "dissolved" vitamins through the bloodstream and into your cells.
- **Water-soluble:** These vitamins (which include vitamin C and all the B vitamins) often act with enzymes to speed up reactions.

# Exploring the Human Excretory System

When the human body breaks down food, it uses as many of the nutrients as possible to fuel cellular processes. Of course, you can only squeeze so much out of any given nutrient. What's left is waste that your body removes via your *excretory system,* which consists of your large intestine and kidneys. The next sections introduce you to these organs and how they remove waste from your body to keep you from getting sick.

## Getting to know your large intestine and how it eliminates solid wastes

After the usable nutrients from food are absorbed into the bloodstream from the small intestine, the leftover material continues on to the large intestine (also called the *colon*). This is where fecal matter (or *feces*) is created. Feces pass out of your large intestine and into the rectum, which acts like a holding tank. When the rectum is full, you feel the need to *defecate* (remove fecal material). This feat, signaling the end of the digestive process, is performed through the *anus.*

As the large intestine converts the leftover material into feces, it absorbs water and some electrolytes from the material and returns that water to the body to prevent dehydration. If too much water is absorbed, constipation occurs; if too little water is absorbed, diarrhea occurs.

### Why you *must* wash your hands

Although the bacteria that produce vitamin K in your intestines are helping to keep you healthy, they can be extremely harmful if they get anywhere else in your body. One of the bacteria that lives in your colon is *Escherichia coli* (or *E. coli*). Any strain of *E. coli* can cause diarrhea and vomiting when ingested; serious *E. coli* contamination can also contribute to *sepsis* (bacteria in the bloodstream, traveling around your body causing infections elsewhere), which can lead to coma and death.

The number one way that *E. coli* gets into food is from dirty hands. When you wipe yourself after defecating, some of the bacteria that were excreted with your feces can easily get onto your hands. If you don't wash your hands to remove any potential bacteria lingering there and then you pick up food with your hands and eat it, you may ingest the bacteria and make yourself sick. All the more reason to never forget to wash your hands after going to the bathroom!

The large intestine absorbs ions (such as sodium) into its cells from the material passing through it. Sodium ions are necessary for many cellular processes, such as the active transport of materials across cell membranes (a process described in Chapter 4). The large intestine also collects (from the bloodstream) ions to be excreted, helping to regulate the amount of ions in your body. If the amount of ions in your body isn't in the normal range, serious effects occur. For example, if your level of sodium and potassium ions (also called *electrolytes*) is abnormal, the ability for muscles to contract properly or for nerves to send impulses correctly is affected, which can interfere with your heartbeat and potentially cause a heart attack.

## Flowing through how your kidneys remove nitrogenous wastes

*Nitrogenous wastes* — unnecessary, excess materials containing nitrogen and resulting from the breakdown of proteins and nucleic acids — are released from the body in urine. In humans, the kidneys are the organs responsible for the production of urine.

You have two kidneys, one on each side of your back, just below your ribs. Like most organs in the human body, the function of a kidney is closely tied to its structure (which we depict in Figure 16-2). As you can see, each kidney has three distinct areas:

- The renal cortex, which is the outer layer
- The renal medulla, which is the middle layer
- The renal pelvis, which becomes a ureter

Each kidney contains more than 1 million *nephrons,* microscopic tubules that make urine. Each nephron contributes to a collecting duct that carries the urine into the renal pelvis. From there, the urine flows down the *ureter,* which is the tube that connects the kidney to the *bladder.*

Urine is spurted from the ureter into the top of the bladder continuously. The bladder holds a maximum of about one pint of urine, but you begin to feel the need to urinate when your bladder is only one-third full. When your bladder is two-thirds full, you start to feel really uncomfortable.

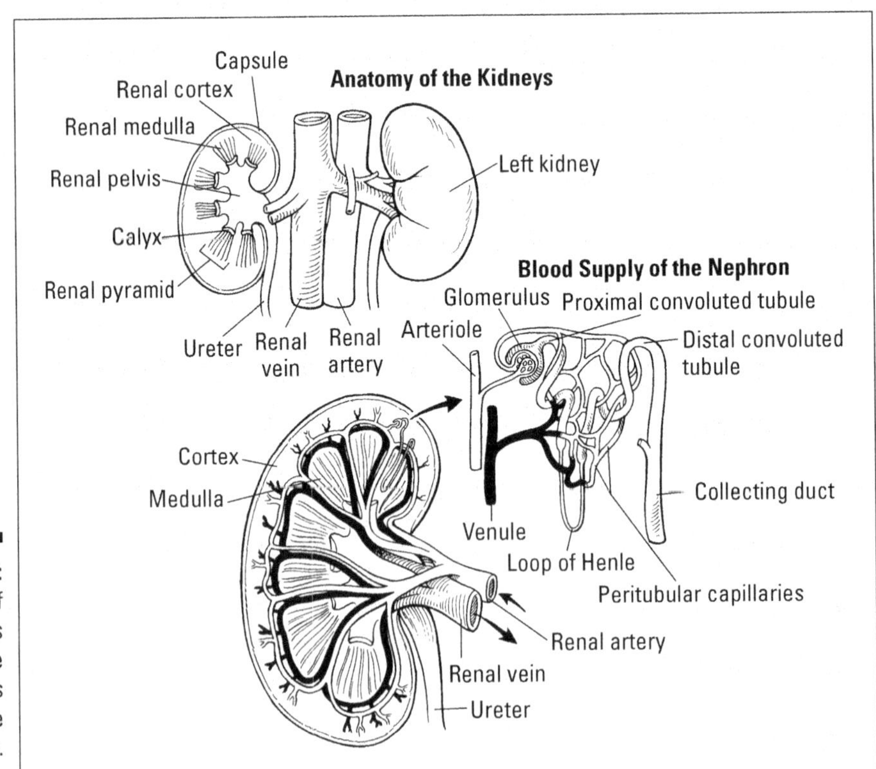


**Figure 16-2:** Structure of the kidneys and the nephrons inside the kidneys.

## Urinary problems

*Incontinence,* the inability to hold urine in the bladder, often becomes an embarrassing problem for men and women later in life. Basically, the need to urinate becomes more urgent and frequent, and sometimes urine leaks out of the bladder uncontrolled. Incontinence is thought to be more prevalent in women than men due to the stress put on the sphincter muscles associated with the bladder during pregnancy and childbirth. Men, however, have an additional urinary issue: enlargement of the prostate gland.

The prostate sits right below the bladder and surrounds the urethra. Its function is to add fluid to semen as the semen passes through the urethra in the penis. As the prostate enlarges (which starts approximately when men reach age 50), it presses on the urethra, squeezing urine back up into the bladder and making urinating painful. If the condition progresses far enough, urine can back up into the kidneys, which can lead to kidney disease. As people age, the number of *nephrons* (the microscopic tubules that make urine) in a kidney declines, and the overall size of the kidney decreases, which can contribute to reduced renal function and lead to serious problems in older people.

Urine leaves the body through the *urethra* (see Figure 16-3), a tube at the bottom of the bladder that opens to the outside of the body. It's held closed by a sphincter muscle. When you want to start urinating, the sphincter muscle relaxes, opening the urethra and letting the urine out.

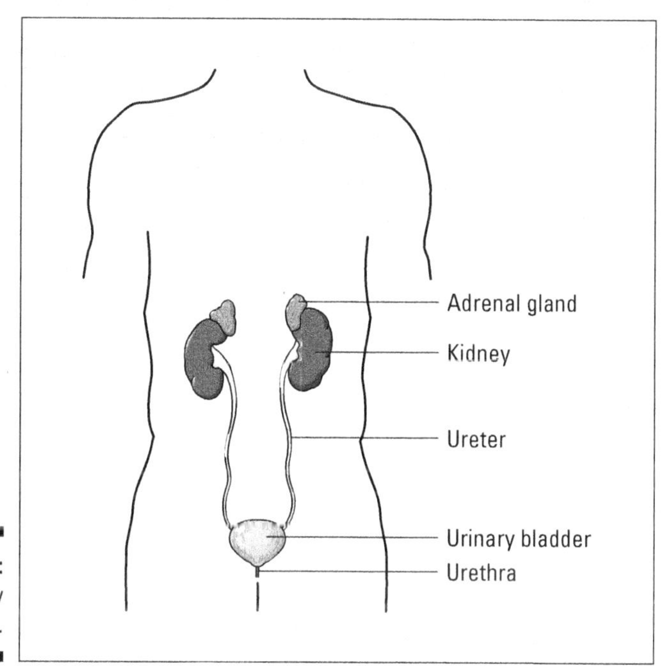


**Figure 16-3:** The urinary system.

*From LifeART®, Super Anatomy 1, © 2002, Lippincott Williams & Wilkins*

# Chapter 17

# Fighting Back: Human Defenses

### In This Chapter

- Distinguishing the good microbes from the bad
- Checking out the components of your innate and adaptive immune systems
- Helping your immune system out when it needs it
- Seeing how age affects your internal defense system

You encounter bacteria and viruses all the time, some of which have the potential to make you very sick. (On the other hand, some bacteria are actually quite beneficial to human existence.) Whether these potential pathogens cause you harm depends on a complicated give and take between their invasion tools and your defenses. You emerge the winner from the vast majority of your microbial encounters due to the combination of your *innate immunity* (a built-in immune system that all humans have) and your *adaptive immunity* (the part of your immune system that develops as you encounter microbes).

In this chapter, we present the structures and cells that keep you safe from microbes and review the options available to you when your body's defenses need a helping hand. We also explain what happens to your immune system as you age.

## Microbial Encounters of the Best and Worst Kinds

*Microbes* (bacteria and viruses) exist on every surface. They're in the air, the water, the soil — even your body. So how can you tell a good microbe from a bad one? The next sections solve the puzzle.

## Good bacteria: Health helpers

The bacteria that normally live in and on your body are your *normal microbiota,* and they play an important role in your general health, in large part because they protect you from disease-causing microbes called *pathogens.*

Your normal microbiota are beneficial to you because they

- Aid the digestive process and assist with blood clotting by releasing vitamin K. (You can't get vitamin K from foods, only from the beneficial bacteria inside your digestive tract. So consider vitamin K the rent the bacteria pay for using your intestines as their home.)
- Take up space and nutrients so pathogens can't easily colonize you.
- Make chemicals called *bacteriocins* that inhibit the growth of other bacteria.

## Bad bacteria: Health harmers

Some bacteria really live up to the bad reputation they've developed over time. These bad bacteria are pathogens, the microbes that cause infectious disease.

To cause disease, bad bacteria must be able to do three things:

- **Enter and colonize the body:** Bad bacteria can enter your body when you breathe or consume food and drink. They can also enter through a wound or be passed along through sexual contact. One of the more infamous pathogens in history is *Yersinia pestis,* the bacterium that caused the Black Death, a wave of bubonic plague that killed about two-thirds of the people living in Europe during the 13th century. This bacterium normally lives and reproduces inside rodents, but if infected rodents and humans are living near each other, the plague bacterium can be transferred to humans by fleas that bite the rodents and then bite the humans.
- **Overcome your defenses:** Your body's defenses are pretty darn good, but some bacteria have tricks to get past them. Other bacteria take advantage of you when you're down. For example, the *Streptococcus pneumoniae* bacteria exist in the throats of normal, healthy people all the time. Most of the time, the bacteria are well-behaved, just hanging about in warm, dark crevices of the throat. But if the host is weakened by a cold or flu, things get ugly. The bacteria get a little power hungry

and begin reproducing rapidly, which can lead to a sinus or ear infection, or even pneumonia.

- **Damage the body:** Pathogens produce toxins and enzymes that damage your tissues. If, for example, food is improperly processed, bacterial toxins may become the secret ingredient in your meals. One good example of this is botulism. This illness is usually caused by improper canning, which allows the bacteria *Clostridium botulinum* to grow in the food and release their toxins. The toxins, not the bacteria, are what make you sick.

## Viruses: All bad, all the time

*Viruses* are the pirate raiders of the microbial world. They're tiny particles containing genetic information that hijack your cells, forcing them to make more viruses. Although, viruses are very different in structure than bacteria, they can also make you sick.

Here's how viruses attack cells (see Figure 17-1 for the visual):

1. **The virus attaches its proteins to a receptor on a cell.**

   You can think of this like inserting your key into the door when you get home. If your key doesn't fit, you can't get in. In #2 in Figure 17-1, you can see the HIV virus attaching to a protein called CD4 that's found on the surface of certain human white blood cells.

2. **The virus shoots its nucleic acid into the cell, taking over the cell.**

   The viral nucleic acid reprograms the cell, turning it into a viral production factory. Instead of doing its job for the body, the cell starts making viral nucleic acid and proteins. The cell even uses its own molecules and energy reserves (ATP) to produce the viral parts. In #3 in Figure 17-1, you can see the viral genetic material of HIV entering the human cell, and in #6, you can see more viral genetic material being made by the cell.

3. **The viral components pull themselves together to form mature viruses.**

   Eventually, this viral replication creates too much of a crowd for the cell to handle and the cell explodes, releasing viral particles to go wreak havoc in other cells in the host's body. The number of viruses that go on attack at this point can range from ten to tens of thousands, depending on the type of virus. In #8 in Figure 17-1, you can see completed viral particles exiting the human cell.

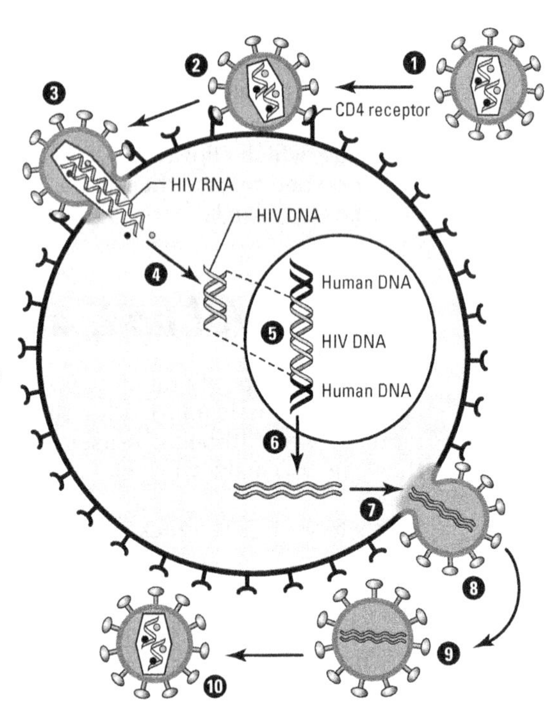


1. **Free Virus.**
2. **Binding and Fusion:** Virus binds to CD4 and coreceptor on host cell and then fuses with the cell.
3. **Penetration:** The viral capsid enters the cell and releases its contents into the cytoplasm.
4. **Reverse Transcription:** The enzyme reverse transcriptase converts the single-stranded viral RNA molecules into double-stranded DNA.
5. **Recombination:** The enzyme integrase combines the viral DNA into the host cell DNA.
6. **Transcription:** Viral DNA is transcribed to produce long chains of viral protein.
7. **Assembly:** Sets of viral proteins come together.
8. **Budding:** Release of immature virus occurs as viral proteins push out of the host cell, wrapping themselves in a new envelope. The viral enzyme protease begins cutting the viral proteins.
9. **Release:** Immature virus breaks free of the host cell.
10. **Maturation:** The viral enzyme protease finishes cutting the viral proteins, and the proteins combine to complete the formation of the virus.

**Figure 17-1:** How viruses attack cells.

Because viruses need a host in order to replicate, they're called *obligate intracellular parasites* (*obligate* means they need it, *intracellular* means they go inside cells, and *parasite* means they use the host's resources for their own benefit and destroy the host in the process).

# Built to Protect You: Innate Human Defenses

Usually you're not aware of all the microbes roaming the world because a) you can't see them and b) your innate immunity keeps most of them from bothering you. *Innate immunity* is the built-in immunity that you have due to the way the human body is constructed. Like the walls of a fortress, your innate defenses can repel all attackers (meaning they're not specifically targeted for one particular pathogen).

You only notice microbes if they manage to get past your innate defenses. When that happens, you need your adaptive immunity to come to your rescue. (See the later section "Learning a Lesson: Adaptive Human Defenses" for more on this part of your immune system.)

Your innate defenses have several ways of fending off the potential pathogens you encounter:

- **Physical barriers:** Your skin and mucous membranes are the barriers that physically block access to your tissues and organs. You can think of these barriers as the walls surrounding your castle.
- **Chemical barriers:** The pH of your stomach acid is a chemical barrier that prevents microbial growth. In other words, your stomach acid is kind of like the boiling oil that castle defenders dump over the walls onto invaders.
- **Dendritic cells:** These cells patrol your body in search of microbes and alert your immune system of impending invasions. Like sentinels patrolling a castle's grounds, they run back to the interior of the castle and alert the commander-in-chief of the army (in this case, the helper T cell) to the presence of invaders.
- **Phagocytes:** These are certain white blood cells that seek and destroy microbes that have successfully entered your body. They actually wrap around invading microbes and eat them alive. Think of phagocytes as your body's hand-to-hand combat specialists.
- **Inflammation:** This is the alarm call that gathers white blood cells to the site of an invasion and helps contain and destroy the microbes, similar to how a warning cry rallies a castle's defenders to fight an invasion.
- **Filters:** The mucus in your nose and throat and the hairs in your nose act as filters that trap microbes and prevent them from getting deeper into your body. But that's not the only filtration system at work in your body. Your lymphatic system screens your body fluids for the presence of microbes and destroys any it finds. All of these filters act a little like a moat around a castle, slowing the enemy down.

We cover each of these innate defense mechanisms in greater detail in the sections that follow.

## Your body's best blockers: Skin and mucous membranes

Your skin and mucous membranes have an important job: serving as your body's first line of defense against bad bacteria and viruses. Both the skin and mucous membranes are *epithelia,* tissues composed of multiple cell layers that are packed tightly together in order to prevent microbes from sneaking in.

Your skin is the largest organ in your body, but its size and tightly packed layers aren't the only characteristics that make it a particularly good microbe repellent. Your skin is good at keeping microbes out for these reasons as well:

- **It's dry.** Microbes need water in order to grow, so many things simply can't grow on your dry skin.
- **It's flaky.** Skin cells regularly fall right off of you, taking microbes with them.
- **It's tough.** Skin cells are reinforced with the protein *keratin* — the same strong protein that forms your hair and fingernails.
- **It's acidic.** The fatty acids from your *sebum* (the oily stuff secreted by the glands in your skin) give your skin a slightly acidic pH, which prevents many microbes from growing.

Just like your skin guards the exterior of your body, your mucous membranes protect your wet interior surfaces. Mucous membranes line your respiratory, gastrointestinal, urinary, and reproductive systems. Although your mucous membranes aren't quite as tough as your skin, they have some unique defenses of their own:

- **They're sticky.** Mucus traps invading microbes as they land on the membranes.
- **They move stuff out.** *Mechanical washing* removes microbes from the surfaces of several mucous membranes. For example, tears wash the eyes, urine flushes out the urethra, and peristalsis moves material through the intestines.

  In the respiratory tract, a blanket of cilia called the *ciliary escalator* moves mucus upward in the throat just like an escalator moves people upward at the mall. The mucus moves to a location where you can cough it out, protecting your lower respiratory tract from infection (see Chapter 4 for more on cilia).

## Tiny but mighty: Molecular defenders

The fluids in your body contain a variety of defensive proteins that help prevent infection. These tiny, invisible defenders bind to microbes, breaking them down and just generally making it difficult for them to get a foothold in your body.

These invisible chemical defenders are found in your tears, saliva, mucus, blood, and tissue fluids. Here are the names of a few of 'em:

- **Lysozyme** is a protein that breaks down one of the chemicals found in bacterial cell walls (*lyse-* is to break, and *zyme* means "enzyme," so a *lysozyme* is an enzyme that breaks down bacteria). It's one of the most common molecular defenders in your body; basically, when bacteria land on you or in you, they encounter lysozyme.
- **Transferrin** in your blood binds iron so microbes don't have enough iron for their growth.
- **Complement proteins** in your blood and tissue fluids bind to microbes and target them for destruction.
- **Interferons** are proteins that are released by cells infected with viruses. They travel to cells all around the infected cell and warn them about the virus. Cells that receive a warning from interferon produce proteins to help protect themselves against viral attack.

## Microbe seeker-outers: Dendritic cells

All around your body, white blood cells called *dendritic cells* look for potentially dangerous microbes with their special cell receptors designed for detecting molecules typical of microbial cells. Because of these special receptors, called *toll-like receptors,* dendritic cells are very good at recognizing foreign microbes. Molecules from bacteria and viruses bind to the toll-like receptors and activate the dendritic cells.

Dendritic cells do two things that are key to your ability to fight infections:

- **They release communicating molecules called cytokines.** Cytokines spread through your body and bind to other cells in your immune system (*cyto-* means "cell," and *kinesis* means "movement," so *cytokines* are literally molecules that travel between cells). The cytokines tell your immune system that microbes have been detected and help activate the cells you need to fight back.
- **They break down microbes into little bits called antigens and then show the antigens to helper T cells.** *Helper T cells* are very important to your pathogen-fighting ability because they recognize foreign antigens and send signals directing other cells of your immune system to fight.

### Invader eaters, big and small: Phagocytes

*Phagocytes* are white blood cells that patrol your body looking for microbes. When they find them, they grab them and eat them alive (*phago-* means "eat," and *-cyte* means "cell," so phagocytes are "eater-cells"). Like dendritic cells (covered in the preceding section), phagocytes activate helper T cells by showing them antigens from the destroyed microbes.

The two types of phagocytes are

- **Neutrophils:** These phagocytes multiply early during an infection and are the first ones to arrive on the scene during inflammation.
- **Macrophages:** These phagocytes live in particular tissues. (*Macro-* means "big," and *phage* means "eat," so *macrophages* are literally "big eaters.")

### Damage control: Inflammation

When microbes do manage to invade, your body responds quickly to try and contain them. The microbes and your own damaged cells trigger a cascade of events that leads to *inflammation,* a local defensive response to cellular damage that's characterized by redness, pain, heat, and swelling. Inflammation fights infection by destroying microbes, confining the infection to one location, and repairing damaged tissue.

Molecules such as histamine that are released during inflammation lead to vasodilation and increased blood vessel permeability.

- **Vasodilation causes blood vessels to widen, allowing more blood to flow to the affected area.** The blood flowing to the infected area delivers clotting elements that trigger blood clots and help confine the infection to one location. It also brings more white blood cells, including phagocytes, to help fight the infection. As a result of the increased blood flow, the infected area becomes warm and red.
- **Increased blood vessel permeability means the walls of the blood vessels loosen up.** This allows cells and materials to leave the blood and enter the tissues where the infection is happening. Phagocytes squeeze through the gaps in the blood vessel walls, crawl to the infection, and start eating the microbes. Fluid leaks from the blood vessels into the tissues and swelling occurs.

## A fluid filterer: The lymphatic system

Your blood constantly delivers materials such as food to your tissues. After fluid leaves the blood and enters the tissues, it needs to be cleaned before it's returned to the blood. The *lymphatic system* (pictured in Figure 17-2) is responsible for checking these fluids for foreign materials, such as microbes, and cleaning them up.

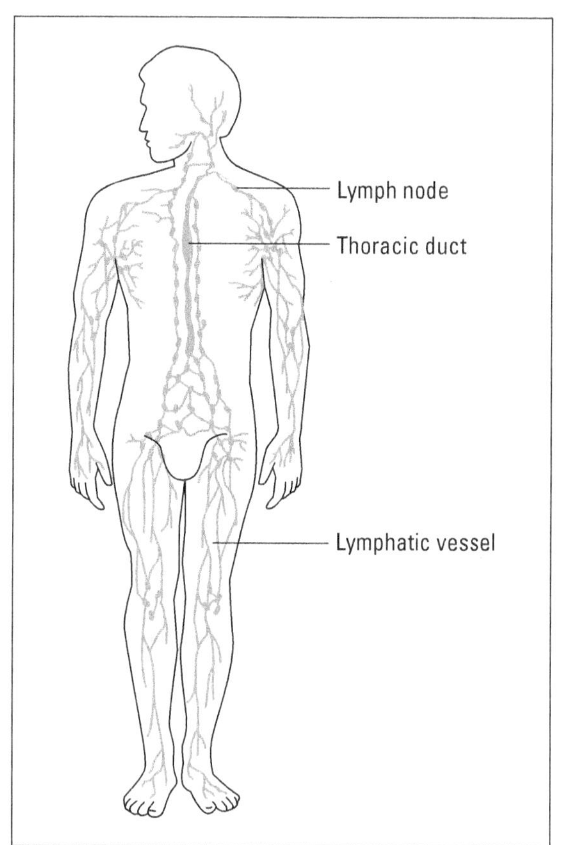


**Figure 17-2:** The lymphatic system.

From LifeART®, Super Anatomy 1, © 2002, Lippincott Williams & Wilkins

Your lymphatic system has two main components:

- **Lymphatic vessels:** These vessels carry fluid, called *lymph*, through a network of *lymph nodes* (which are lymphatic organs) and then back to the circulatory system. (***Note:*** Lymphatic vessels form a circulatory system that's similar to, but separate from, your blood vessels.) Fluid from your tissues drains into your lymphatic vessels and then flows in a one-way direction toward your heart. As lymph passes through the lymph nodes, defensive proteins and white blood cells clean it of any foreign material, including microbes. Fluid from the lymph then reenters the bloodstream near the heart.

- **Lymphatic organs:** Lymph nodes, the spleen, the tonsils, and the thymus are all lymphatic organs. They're full of white blood cells that help fight infection by destroying foreign material. Patches of lymphatic tissue are also scattered around in different organs of your body.

The spleen is a little different from the other lymphatic organs because it's connected to the circulatory system, so it filters and cleans blood rather than lymph. It's considered a lymphatic organ, though, because it contains so many white blood cells and because its job is to clean up foreign material.

## Learning a Lesson: Adaptive Human Defenses

Your body's innate defenses are incredible, and they prevent infection by most of the microbes that you encounter in your life. But every now and then, a microbe comes along that gets around your innate defenses and into your body. When your innate defenses are breached, it's time for the troops of your adaptive immunity to rally and fight back.

Your *adaptive immunity* gets its name because it adapts and changes as you go through life and are exposed to specific microbes. If, for example, you're infected with *E. coli,* only those white blood cells that recognize particular molecules on *E. coli* will be activated. If you face a different infection, say the bacteria *Staphylococcus aureus,* only the set of white blood cells that recognizes specific molecules on *S. aureus* will be activated. In other words, when your adaptive defenses come to your rescue, exactly the right team of white blood cells is activated to fight each pathogen. That means your adaptive defenses learn to recognize specific pathogens after you encounter them.

One of the awesome features of your adaptive immunity is that it can remember a pathogen it has encountered before. This *immunologic memory* allows your immune system to respond much more effectively when you reencounter a particular pathogen.

Certain cells of your immune system, called *memory cells,* remain in a semi-activated state after your first encounter with a microbe. These memory cells and their descendants hang around for a long time after they're activated in the first battle. When the same pathogen shows up again, these cells multiply quickly and efficiently destroy the pathogen before you even realize it came back. Memory cells are therefore the reason why you can get some illnesses just once and be protected from getting them again.

We describe the other components of your adaptive immune system in the following sections.

## Commanders-in-chief: Helper T cells

*Helper T cells* are white blood cells that coordinate your entire adaptive immune response. (They're called T cells because they mature in the *thymus,* which is one of your lymphatic organs.) Helper T cells receive signals from the white blood cells of your innate defenses, such as dendritic cells and phagocytes, and relay those signals to the fighters of your adaptive defenses: the B cells and cytotoxic T cells (more on these in the next sections).

Helper T cells are also called *CD4 cells* because they have a protein on their surface called CD4.

Dendritic cells and phagocytes (which are both considered *antigen-presenting cells*) activate helper T cells by showing them the antigens from the microbes they've found. Here's how the process works:

1. **Antigen-presenting cells attach pieces of the foreign antigen to proteins that they display on their surface.**

   By putting antigens on their surfaces, the antigen-presenting cells can show the antigen to the helper T cells. They're basically holding out the antigen to the helper T cells and saying, "Look what I've found!"

2. **Antigen-presenting cells also release cytokines, signaling that they've detected a foreign antigen.**

   You can think of this as the antigen-presenting cells telling the helper T cells "This looks dangerous!"

3. **Helper T cells bind to the displayed antigen using a receptor called a T cell receptor.**

   Only T cells whose T cell receptors are the right shape can bind a specific antigen. So, only the T cells that are the right T cells to fight this antigen will be activated.

Helper T cells also receive signals from antigen-presenting cells. Receptors on the helper T cells bind the signals, acting like ears so the T cells can "hear" the alarm call of the antigen-presenting cells.

After T cells bind to the antigens on antigen-presenting cells and receive the cells' signals, they activate and begin releasing signals to the other cells of the adaptive immune system.

## Soldiers on the march: B cells and antibodies

*B cells* are white blood cells that become activated when they detect foreign antigens with their B cell receptors or receive signals from helper T cells. They're activated to form two types of cells: plasma cells and memory cells (for more on memory cells, see the earlier "Learning a Lesson: Adaptive Human Defenses" section).

Plasma cells produce *antibodies,* defensive proteins that bind specifically to antigens. The antibodies produced by plasma cells are released into the blood, where they can circulate around the body. Anything in the body that's tagged with antibodies — such as invading pathogens — is marked for destruction by the immune system.

Following are the reasons why antibodies make it easier for the immune system to fight infection:

- Phagocytes can bind to antibodies, so anything with antibodies bound to it is easier for the phagocytes to grab and eat.
- Antibodies stick to pathogens and drag them into clumps. This makes phagocytes more efficient because it takes fewer "mouthfuls" to deal with the problem.
- Antibodies stick to the surfaces of viruses, preventing them from binding to new host cells.

Each plasma cell makes antibodies that are specific to one antigen, so you can think of plasma cells like the archers of the immune system. Each archer is trained to hit just one type of target. When you have an infection, the helper T cells call out just the set of archers needed to fight that particular pathogen.

Each adaptive immune response is tailored specifically to fight the invading pathogen. This specificity occurs because B cells (and T cells) activate only when their receptors recognize a specific foreign antigen. So, out of the thousands of different B (and T) cells your body can produce, only a small subset reacts to each pathogen.

## Cellular assassins: Cytotoxic T cells

If the microbes try to hide inside your cells so the antibodies can't find them, that's when cytotoxic T cells come into play. *Cytotoxic T cells* are white blood cells that are experts at detecting infected host cells. When they detect foreign antigens on the surface of an infected host cell, they order the cell to commit suicide — a necessary sacrifice in order to destroy the hidden microbes.

Cytotoxic T cells are also called *CD8 cells* because they have a protein on their surface called CD8.

# Giving Your Defenses a Helping Hand

As powerful as the human immune system is, microbes are amazingly clever and diverse, which means sometimes you may encounter a pathogen with the ability to get around all of your defenses. This is where science and medicine come in. Scientists study microbes and how they work, searching for ways to prevent pathogens from infecting and damaging the body. Doctors study how the body works and how to recognize the signs of different illnesses so they know which tools to use to help you fight off disease. Together, scientists and doctors have found ways to give the human immune system a helping hand when necessary. The next sections introduce you to these immune system helpers.

## Killing bacteria with antibiotics

*Antibiotics* are molecules made by microbes that kill bacteria. The first and most famous antibiotic, penicillin, is produced by a mold that looks a lot like the green stuff you see on old bread in your kitchen. Many other antibiotics are produced by bacteria that live in the soil.

The structures and enzymes that antibiotics target are unique to bacterial cells, so they have little effect on human cells (see Chapter 4 for more on the differences between bacterial and human cells).

Antibiotics worked so well after they were first discovered that people thought they'd won the war against bacteria. Funding for research on new antibiotics decreased because people thought they had enough weapons in their arsenal. But while people were celebrating their victory, the microbes continued to evolve. Today, humans are faced with a new microbial problem: *antibiotic-resistant bacteria.* These bacteria can't be killed by some, or even all, of the existing antibiotics.

The conundrum modern doctors now face is that using antibiotics increases the chances that antibiotic-resistant strains of bacteria will develop. It's purely natural selection (see Chapter 12 for more on this concept). In other words, when antibiotics are used, the most susceptible bacteria die first, leaving the more resistant bacteria to survive. The resistant bacteria multiply, creating new hordes that are more resistant to the antibiotic than the last horde. Repeat this cycle a few times, and the antibiotic no longer works at all.

Every year, nearly 100,000 Americans die from hospital-acquired infections (called *nosocomial infections*) related to antibiotic-resistant bacteria. And that's just part of the problem. Infections that humans thought they had under control, including such nasty diseases as tuberculosis and bubonic plague, are rearing their ugly heads in developing countries around the world.

Scientists and doctors are teaming up again to fight the threat of antibiotic-resistant bacteria. Scientists are searching for new antibiotics and new bacteria-fighting strategies, while doctors are being careful about how they prescribe antibiotics. By saving antibiotics for when they're really needed, doctors can slow down natural selection and help keep antibiotics working for as long as possible.

## Using viruses to kill bad bacteria

It may seem strange to think of a helpful virus, but that's exactly what a *bacteriophage* (or *phage* for short) is: a virus that attacks only bacterial cells.

Bacteria-blasting viruses were first discovered in the early years of the last century at the Pasteur Institute in Paris. Canadian microbiologist Félix d'Hérelle came across the little creatures while looking for a means of treating dysentery in Paris. He saw phages take on and completely destroy a whole colony of much larger bacteria. Logically enough, he hoped the microbes would help eliminate some of the world's worst bacterial infections.

Until about 1940, these tiny microbes (they're only about a fortieth of the size of bacteria) were the miracle cure for many bacterial infections. Then antibiotics came on the scene, and the medical world turned its back on these little creatures. Because you have to find just the right bacteria to appeal to the palate of a specific phage or else the deal is off, antibiotics — which have a much broader spectrum — seemed like a better solution to the problem of infection.

Yet the rise of antibiotic-resistant bacteria (see the preceding section for more on this) has caused modern doctors and scientists to once again consider the beneficial features of phages. Here are the prominent ones:

- Phages are among the most abundant creatures on Earth.
- Phages survive and reproduce in the same neighborhoods that bacteria call home. They blissfully swim around in piles of sewage and hide in cozy little corners in your body.

- Phages reproduce rapidly. After phages invade a bacterial cell, as many as 200 new phages are produced per hour. At that rate, it doesn't take long before the rambunctious little phages burst right through the cell wall. Then it's bye-bye bacterium. But that's hardly the end of it. The infant phages move on to nearby bacteria, making short work of every cell in their path. Before long, a whole colony of bacteria is history.

Scientists are currently working to solve the problems that remain with phage therapy, which include the time it takes to identify which phages like to knock off which bacteria. Many scientists advise conservative use of phage therapy to help prevent the development of phage-resistant bacteria, but they also say that phage therapy could be a valuable treatment-of-last-resort, to be used when even the most powerful antibiotics fail.

## Fighting viruses with antiviral drugs

Controlling viruses is even more difficult than controlling bacteria. Viruses aren't cells, and they have very few molecules that can be targeted by drugs. They're difficult to see and isolate, and they're equally difficult to classify. And here's another difficulty in the development of antiviral therapies: Because the host cell and viruses become so integrated, it's sometimes hard to target the virus without destroying or upsetting the mechanics of the cell itself. So, in many cases, all doctors can do is treat the symptoms of viral illnesses rather than the illnesses themselves.

The medical world does have a couple options for fighting viruses, though:

- Interferons and other cytokines can be used to stimulate the immune system's response to viruses. These human proteins can now be made in a lab by scientists who genetically engineer bacteria (flip to Chapter 9 for more on genetic engineering).
- Antiviral drugs that target specific viral proteins can be used to treat a number of viral diseases, including the flu, herpes, and HIV.

The term *antibiotic* specifically applies to drugs that fight bacterial cells. When talking about drugs for viral infections, be sure to call them *antiviral* drugs and not antibiotics.

## Getting ahead of the game with vaccines

*Vaccines,* solutions containing pieces of microbes that are introduced into the body, prevent diseases by generating immunologic memory. When you get a vaccine, the antigens from the pathogen enter your body and are processed by your immune system. The antigen-presenting cells pick up the pieces and show them to the helper T cells. The helper T cells then activate the B cells and cytotoxic T cells. In all the excitement of the adaptive immune response, the memory B and T cells are developed. They remember the antigens and are ready to defend you if the real pathogen ever shows up. If it does, your immune system kicks into gear so fast and hard that the pathogen is blown away before you even realize it was ever there.

## Vaccine safety

One issue that challenges the success of vaccines today is people's fears about vaccine safety. Because of these fears or mistrust of vaccines, some people are choosing not to vaccinate their children, a decision that ultimately puts the children at greater risk for infectious disease.

Following are a few points to consider about the safety of vaccination:

- **The risks from a vaccine are less than the risks from the disease.** All vaccines have risks and can cause side effects. However, in order for a vaccine to be licensed by the U.S. Food and Drug Administration, the side effects must be far less severe than the effects of the disease, and the risk of having side effects must be much lower than the risk of getting the disease.
- **Many diseases that are perceived as mere nuisances can actually have extremely serious complications.** Measles, for example, which some people think of as a relatively harmless childhood disease, is the sixth most common killer of children worldwide. Measles infection can result in complications such as *encephalitis* (swelling of the brain) and pneumonia. According to the WHO, measles killed 777,000 children during 2000 alone. This statistic is especially tragic when you consider that measles is a vaccine-preventable disease.
- **Many people in rich nations have no first-hand knowledge of the full impacts of infectious disease.** Most younger people who live in rich nations such as the United States or countries in Europe have grown up in a time when vaccinations were easily available. Few people who had polio are still alive in these countries, and hardly anyone remembers the days when people ended up in iron lungs because polio had paralyzed the muscles they needed to breathe. Because of a lack of knowledge, the fear of infectious diseases has declined in these countries, leading people to question the need for vaccinations.
- **The Internet spreads rumors like wildfire.** The Internet brings a world of information right into your home. The problem with that information, however, is that it hasn't all been checked for its accuracy. Books are checked by editors, and scientific and medical articles are carefully reviewed by groups of scientists and doctors before they're published. All you need to put information on the Internet, however, is the cash for a domain name and host server. An official-looking Web site can fool people about the reliability of its information, so always check the source of your information.

***Tip:*** Two organizations that have excellent information on vaccine safety are the Centers for Disease Control and Prevention (`www.cdc.gov`) and The Children's Hospital of Philadelphia (`www.chop.edu/service/vaccine-education-center`).

Vaccines are the only tool people have to control the spread of some viral diseases. Following are the different types of vaccines:

- **Inactivated vaccines** contain killed pathogens. The parts of the pathogen are included, but the pathogen can't reproduce.
- **Attenuated (live) vaccines** contain weakened pathogens. The pathogens have been altered in the lab so they have most of their original form, except for the parts that let them make you sick.
- **Subunit vaccines** contain just bits of the pathogen, such as the antigens most recognized by the body.

Smallpox has been virtually wiped off the face of the Earth by vaccination. The World Health Organization (WHO), which works to control the spread of infectious disease around the world, is now working to eradicate polio through vaccination and also plans to target measles. Eradication of smallpox, polio, and measles is possible because humans are the only host species for these viruses, which means the viruses can't multiply and mutate in animal hosts. So when you get vaccinated against these diseases, your immunologic memory takes over from there and you're good to go.

Flu viruses are a bit different. They mutate so rapidly that new vaccines are required each year. When a virus mutates, it can continually adapt to new intracellular environments and escape from the host's immune response. Mutations can change everything from the strength of a virus to its ability to latch on to new kinds of cell types or infect new animal hosts. Case in point: The swine flu pandemic of 2009 was caused by a strain of H1N1 influenza that had a unique mixture of genes from three different strains of flu, some of which hadn't infected humans before. Because the H1N1 influenza had a unique combination of genes, it had a unique combination of antigens, and no humans had existing immunologic memory. This lack of immunity allowed the virus to spread far and wide, causing the pandemic.

# Aging and Ailing: Changes in the Immune System

As your body grows, develops, and ages, your immune system changes. During the busy, trying time of puberty, the thymus starts to get smaller and smaller until it's virtually nonexistent later in adulthood. Without the thymus, T cells don't get differentiated as well or as often. Production of B cells starts to wane as well. As the cells in the bone marrow that are precursors to the T cells and B cells go through a slowdown in production, the bone marrow becomes less able to put cells through cell division.

Weakening of the immune system may explain why people toward the high end of the average life span (78 years for people living in the United States) become more prone to infection.

In a rather nasty paradox, older people also have an increased risk of *auto-immune diseases,* which are diseases in which the body's healthy cells are attacked by the cells of its own immune system. One example of an autoimmune disease that's much more common in older people than younger folk is arthritis. In people with *arthritis,* the cells of the immune system attack the cells lining the joint spaces, causing inflammation and deterioration (not to mention pain and swelling).

## Chapter 18

# The Nervous and Endocrine Systems, Messengers Extraordinaire

### In This Chapter

- Diving into the ins and outs of nervous systems
- Figuring out how the brain and senses work
- Seeing how a nerve impulse travels through your nervous system
- Discovering how the endocrine system uses hormones to regulate your inner workings

With all the metabolic processes and reactions going on in living things, organisms need to be able to exert some control in order to avoid chaos. Enter the nervous and endocrine systems. The nervous system, which consists of a brain and nerves, is responsible for picking up information from the organism's sense organs, interpreting that information, and coordinating a response. The endocrine system releases chemical hormones that travel throughout the body and regulate metabolic cycles. In this chapter, we fill you in on the structure and function of both systems and explore how the human body responds to their signals.

## The Many Intricacies of Nervous Systems

Animals are the only living things on Earth with complex nervous systems that first receive and interpret sensory signals from the environment and then send out messages to direct the animal's response. The complexity of an animal's nervous system depends on its lifestyle and body plan.

- Animals whose bodies don't have a defined head or tail have *nerve nets,* which are weblike arrangements of nerve cells that extend throughout the body. A starfish is a good example of an animal with a nerve net.

- Animals with a defined head (such as worms, insects, reptiles, mammals, and birds) possess a two-part nervous system:
  - The *central nervous system* (CNS) consists of the animal's brain and central neurons. It's housed in the head.
  - The *peripheral nervous system* (PNS) consists of all the nerves that travel to the rest of the animal's body.

The tendency in animals to have neurons concentrated in the head end of the body is called *cephalization.* The trend toward having a central nervous system is referred to as *centralization.*

In the following sections, we give you more detail about the CNS and PNS so you can tell the two apart aside from where they're located. We also get you acquainted with the worker bees of the nervous system — neurons — and explain how your neurons can bypass your brain on special occasions.

## Distinguishing between the CNS and PNS

In all animals with a backbone, including you, the CNS (pictured in Figure 18-1) consists of a brain and a spinal cord. The *brain* contains centers that process information from the sense organs, centers that control emotions and intelligence, and centers that control homeostasis of the body (see Chapter 13 for more on homeostasis). The *spinal cord* controls the flow of information to and from the brain; it sits within a liquid called *cerebrospinal fluid* that guards the CNS against shocks caused by movement and helps supply nutrients and remove wastes.

Both the brain and the spinal cord are highly protected. One layer of protection is the blood-brain barrier, which is created by the capillaries surrounding the brain. These capillaries are highly selective about what they allow to enter the brain or cerebrospinal fluid. The second layer of protection is the meninges, two layers of connective tissue that surround the brain and spinal cord.

From there the nervous system branches off into the PNS (also shown in Figure 18-1), which is divided into two systems:

- **Somatic nervous system:** This part of the PNS carries signals to and from the skeletal muscles. It controls many of an animal's voluntary responses to signals in its environment. In your case, the movements you make when walking, throwing a baseball, or driving a car are all controlled by your somatic nervous system.
- **Autonomic nervous system:** This part of the PNS controls the (mostly involuntary) internal processes in the body, such as heartbeat and digestion. It has two divisions that work opposite each other to maintain homeostasis:
  - The *sympathetic nervous system* automatically stimulates the body when action is required. This is the part of the nervous system

responsible for the *fight-or-flight response,* which stimulates a surge of adrenaline to give the body quick energy so it can escape danger. The sympathetic nervous system also quickens the heart rate to move blood through the blood vessels faster and releases sugar from the liver's glycogen stores into the blood so fuel is readily available to the cells.

- The *parasympathetic nervous system* stimulates more routine functions, such as the secretion of digestive enzymes or saliva. In contrast to the sympathetic nervous system, the parasympathetic nervous system slows down the heart rate after the fight-or-flight response is no longer needed.

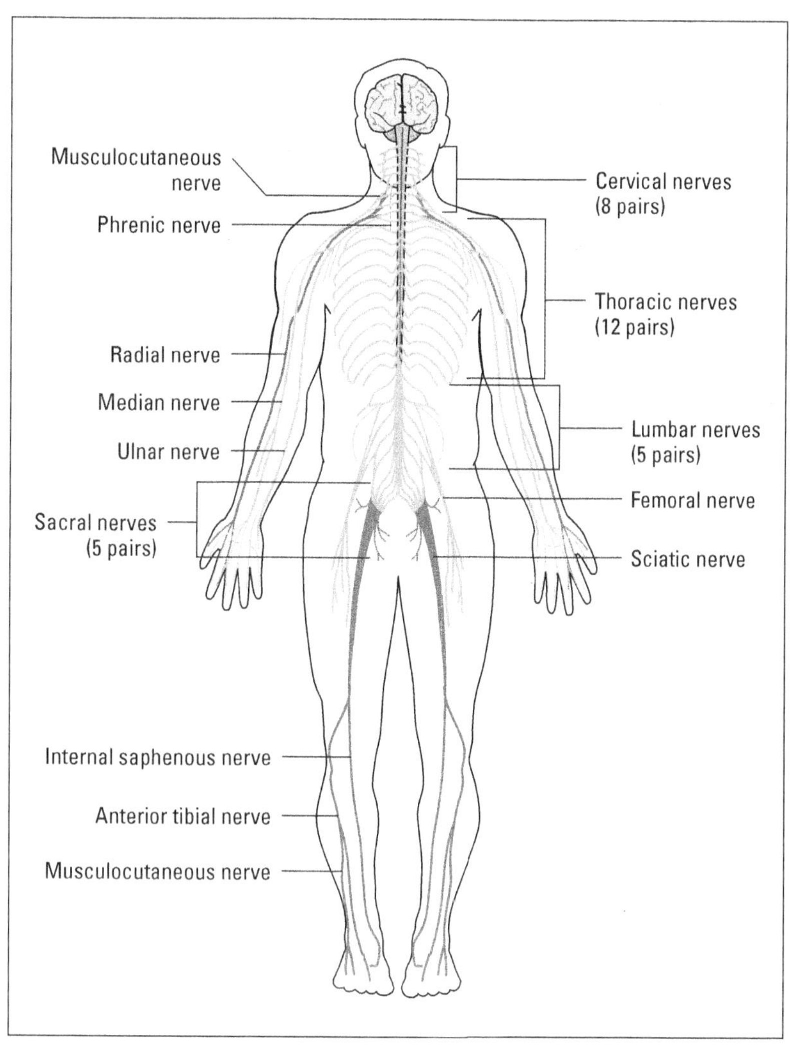


**Figuro 18 1:** The human nervous system.

From LifeART®, Super Anatomy 1, © 2002, Lippincott Williams & Wilkins

## Branching out to study neuron structure

The nervous system contains two types of cells: neurons and neuroglial cells. *Neurons* are the cells that receive and transmit signals; *neuroglial cells* are the support systems for the neurons (in other words, they protect and nourish the neurons).

Neurons have a rather elongated structure, as you can see in Figure 18-2.

- Each neuron contains a *nerve cell body* with a nucleus and organelles such as mitochondria, endoplasmic reticulum, and a Golgi apparatus (see Chapter 4 for more on cells and organelles).
- Tiny projections called *dendrites* branch off the nerve cell body at the receiving end of the neuron. The dendrites act like tiny antennae that pick up signals from other cells.
- The opposite side of the neuron extends into a long, thin, branching fiber called an *axon*. The axon is insulated by a myelin sheath made up of segments called Schwann cells.

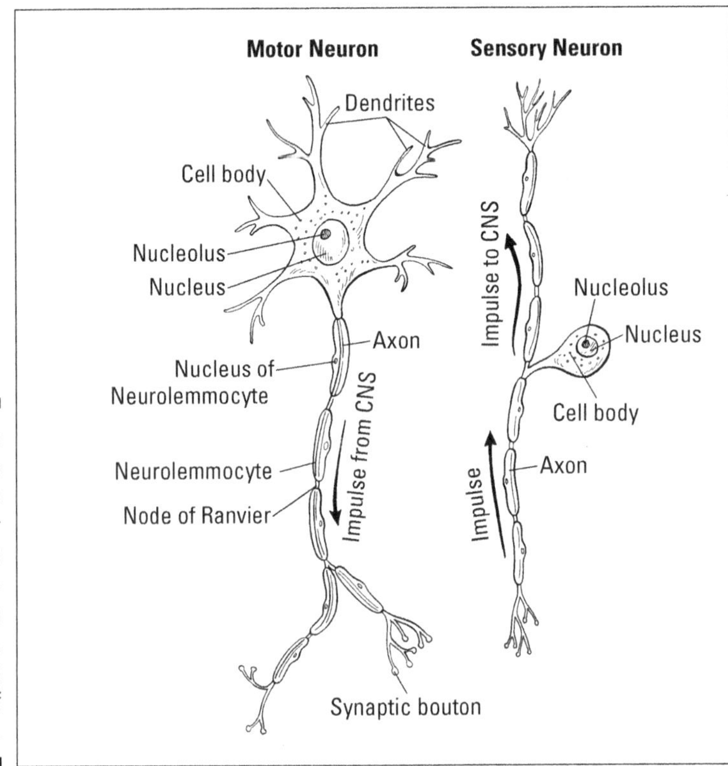


**Figure 18-2:** The basic structure of a motor neuron (left) and a sensory neuron (right), including the path of an impulse.

Nerve impulses enter a neuron through the dendrites. They then travel down the branches of the dendrites to the nerve cell body before being carried along the axon. When the impulses reach the branches at the end of the axon, they're transmitted to the next neuron. Impulses continue to be carried in this fashion until they reach their final destination.

## Processing signals with the three types of neurons

The three major functions of a nervous system are to collect, interpret, and respond to signals. Different types of neurons carry out each of these functions.

- **Sensory neurons collect sensory information and bring it to the CNS.** Also called *afferent neurons,* sensory neurons are designed primarily to receive initial stimuli from sense organs — the eyes, ears, tongue, skin, and nose. However, they're also responsible for receiving internally generated impulses regarding adjustments that are necessary for the maintenance of homeostasis. For example, if you touch the tip of a knife, the sensory neurons in your finger will transmit impulses to other sensory neurons until the impulse reaches an interneuron.
- **Interneurons within the CNS integrate the sensory information and send out responding signals.** Also called *connector neurons* or *association neurons,* interneurons "read" impulses received from sensory neurons. When an interneuron receives an impulse from a sensory neuron, the interneuron determines what (if any) response to generate. If a response is required, the interneuron passes the impulse on to motor neurons. To continue with the example from the preceding bullet, the interneurons in the cerebral cortex of your brain will process the incoming sensory information and send out responding signals.
- **Motor neurons carry the responding signals from the CNS to the cells that are to carry out the response.** Also called *efferent neurons,* motor neurons stimulate effector cells that generate reactions. To conclude our example, responding signals from your brain travel through motor neutrons until they reach your muscles, signaling your muscles to contract and pull your finger away from the sharp knife.

## Acting without thinking

Sometimes the nervous system can work without the brain, as in a reflex arc. A *reflex arc* gives sensory nerves direct access to motor nerves so information can be transmitted immediately.

When you touch a hot stovetop, your sensory nerves detect the excessive heat and instantly fire off a message to your motor nerves that says, "Pull your hand away!" The motor nerves call the proper muscles into action to move your hand away before you can even "think" about it. Your brain comes into play after the fact when you sense the pain of the burn or think to yourself how silly it was to touch a hot stovetop in the first place.

# What a Sensation! The Brain and the Five Senses

The brain is the master organ of the body in animals with a central nervous system because it takes in all the information received by the animals' sense organs and produces the appropriate responses. The brain consists of three main parts: the forebrain (uppermost part), the midbrain (middle brain), and the hindbrain (lowermost part). These three parts are further organized into four major regions that we present in the following list:

- **Cerebrum:** Also called the *telencephalon,* the *cerebrum* is the largest part of the brain and is responsible for consciousness. It's located at the uppermost part of the brain and divided into left and right halves, which are called *cerebral hemispheres.* Each cerebral hemisphere has four lobes named for the bones of the skull that cover them: frontal, parietal, temporal, and occipital. Specific areas of the lobes are responsible for certain functions, such as concentration, speech recognition, memory, and so on.
- **Diencephalon:** Found at the center of the brain, the *diencephalon* is a structure that consists of the thalamus and hypothalamus. The *thalamus* processes information going to and from the spinal cord, and the *hypothalamus* controls homeostasis by regulating hunger, thirst, sleep, body temperature, water balance, and blood pressure. At the base of the hypothalamus is the *pituitary gland,* which helps maintain homeostasis in the body by secreting many important hormones.
- **Cerebellum:** The *cerebellum,* which is found at the base of the brain, coordinates muscle functions such as maintaining normal muscle tone and posture.
- **Brain stem:** Located below the cerebellum, the *brain stem* is made up of three structures: the midbrain, the pons, and the medulla oblongata. The brain stem controls critical functions such as breathing and your heartbeat.

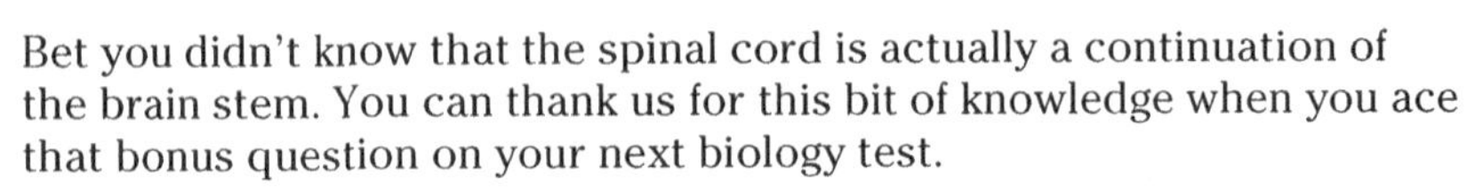
Bet you didn't know that the spinal cord is actually a continuation of the brain stem. You can thank us for this bit of knowledge when you ace that bonus question on your next biology test.

The human sense organs — eyes, ears, tongue, skin, and nose — help to protect the body. They're filled with receptors that relay information through sensory neurons to the appropriate places within the nervous system. Each sense organ contains different receptors.

- **General receptors** are found throughout the body. They're present in skin, visceral organs (*visceral* meaning in the abdominal cavity), muscles, and joints.
- **Special receptors** include *chemoreceptors* (chemical receptors) found in the mouth and nose, *photoreceptors* (light receptors) found in the eyes, and *mechanoreceptors* (movement receptors) found in the ears.

Table 18-1 compares the various types of receptors found in an animal's nervous system.

**Table 18-1 Types of Receptors & Their Functions**

| *Receptor* | *Location* | *Function* |
|---|---|---|
| Chemoreceptor | Taste buds, cilia in nasal cavity | Detect chemicals in food and air |
| Mechanoreceptor | Cilia in ear | Detect movement of ear drum and *ossicles* (ear bones) |
| Osmoreceptor | Hypothalamus | Detect concentration of solutes in the bloodstream |
| Photoreceptor | Retina of eye | Detect light |
| Proprioceptor | Muscles | Detect positioning and movement of limbs |
| Stretch receptor | Lungs, tendons, ligaments | Detect expansion or elongation of muscle tissue |

We delve into the five senses in greater detail in the sections that follow, with a focus on how these senses operate in humans.

## Oooh, that smell: Olfaction

If you walk in the door of your home and smell an apple pie baking and peppers and onions sautéing, how do you know that the apple pie is apple pie and that the pepper and onions are in fact peppers and onions and not eggplant and zucchini? The key lies in your *olfactory cells* (you can remember the name of these cells by thinking that they "smell like an old factory"); they're responsible for your sense of smell.

Olfactory cells line the top of your nasal cavity. They have cilia on one end that project into the nasal cavity (see Chapter 4 for more on cilia). On the other end, they have olfactory nerve fibers that pass through the ethmoid bone and into the olfactory bulb, which is directly attached to the cerebral cortex of your brain.

As you breathe, anything that's in the air you take in enters your nasal cavity. You don't "smell" air or dust or pollen, but you do smell chemicals. The olfactory cells are chemoreceptors, which means they have protein receptors that can detect subtle differences in chemicals.

As you breathe upon walking into your kitchen, the chemicals from the apple pie, peppers, and onions waft into your nasal cavity. There, the chemicals bind to the cilia, generating a nerve impulse that's carried through the olfactory cells, into the olfactory nerve fiber, up to the olfactory bulb, and directly to your brain. Your brain then determines what you're smelling. If the scent is something you've smelled before and are familiar with, your brain recalls the information that has been stored in your memory. If you're sniffing something that you haven't experienced before, you need to use another sense, such as taste or sight, to make an imprint on your brain's memory.

## Mmm, mmm, good: Taste

The senses of smell and taste work closely together. If you can't smell something, you can't taste it either. That's because taste buds and olfactory cells (described in the preceding section) are chemoreceptors designed to detect chemicals. Your tongue is covered with *taste buds,* clusters of cells that specialize in recognizing tastes. The *taste receptor cells* within your taste buds allow you to detect five different types of taste: sweet, sour, bitter, salty, and umami (savory).

A lot of people think that taste buds are the little bumps on the tongue, but they're wrong. Those little bumps are actually papillae; the taste buds exist in the grooves between each papilla.

Foods contain chemicals, and when you put something into your mouth, the taste receptor cells in your tongue can detect what chemicals you're ingesting. Each taste bud has a pore at one end that allows the chemicals in the food to enter the taste bud and reach the taste receptor cells. The taste receptor cells are connected at the other end to sensory nerve fibers that carry the taste signals to the brain.

The sense of taste allows you to enjoy food, which you must ingest to live, but it also serves a higher function. When your taste buds detect chemicals, they send the signal to your brain that sets in motion the production and release of the proper digestive enzymes necessary for breaking down the food you're ingesting. This function allows your digestive system to work optimally to retrieve as many nutrients as possible from food.

## Now hear this: Sound

Your ears are the sense organ for your sense of hearing, which kicks into high gear when sound waves are shuttled through your ear canal and into your middle ear — the place where your eardrum is located. When a sound wave hits your eardrum, the eardrum moves tiny bones (specifically the malleus, incus, and stapes). The movement of these bones is picked up by the mechanoreceptors in your inner ear, which exist on cilia-containing hair cells between the end of the semicircular canals and the vestibule. When the cilia move, they create an impulse that's sent through the cochlea to the eighth cranial nerve, which carries the impulse to the brain so it can interpret the information as a specific sound.

Your ears do something else for you that's not in any way related to your sense of hearing: They help you maintain your *equilibrium.* In other words, they keep you from falling over because they contain fluid within the semicircular canals of the inner ear. When this fluid moves, that movement is ultimately detected by the cilia. The cilia transmit impulses to your brain about angular and rotational movement, as well as movement through vertical and horizontal planes, all of which helps your body keep its balance.

When the fluid in your ear doesn't stop moving, you can develop motion sickness.

## Seeing is believing: Sight

Vision is perhaps the most complex of all the senses because it depends upon the correct structure and function of all the parts of your eye.

- The colored part of the eye is the *iris,* which is actually a pigmented muscle that controls the size of your pupil. The *pupil* is the "black hole" at the very center of your iris; it dilates to allow more light into the eye and contracts to allow less light into the eye.
- The *cornea* covers and protects the iris and pupil.
- The *lens* of the eye is located just behind the pupil. It's connected to the iris by a small muscle called the *ciliary muscle,* which changes the shape

of the lens to help you see things that are both far and near. Specifically, the lens flattens so you can see farther away, and it becomes rounded so you can see things that are near to you. The process of changing the shape of the lens is called *accommodation.* (People lose the ability of accommodation as they grow older, prompting the need for glasses.)

- The *vitreous body* is located behind the lens and is filled with a gelatinous material called *vitreous humor.* This substance gives shape to the eyeball and also transmits light to the very back of the eyeball, where the retina lies.
- The *retina* contains two types of photoreceptors that detect light:
  - *Rods* detect motion and can function in low light.
  - *Cones* detect fine detail and color and work best in bright light. The three types of cones each detect a different color — red, blue, or green. (Color blindness occurs when one type of cone is lacking. For example, if you're missing your red cones, you can't see red.)

Light strikes the rods and cones, generating nerve impulses that travel to two types of neurons: first to the bipolar cells and then to the ganglionic cells. The axons of the ganglionic cells form the *optic nerve,* which carries the impulses directly to the brain.

Approximately 150 million rods are in a retina, but only 1 million ganglionic cells and nerve fibers are there, which means many more rods can be stimulated than there are cells and nerve fibers to carry the impulses. Your eye must therefore combine "messages" before the impulses can be sent to the brain.

## A touchy-feely subject: Touch

Your eyes, ears, nose, and tongue all contain special receptors. Your skin, on the other hand, contains only general receptors, but it possesses a combination of general receptors that allow you to distinguish between different types of touches, from fine touches and pressure to vibration and painful sensations:

- Pain receptors make it possible for you to feel pain, like what you experience when you slam your finger in a car door.
- Mechanoreceptors allow you to feel pressure or fine touches, like a hug, a handshake, or a gentle stroke.
- Thermoreceptors help you feel temperature, like heat on a hot summer day.

These three types of general receptors are interspersed throughout your skin's surface, but they're not distributed evenly. For one thing, you have more of some types of receptors than others (as in you have many more receptors for pain than you do for temperature). For another thing, some parts of your skin have more receptors than other parts. Case in point: You have many more receptors in the skin covering the tips of your fingers than you do in the skin on your back.

When your general receptors are activated, they generate an impulse that's carried to your spinal cord and then up to your brain.

# Following the Path of Nerve Impulses

Nerve impulses pass through a neuron in about seven milliseconds — that's faster than a lightning strike! The sections that follow explain how a nerve impulse moves through a neuron and how it then makes its way from one neuron to the next.

## Traveling from one end to the other

Before a nerve impulse can enter the brain, it must first pass from one side of a neuron to the other. When a neuron is inactive, just waiting for a nerve impulse to come along, the electrical difference across the nerve cell membrane is called its *resting potential.* In this state, the neuron is polarized with sodium on the outside of its membrane and potassium on the inside. (A neuron is *polarized* when the electrical charge on the outside of its membrane is positive and the charge on the inside is negative.)

The outside of a polarized neuron contains excess sodium ions ($Na^+$); the inside of it contains excess potassium ions ($K^+$). How can the charge inside the cell be negative if positive ions are present, you ask? Well, the cell pushes out greater numbers of $Na^+$ than the number of $K^+$ it pulls in. So, the cell becomes more positive outside and less positive — or negative — inside.

Following is a description of the process that causes a resting neuron to send a nerve impulse and then return to a resting state; Figure 18-3 gives you the visual:

1. **Sodium ions move inside the neuron's membrane, causing an action potential.**

   An *action potential* is the wave of electrical activity that represents the nerve impulse. When a stimulus reaches a resting neuron, the gated ion channels on the resting neuron's membrane open suddenly, allowing the $Na^+$ that was on the outside of the membrane to rush into the cell. As this happens, the neuron becomes *depolarized* (more positive ions go charging to the inside of the membrane), making the inside of the cell positive as well.

Each neuron has a threshold level. The *threshold* is the point at which there's no holding back. After the stimulus goes above the threshold level, more gated ion channels open, allowing more $Na^+$ inside the cell. This causes complete depolarization of the neuron, creating an action potential. In this state, the neuron continues to open $Na^+$ channels all along the membrane, creating an all-or-none phenomenon. *All-or-none* means that if a stimulus doesn't exceed the threshold level and cause all the gates to open, no action potential results; after the threshold is crossed, there's no turning back — complete depolarization occurs, and the stimulus is transmitted.

When an impulse travels down an axon covered by a myelin sheath, the impulse must move between the uninsulated gaps that exist between each Schwann cell. These gaps are called the *nodes of Ranvier.* During the action potential, the impulse undergoes *saltatory conduction* (think salt, as in the sodium ions that allow this to happen) and jumps from one node of Ranvier to the next node of Ranvier, increasing the speed at which the impulse can travel.

2. **Potassium ions move outside the membrane, and sodium ions stay inside the membrane, repolarizing the cell.**

   After the inside of the neuron is flooded with $Na^+$, the gated ion channels on the inside of the membrane open to allow the $K^+$ to move to the outside of the membrane. With $K^+$ moving to the outside, the cell's balance is restored by repolarizing the membrane. (The result is a polarization that's opposite of the initial polarization that had $Na^+$ on the outside and $K^+$ on the inside.) Just after the $K^+$ gates open, the $Na^+$ gates close; otherwise, the membrane wouldn't be able to repolarize.

3. **The neuron becomes hyperpolarized when there are more potassium ions on the outside than there are sodium ions on the inside.**

   When the $K^+$ gates finally close, the neuron has slightly more $K^+$ on the outside than it has $Na^+$ on the inside. This causes the cell's potential to drop slightly lower than the resting potential, and the membrane is said to be hyperpolarized. This period doesn't last long, though. After the impulse has traveled through the neuron, the action potential is over, and the cell membrane returns to normal.

4. **The neuron enters a refractory period, which returns potassium to the inside of the cell and sodium to the outside of the cell.**

   The *refractory period* is when the $Na^+$ and $K^+$ are returned to their original sides (that means $Na^+$ on the outside and $K^+$ on the inside). A protein called the *sodium-potassium pump* returns the ions to their rightful sides of the neuron's cell membrane. The neuron is then back to its normal polarized state, which is where it stays until another impulse comes along.

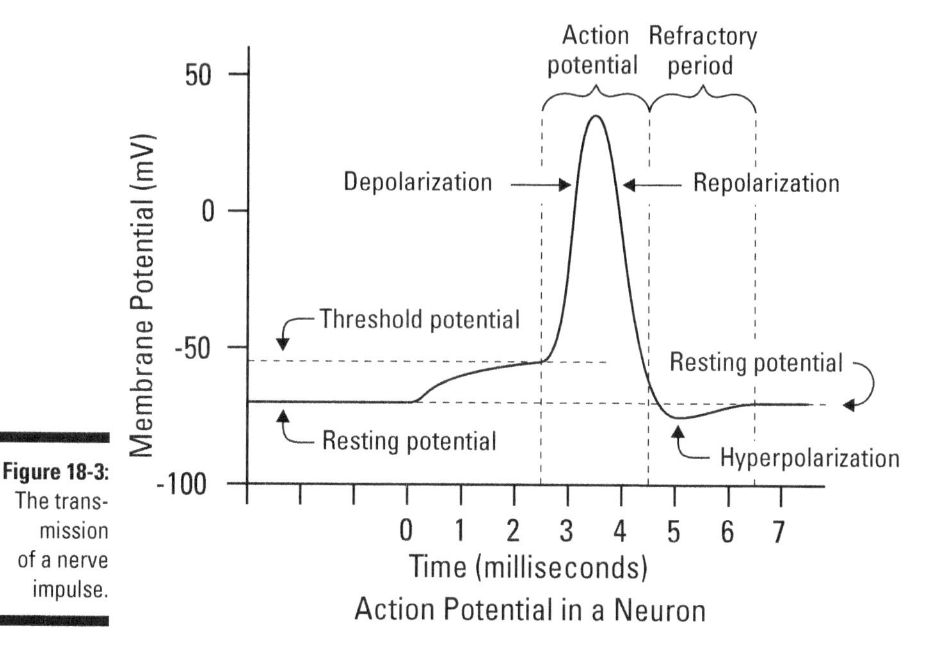


**Figure 18-3:** The transmission of a nerve impulse.

# Jumping the gap between neurons

Nerve impulses can't move directly from one neuron into another one because neurons don't actually touch each other. Gaps exist between the axon of one neuron and the dendrites of the next neuron. Nerve impulses must therefore move through that gap to continue on their path through the nervous system.

The gap between two neurons is called a *synaptic cleft* or *synapse.*

In many invertebrate animals and in the brains of vertebrate animals, impulses are carried across synapses by electrical conduction. However, outside of the brain, such as when a neuron sends a signal to a muscle cell, nerve impulses are conducted across synapses by a series of chemical changes, which occur as follows:

1. **Calcium gates open.**

   At the end of the axon from which the impulse is coming (called the *presynaptic cell* because the axon precedes the synapse), the membrane depolarizes, allowing the gated ion channels to open so they can let in calcium ions ($Ca^{2+}$).

2. **Synaptic vesicles release a neurotransmitter.**

   When the $Ca^{2+}$ rushes into the end of the presynaptic axon, synaptic vesicles connect to the presynaptic membrane. The synaptic vesicles then release a chemical called a *neurotransmitter* into the synapse.

3. **The neurotransmitter binds with receptors on the postsynaptic neuron.**

   The chemical that serves as the neurotransmitter diffuses across the synapse and binds to proteins on the membrane of the neuron that's about to receive the impulse. The proteins serve as the receptors, and different proteins serve as receptors for different neurotransmitters — that is, neurotransmitters have specific receptors.

4. **Excitation or inhibition of the postsynaptic membrane occurs.**

   Whether excitation or inhibition occurs depends on what chemical served as the neurotransmitter and the result that it had. For example, if the neurotransmitter causes the $Na^+$ channels to open, the membrane of the postsynaptic neuron becomes depolarized, and the impulse is carried through that neuron. If the $K^+$ channels open, the membrane of the postsynaptic neuron becomes hyperpolarized, and the impulse is stopped dead.

   If you're wondering what happens to the neurotransmitter after it binds to the receptor, here's the story: After the neurotransmitter produces its effect, whether that effect is excitation or inhibition, it's released by the receptor and goes back into the synapse. In the synapse, enzymes degrade the chemical neurotransmitter into smaller molecules. Then the presynaptic cell "recycles" the degraded neurotransmitter, sending the chemicals back into the presynaptic membrane so that during the next impulse, when the synaptic vesicles bind to the presynaptic membrane, the complete neurotransmitter can again be released.

Table 18-2 lists some common neurotransmitters and their functions.

**Table 18-2 Characteristics of Common Neurotransmitters**

| *Neurotransmitter* | *Source* | *Function* |
|---|---|---|
| Acetylcholine | Secreted at gaps between the neurons and muscle cells | Stimulates or inhibits contraction of muscles, depending on receptor |
| Dopamine | Created from amino acids | Affects movement, emotion, and feelings of pleasure, and plays an important role in drug addiction |
| Epinephrine | Created from amino acids | Responsible for fight-or-flight response |
| Norepinephrine | Released by postganglionic axons | Increases blood pressure |
| Serotonin | Produced through enzymatic reaction involving tryptophan | Regulates sleep, calms anxiety, and affects sexual behavior |

# The Endocrine System: All Hormones Are Not Raging

The *endocrine system* (shown in Figure 18-4) is the system that handles hormone production and secretion within an organism. It keeps a check on cellular processes and components of the bloodstream and can make adjustments as necessary.

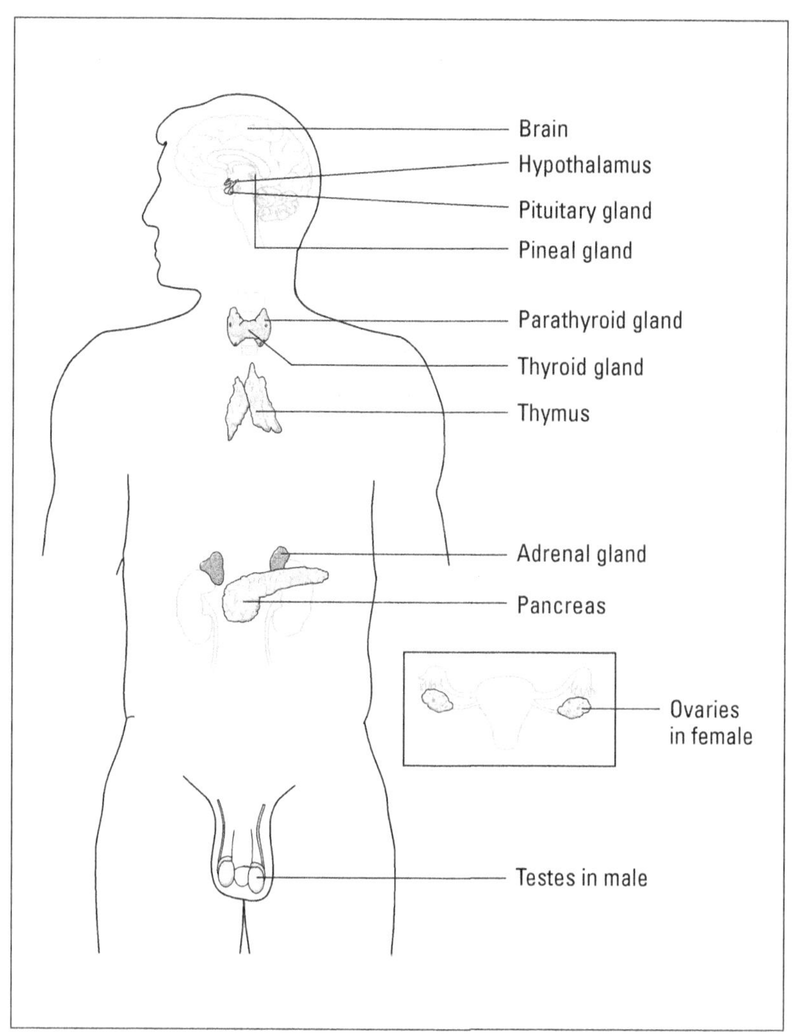


**Figure 18-4:** The ondocrinc system.

*From LifeART®, Super Anatomy 1, © 2002, Lippincott Williams & Wilkins*

The endocrine system contains organs called *endocrine glands* that produce *hormones* — chemical messengers that coordinate the activities of specific cells in certain areas of the body. Hormones are carried in the bloodstream to a target tissue elsewhere in the body, where they must be absorbed into the tissue before they can have an effect.

The word *endocrine* stems from a Greek word meaning *within*. Endocrine glands secrete their products into the bloodstream, which remains within the body. On the other hand, *exocrine glands* secrete products to the outside of the body. Examples of exocrine gland secretions are sweat and saliva.

The next sections fill you in on how hormones operate and get you acquainted with some of the important jobs they perform.

## Seeing how hormones work

Hormones are long-distance messengers, carrying their message through the bloodstream to *target cells,* the cells that respond to the hormone, throughout the body. In order for a cell to respond to a particular hormone, it must have *receptors* (molecules that bind to signaling molecules like hormones and cause changes in the behavior of cells) for that hormone. (Cells that don't have receptors for a particular hormone don't respond to that hormone.) When the hormone binds to the receptor on target cells, the receptor is activated and causes a change in the behavior of the cell.

Hormones in vertebrates can be divided into two groups:

- **Peptide hormones,** such as insulin, are short chains of amino acids; you can think of them as very small proteins. Peptide hormones are *hydrophilic* (water loving), so they don't pass easily through cell membranes.

  The receptors for peptide hormones are embedded in the plasma membranes of target cells. Peptide hormones bind to their receptors at the cell surface, activating the receptor and causing a change in the part of the receptor that extends into the inside of the cell. The internal part of the receptor interacts with molecules inside the cell to cause a change in behavior. Because the message from the hormone is passed through the plasma membrane, this signaling process is called *signal transduction.*

- **Steroid hormones,** such as testosterone and estrogen, are lipids, so they're hydrophobic (water fearing) and can pass easily through the hydrophobic layer of the plasma membrane and enter cells. Thus, the receptors for steroid hormones are located inside the cell.

  Once inside the cell, steroid hormones pass through the cytoplasm of the cell and diffuse into the nucleus. Inside the nucleus, steroid hormones bind to receptor proteins, forming an activated complex. The activated hormone-receptor complex directly causes a change in the behavior of the cell, often by acting as transcription factors that turn on the transcription of certain genes. After these genes are transcribed and translated, the newly made proteins perform the function that represents the cell's response to the hormone (see Chapter 8 for more on the transcription and translation of proteins).

## Surveying the general functions of hormones

Hormones play several important roles, whether they come from a plant, an invertebrate animal, or a vertebrate animal. Following are just a few examples of the many functions that hormones serve within an organism:

- **Assuring that growth occurs properly:** In humans, growth hormones must be secreted at normal levels by the pituitary gland throughout childhood and adolescence. The extremes of having too many or too few growth hormones are obvious — giants or midgets, respectively. In invertebrate animals, such as insects, growth hormones are responsible for *molting,* which is the shedding of the outer layer (the exoskeleton). Hormones also regulate growth in plants, and you can find out more about that in Chapter 21.
- **Ensuring that development and maturation occur properly and on time:** In insects, *metamorphosis* — the process of changing body forms during developmental stages — is controlled by a substance called juvenile hormone. (Metamorphosis is the process that changes a larva or caterpillar into a pupa and then into a moth or a butterfly.) In vertebrates, hormones trigger the transition from juvenile to adult forms and cause the onset of sexual maturity. Plant hormones regulate developmental events such as seed germination and flowering.
- **Making sure that reproduction occurs at the best possible time:** For humans, who have steady supplies of food year-round and sheltered environments in which to live, reproduction can occur whenever the urge hits (see Chapter 19 for the specifics on human reproduction). But for other animals and plants, reproduction needs to occur during certain seasons of the year when climate and food supplies are optimal. Consequently, hormones stimulate these organisms' reproductive urges only when the climate and food conditions are just right.

So many different hormones are found in animals, that we can't possibly tell you about them all. However, we do want to give you a peek into the functions of specific hormones, which is why Table 18-3 highlights some of the many hormones found in mammals like you, along with the glands that produce them and their major functions.

**Table 18-3 Several Important Mammalian Hormones**

| *Hormone* | *Gland* | *Function* |
|---|---|---|
| Adrenaline (epinephrine) | Adrenal gland | Increases metabolism and glucose in the blood; constricts some blood vessels |
| Aldosterone (mineralocorticoids) | Adrenal gland | Regulates the balance of salt and water in the body by causing kidneys to reabsorb sodium and release potassium |
| Antidiuretic hormone | Pituitary gland | Signals kidneys to retain water |
| Estrogen | Ovaries | Stimulates growth of the uterine lining; triggers and maintains secondary sex characteristics in females |
| Follicle-stimulating hormone (FSH) | Pituitary gland | Stimulates production of eggs and sperm |
| Glucagon | Pancreas | Raises glucose (sugar) in the blood |
| Growth hormone | Pituitary gland | Stimulates growth of bones; promotes metabolic functions |
| Insulin | Pancreas | Lowers glucose (sugar) in the blood |
| Luteinizing hormone (LH) | Pituitary gland | Stimulates ovaries and testes |
| Melatonin | Pineal gland | Regulates sleep and wake cycles |
| Oxytocin | Pituitary gland | Triggers contraction of the uterus and mammary glands (to release milk) |
| Progesterone | Ovaries | Supports growth of the uterine lining |
| Prolactin | Pituitary gland | Stimulates milk production |
| Testosterone | Testes | Stimulates sperm formation; triggers and maintains secondary sex characteristics in males |
| Thyroid-stimulating hormone | Pituitary gland | Stimulates the thyroid gland |
| Thyroxine ($T_4$) | Thyroid | Stimulates and maintains metabolism |

# Chapter 19

# Reproduction 101: Making More Animals

### In This Chapter

- Becoming aware of how asexual reproduction occurs in animals
- Getting an up-close-and-personal look at sexual reproduction
- Seeing how the birds and bees (and other animals) do it
- Figuring out how a cell "knows" where to go and what to do
- Exploring the complexities of development and aging

This chapter is your chance to find out all about how babies are made. It covers the nuts and bolts of how animals reproduce (we explain how plants reproduce in Chapter 20). Get ready to discover what happens inside a female's body to prepare for reproduction, how different types of animals (including humans) actually mate, how offspring develop prior to birth, and how gender is determined in humans.

## This Budding's for You: Asexual Reproduction

Asexual reproduction allows organisms to reproduce rapidly and without a partner, which makes asexual organisms essentially just fresher, younger versions of their original selves. Also, asexual organisms don't really die; instead, they just bud off into new versions of themselves and continue on.

The basic cellular process that makes asexual reproduction possible is *mitosis*, the type of cell division that produces exact copies of parent cells (see Chapter 6 for details).

Asexual reproduction occurs by several different methods in a variety of animals:

- **Budding** happens when a small outgrowth begins on the original organism. That outgrowth gradually becomes larger and eventually separates to create a new individual. Several species of invertebrates, including the hydra, produce offspring by budding.
- **Fission** occurs when the original organism grows larger and then splits in two. Sea anemones are an example of an invertebrate that reproduces asexually by fission.
- **Fragmentation** happens when small pieces of the original organism break off and then grow into complete individuals. Starfish are among the animals that use fragmentation to reproduce.

For organisms that are widely separated from others of their kind and for organisms that are doing well in a particular environment, asexual reproduction poses quite the advantage. Yet what makes asexual reproduction an asset for some species — the fact that it doesn't allow for change — also makes it a disadvantage. If a disease strikes or the environment changes and all the organisms are identical, they'll all be affected equally. If the disease can easily kill the organisms, for example, they'll all die. If they were the only organisms of their kind, then an entire species would be wiped out in one fell swoop. Species ultimately have a better chance of surviving changes if their members have some differences from one another.

# The Ins and Outs of Sexual Reproduction

Whereas asexual reproduction involves the entire organism (one whole organism splits into other whole organisms), sexual reproduction starts at the cellular level and involves two parents. The new organisms that result from sexual reproduction develop and grow over time.

In animals, sexual reproduction begins with an egg and a sperm, each of which is a single cell. Mating combines the two single cells to produce an entirely new organism that contains either eggs or sperm. Therefore, the new organism can continue the life cycle and contribute its genetic material to yet another generation of the species.

The following sections delve into the details of sexual reproduction, from the cells and mating rituals that make it possible to how humans go about creating new life. (We focus on humans rather than other species because that's obviously what's most applicable to you.)

## Getting to know gametes

*Gametes* are the sex cells of sexually reproducing organisms. Two kinds of gametes exist: sperm and eggs. Each sperm and each egg contain half the number of chromosomes that are normally present in the whole organism. When a sperm and an egg join, the organism has all the chromosomes it needs.

*Gametogenesis* is the process that produces gametes. During gametogenesis, meiosis occurs to reduce the number of chromosomes in the cells by half (see Chapter 6 for a breakdown of meiosis). Human cells, for example, have 46 chromosomes, so human gametes have only 23. The entire process of gametogenesis is controlled by *hormones,* those protein-containing substances that start, stop, and alter many metabolic processes (flip to Chapter 18 for more on hormones). We explain exactly how gametogenesis works in the sections that follow.

***Note:*** We focus our explanation on humans, but the process of gametogenesis works pretty much the same in all animals that undergo sexual reproduction. The differences lie in how the animals mate and transfer their gametes.

### Spermatogenesis: Making little swimmers

*Sperm* are the male gametes, and *spermatogenesis* is the metabolic process that produces them. Beginning at puberty, human males start producing millions of sperm each day. The sperm only survive inside the male for a short time, which is why they need to be produced continually.

The process of spermatogenesis begins in *spermatogonia,* cells containing 46 chromosomes. These cells line the walls of the seminiferous tubules. When a human male's hormones pull the trigger to initiate spermatogenesis, the spermatogonia go through mitosis to produce cells called *primary spermatocytes* that also contain 46 chromosomes. Here's how the rest of the process goes down from there:

1. **Each primary spermatocyte undergoes meiosis to produce two secondary spermatocytes.**

   *Secondary spermatocytes* are cells that contain 23 replicated chromosomes (we cover replication in Chapter 6).

2. **Each secondary spermatocyte completes meiosis, producing four spermatids, each of which contains 23 chromosomes.**

3. **Spermatids develop into spermatozoa.**

   *Spermatozoa* is just the technical name for what you know as a sperm. To become a full-fledged sperm, the spermatid must mature to the point where it has a tail, middle piece, and a head. The tail is a flagellum that moves the sperm through body fluids — in other words, it's what allows the sperm to swim. The middle piece contains many mitochondria, which supply the energy for the sperm's tail to move, and the head contains the cell's 23 chromosomes inside its nucleus.

### Oh, oh, oogenesis: Making eggs

*Oogenesis* is how human females produce *eggs,* the female gamete. A human female is born with all the eggs she'll ever have (which means oogenesis occurs when the female is a developing fetus). These eggs lie dormant from birth until puberty, at which time the female's hormones kick-start the eggs into the menstrual cycle, which continues monthly from puberty until the woman begins menopause.

The process of oogenesis begins in cells called *oogonia* (singular: *oogonium*) that contain all 46 human chromosomes. These cells grow in size and mature into *primary oocytes.* Here's a breakdown of the rest of the process:

1. **Primary oocytes begin going through meiosis but pause early in the first half of the process.**

   Each primary oocyte remains paused in meiosis until the female enters puberty.

2. **After the hormones of sexual development start flooding the female's ovaries during puberty, the menstrual cycle begins, triggering ovulation and the completion of meiosis in the primary oocyte.**

   *Ovulation,* the release of an egg from an ovary, occurs at the midpoint of the menstrual cycle. To prepare for release and possible *fertilization,* the joining of egg and sperm, the primary oocytes continue on through meiosis, completing the first meiotic division to produce two unequal cells: a *secondary oocyte* (also called a *daughter cell*), which receives most of the cytoplasm (so it's large), and the *first polar body,* which has minimal cytoplasm (so it's tiny). Both the secondary oocyte and the polar body contain 23 replicated chromosomes.

3. **The secondary oocyte and first polar body complete the second division of meiosis.**

   The secondary oocyte undergoes another unequal division of cytoplasm, producing a larger egg and a smaller second polar body. The first polar body divides equally into two polar bodies. The egg and the three polar bodies each contain 23 chromosomes.

   The splitting of the cytoplasm, called *cytokinesis,* is unequal so that the egg ends up full of cytoplasm. The egg gets most of the cytoplasm so it can hold plenty of the nutrients and organelles necessary for a developing embryo. Meiosis in one original oogonium produces just one functional egg. The three polar bodies also produced just wither away.

In human females, the meiotic division that pauses in the oocyte can remain paused for 40 years or more! From before a woman is born until her ovulation ends during menopause, oocytes are in "hang time," just waiting to develop into an egg and get fertilized. Of the thousands of oocytes that a female is born with, only about 500 develop into eggs during the woman's lifetime.

## Mating rituals and other preparations for the big event

Mating in humans can take place whenever a man and a woman are in the mood to do so. Most other animals, however, must follow more rigid reproductive cycles.

Pretend for a moment that you're a female oyster living in the ocean. Every season you release 60 million eggs into the water (no, we're not exaggerating). With that many eggs, your mating process is really left to chance. You just have to hope that one of your eggs happens to bump into a male oyster's sperm so fertilization can occur. Leaving the continuation of the oyster species to chance is one reason why oysters release such huge numbers of eggs in the first place. Obviously, most of the eggs never bump into Mr. Right; if they did, the ocean would be overflowing with oysters, and pearls wouldn't be so expensive. But what happens if you release all those eggs just because you feel like it and the male oysters in the area aren't in the mood and don't release any sperm? Then you're left alone in your shell with nothing to show for it. Animals have reproductive cycles and specific mating seasons to ensure this type of scenario doesn't happen.

In the next few sections, we explain how animals decide when it's time to start looking for a mate and how they attract one. We also fill you in on the details of human reproductive cycles because they play a huge part in the continuation of mankind.

### Syncing up for mating season

Animals of the same species need to be in sync to have successful sexual reproduction. In other words, they need to be attempting to mate at the same time.

Most species mate when the conditions are optimal. Often that's the time of year when the offspring has the best chance for survival. For example, in deer, the *gestation period* (the length of time the fetus is developing inside a female mammal) is approximately five or six months. The best time for a *fawn* (a baby deer) to be born is in the spring because food is plentiful, temperatures are a bit warmer, leaves are on the trees, and shrubs can provide cover. A spring birth also gives the fawn the longest period of time to develop before conditions get harsh the next winter. So, backing up six months from spring puts the mating season for deer around October or November. And that's exactly when you can see bucks competing for does. The strongest buck, which is supposedly the one with the strongest genes, gets to mate and pass on its genetic material to continue the species.

Sometimes, however, mating is triggered by a specific environmental signal. For instance, *aquatic* (water-living) animals that live in the desert reproduce only when the scarce desert rainfall produces a temporary pond. During the dry season, these animals are in *diapause,* which is a dormant state. In diapause, the metabolism of the animal is very low, and extreme heat and dry weather don't affect it. When rain does come, desert-dwelling aquatic animals become active immediately, breed quickly, and have offspring that develop as fast as they can before the pond dries up. Then, the new generation gets to sit in the desert in diapause waiting for the next thunderclouds to appear in the sky.

### Attracting a partner

The birds and bees may reproduce sexually, but they don't fall in love. Nor do they suffer angst over whether the other bee will remain committed or worry about whether their partner will be faithful. Bees "do it" solely for the purpose of creating more bees.

Although love may not be a requirement for sexual reproduction, attraction most certainly is. Animals generally rely on one or both of the following to help them attract a mate:

- **Mating rituals:** *Mating rituals* are behaviors that animals use to attract a mate. During mating rituals, animals (usually males) "show off" their best features in the hope of proving they'd be good partners. One male may fight other males to show that he's the strongest; another male may show that he controls access to lots of resources such as food or nesting space. Consider doves. When a male dove is trying to win a mate, he struts around bowing and cooing to the female, trying to win her over. But before they mate, they build a home (sound familiar?). The male and female doves work together to first choose a place for their nest and then build it. During the period of time that they're building the nest, they take a break and *copulate* (that's the formal way of saying they had sex). A few days later, the female lays two eggs in the new nest, and when the chicks hatch, both parents feed them. When the chicks are old enough to start feeding themselves (in about two to three weeks), the adults repeat the reproductive cycle and start courting all over again.

- **Secondary sex characteristics:** These develop as an animal matures, and they're far less obvious than *primary sex characteristics,* which are the male and female reproductive organs. Secondary sex characteristics in humans include hair growth and distribution (beards in males), deepening of the voice (in males), increase in muscle mass (males), increase in amount and distribution of fat (in females), and development of breasts (in females). Male deer grow antlers, male lions grow manes, and male peacocks develop a fan of beautiful tail feathers.

### Human reproductive cycles

Even though humans are capable of reproducing year-round, reproductive cycles are still involved. Sperm can fertilize an egg any given day of the week, but eggs can only be fertilized a few days out of every month. Human sexual reproduction is therefore controlled by the monthly *ovarian cycle* (the development of the egg in the ovary) and *menstrual cycle* (the periodic series of changes associated with menstruation), both of which are controlled by hormones (for more on hormones, see Chapter 18).

As described in the earlier "Oh, oh, oogenesis: Making eggs" section, an oocyte needs to complete meiosis and mature into an egg before it can be released by the ovary. Believe it or not, the brain runs this process.

The hypothalamus, which is found at the center of the brain, keeps a check on how much of the hormones estrogen and progesterone are floating through the bloodstream. When the levels decline, the hypothalamus secretes a hormone called *gonadotropin-releasing hormone* (GnRH) that heads straight for the pituitary gland (which is also found in the center of the brain) and stimulates part of the pituitary to secrete follicle-stimulating hormone (FSH) and luteinizing hormone (LH). FSH and LH begin the ovarian cycle by kick-starting meiosis again and continuing development of the *follicles* (another name for the oocytes that are suspended in meiosis) so they can release an egg. As the follicles grow, they release the hormone *estrogen.*

When the hypothalamus detects rising levels of estrogen in the bloodstream, it releases even more GnRH, causing the pituitary gland to release a large amount of LH about midway through the ovarian cycle. LH stimulates the release of the egg from the follicle in the ovary — in other words, LH triggers ovulation. It also triggers the remaining follicle cells to develop into a mass of cells called the *corpus luteum,* which secretes estrogen and progesterone for the rest of the ovarian cycle (about two more weeks). These hormones prepare the body for a possible pregnancy by spurring the tissues lining the uterus to develop thicker blood vessels, which brings more nutrients into the uterus.

The hypothalamus can detect when the levels of estrogen and progesterone have reached the point where the lining of the uterus is ready for implantation. If fertilization

- **Has occurred:** The fertilized egg implants in the lining of the uterus and an embryo begins developing. That embryo immediately starts to secrete the hormone *human chorionic gonadotropin* (hCG). The presence of hCG ensures that estrogen and progesterone production continue so the lining of the uterus remains nourished by larger blood vessels. After the *placenta* (a blood-filled, nutrient-rich temporary organ) has formed, the embryo gets its nutrients and blood supply through the umbilical cord connecting the embryo to the placenta, which is connected to the mother's blood supply. Therefore, the production of hCG by the embryo declines after the placenta is up and running.

  Pregnancy tests detect the presence of hCG. Because hCG is produced solely by fertilized eggs, only women with a fertilized egg in their bodies should have detectable levels of hCG.

- **Hasn't occured:** No hCG is produced, and the hypothalamus tells the pituitary gland to stop producing FSH and LH. The lack of FSH and LH stops the production of estrogen and progesterone, which causes the lining of the uterus, sometimes referred to as the *endometrium,* to stop receiving all that extra nourishment. The endometrium then starts to disintegrate and eventually sloughs off and is carried out of the body by the menstrual flow. The first day of menstrual flow is the first day of the menstrual cycle (pictured in Figure 19-1). The drop in LH also causes the corpeus luteum to disintegrate, ending the ovarian cycle.

The ovarian cycle and the menstrual cycle occur simultaneously and are synchronized to each other by hormones, but each cycle consists of different events.

- The ovarian cycle includes the development of the follicle, the secretion of hormones by the follicle, ovulation, and the formation of the corpus luteum. It occurs in the ovary, takes about 28 days to go from beginning to end, and is controlled by GnRH, FSH, LH, and estrogen.
- The menstrual cycle includes the thickening of the endometrium to prepare for possible implantation and the shedding of the endometrium if there's no implanted embryo. It occurs in the uterus, takes about 28 days, and is controlled by the levels of progesterone and estrogen.

Many more eggs are fertilized in a woman's lifetime than she may realize. Not every fertilized egg results in a bouncing baby boy or girl. If the hormone levels aren't right from the start, a fertilized egg may never implant or may implant but not secrete enough hormones to maintain the pregnancy. If an embryo doesn't produce a sufficient amount of hCG, the pregnancy fails to continue, and the embryo aborts itself (a *spontaneous abortion* is another term for a miscarriage). An unusually heavy menstrual period that started a few days late is often really the spontaneous abortion of a fertilized egg that didn't work out.

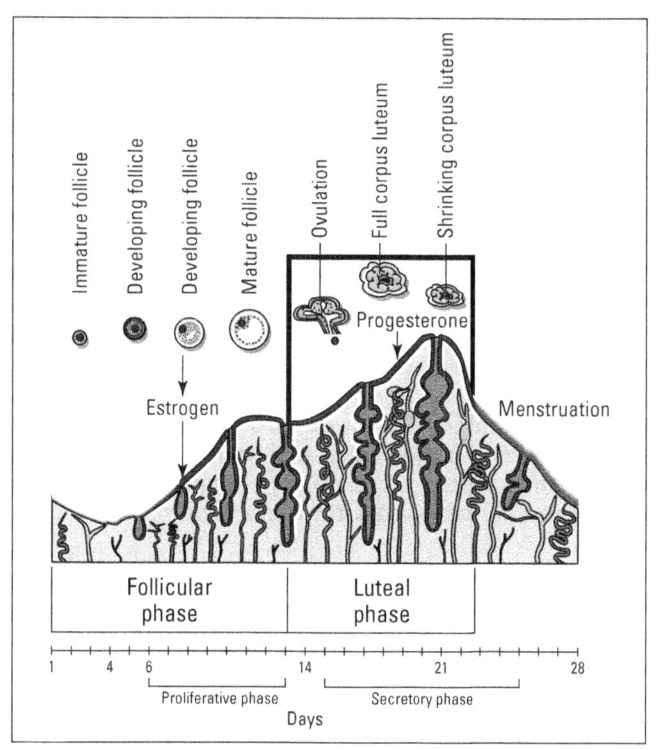


**Figure 19-1:** The menstrual cycle.

From LifeART®, Super Anatomy 3, © 2002, Lippincott Williams & Wilkins

## Talk about overstaying your welcome

After a follicle develops and releases an egg, the empty follicle is called a *corpus luteum,* which is Latin for "yellow body." If the egg becomes fertilized and implants in the uterus, the corpus luteum hangs around to help out with the beginning stages of pregnancy. It secretes progesterone for a few weeks until the placenta is fully developed and can secrete progesterone on its own. (The progesterone helps keep the lining of the uterus rich with blood and nutrients for the developing embryo.) Sometimes the corpus luteum sticks around for a few months, but in most cases, it eventually shrinks and withers away during the pregnancy.

Notice how we said "in most cases"? About 10 percent of the time, the corpus luteum hangs out in the ovary far longer than it should. Sometimes it remains even if the woman isn't pregnant. When that happens, the corpus luteum can turn into a cyst, which is aptly called a *corpus luteum cyst.* Usually, this cyst on the ovary isn't a problem, unless it continues to grow, twist, or rupture. Only then does it need to be removed surgically.

## How humans mate

The most important thing you should know about mating is that members of different species can't successfully reproduce with each other. After all, the whole point of sexual reproduction is to create a new generation that contains the genetic information from the previous generations. Interspecies sexual reproduction doesn't work because different species contain different numbers of chromosomes, and those chromosomes contain differing genes. For instance, humans carry 46 chromosomes in each cell, whereas chimpanzees have 48 per cell. If a human and a chimp were to mate, the cell divisions wouldn't be equal, and a theoretical offspring (we won't even go there) probably wouldn't be able to survive.

Eggs are actually surrounded by a layer of proteins on top of the plasma membrane that contains receptor molecules made solely for receiving sperm of the same species. In human eggs, the zone that prevents fertilization by a different species is called the *zona pellucida.* Only human sperm can crack the code to get into the egg.

In the following sections, we stick to how humans make babies. (If your parents have yet to have this conversation with you, you can let them know they're off the hook.)

### Human reproductive systems

Before you dive into the details of human sexual reproduction, it helps to know a little bit about the organ systems involved. As you can see in Figure 19-2, the male reproductive system is made up of the penis, the testes, and the seminiferous tubules. Sperm is made in these tubules through the process of spermatogenesis (which we explain earlier in this chapter).

The female reproductive system consists of the vagina, the uterus, the ovaries, and the fallopian tubes (see Figure 19-3). A woman's egg cells are produced in her ovaries through the process of oogenesis (which we walk you through earlier in this chapter).

### Sexual intercourse

In order to prepare the body for *sexual intercourse* (the insertion of the man's penis into the woman's vagina), men and women engage in activities that increase *arousal* (responsiveness to sensory stimulation). When a man is sexually aroused, his penis becomes erect because the erectile tissue within the penis fills with blood. This erection allows the penis to stiffen so it can remain inside a woman's vagina during intercourse. In females, the *clitoris* is the sexually sensitive organ. It has erectile tissue and a glans tip, just like a penis does. When a woman becomes aroused, the erectile tissue within the clitoris (which is located toward the front end of the labia just under the pubic bone) fills with blood, and the increased pressure causes drops of fluid to be squeezed out of the tissue. This lubrication prepares the vagina for sexual intercourse so the erect penis can be inserted into it easily.

When the penis is fully inside the vagina, the tip of the penis is as close as possible to the woman's cervix. The *cervix* is the bottom end of the uterus, which extends down into the vagina. Sperm must travel through the cervix to enter the uterus. During sexual intercourse, sperm travel from the man's *epididymides* (tubules in the scrotum that store produced sperm) to his *vas deferens,* tubes that carry sperm from the scrotum to the urethra so they can be ejaculated.

The actions that occur during sexual intercourse serve to bring the man and woman to the climax of stimulation, which is followed shortly by orgasm.

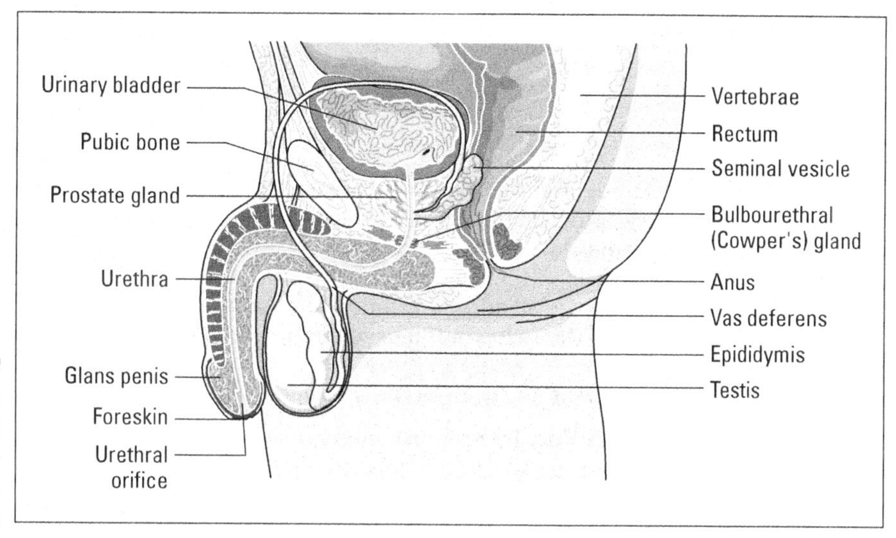


**Figure 19-2:** The male reproductive system.

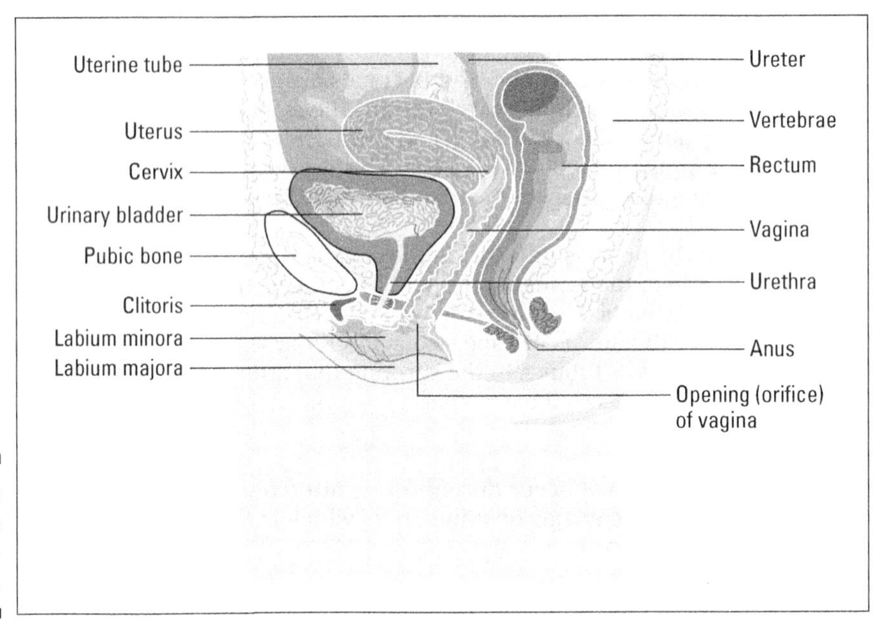


**Figure 19-3:** The female reproductive system.

*From LifeART®, Super Anatomy 1, © 2002, Lippincott Williams & Wilkins*

### Orgasm

Believe it or not, *orgasm,* the highly pleasurable climax of sexual intercourse, serves a physiological purpose. As the sexual stimulation of a male intensifies, the sperm move from the vas deferens into the urethra and secretions from three glands — the seminal vesicles, the prostate gland, and the bulbourethral gland — all add their fluids to create *semen* (seminal fluid). The semen contains the following "ingredients" that help promote fertilization:

- **Fructose:** This sugar gives the sperm energy to swim upstream.
- **Prostaglandins:** These hormones cause contractions of the uterus that help propel the sperm upward.
- **A pH of 7.5:** This relatively high pH provides the basic solution in which sperm can live and helps neutralize the acidic conditions of the vagina, which would otherwise kill the sperm.

Orgasm occurs at the height of sexual stimulation and is signaled by muscular contractions and a pleasurable feeling of release. The muscular contractions cause semen to be expelled from the penis, which is called *ejaculation.* The average amount of semen expelled during one ejaculation is less than 1 teaspoon, but it contains more than 400 million sperm.

When orgasm occurs in a male, a sphincter muscle closes off the bladder to prevent urine from entering the urethra. Shutting out urine allows the urethra to be used solely for ejaculation at that time. (In males, the urethra is shared by both the urinary tract and the reproductive tract; in females, the urethra is solely part of the urinary tract.)

In females, the height of sexual stimulation also causes intense muscular contractions and a pleasurable feeling of release. The fluid released inside the vagina helps create a watery environment that the sperm can swim in. The muscular contractions of the uterus slightly open the cervix, which allows sperm to get inside the uterus and also helps "pull" sperm upward toward the fallopian tubes.

### Fertilization

After ejaculation, sperm have quite a bit of swimming to do before they find the egg. They have to travel from wherever they're deposited in the vagina, through the muscular cervix, up through the entire uterus, and finally into the fallopian tubes, which is where *fertilization* (the joining of sperm and egg) actually occurs.

Because a human egg lives no longer than 24 hours after ovulation and human sperm live no longer than 72 hours, intercourse that occurs in the three-day period prior to ovulation or within the day after ovulation is the only chance of fertilization during a given month.

If the sperm does find its way to the egg, it must penetrate the egg in order to supply it with its 23 chromosomes. However, human eggs have several layers of cells and a thick membrane surrounding them. To get through all that, the sperm produces enzymes in a structure near its nucleus called the *acrosome*. The acrosomal enzymes digest the protective layers of the egg (so the sperm essentially "eats" its way into the egg). But the sperm isn't alone in its efforts. The egg helps the sperm get inside by going through physical and biochemical changes. After the sperm has successfully joined with the egg, the two gametes create a cell that contains the full human chromosome count of 46.

## How Other Animals Do It

Humans obviously aren't the only animals that mate and undergo sexual reproduction. All kinds of animals reproduce sexually. The question is: How do they do it?

Following is a look at the mating styles of birds, bees, worms, and sea urchins, as well as explanations of how the species' zygotes form (a *zygote* is the fertilized cell that results when two gametes join together). ***Note:*** We highlight these animals' mating styles and zygote formation because they give an overview of the different strategies for sexual reproduction that occur in the animal kingdom.

- **Bees:** The process by which bees reproduce is called *parthenogenesis,* meaning "virgin production." (*Partheno-* is Greek for "virgin" [as in the Parthenon], and *genesis* means "production.") A queen bee receives all the sperm she'll ever be impregnated with during her nuptial flight when she mates with the *drones* (male bees that never have a full chromosome count). The queen never mates with the drones again because she doesn't have to — she has stored their sperm cells inside her body, leaving her in total control of when fertilization occurs.

  When the queen lays eggs and releases the sperm, the eggs become fertilized. Those fertilized eggs develop into females, many of which are worker bees (these bees are born diploid but never produce gametes, which means they can't reproduce). The other fertilized eggs develop into new queens. When the queen lays eggs but withholds sperm and prevents fertilization, the unfertilized eggs develop into male drones. The cells in drones' testes develop into gametes that are received by the new queen, which technically is one of their sisters. (If you're confused by what *haploid* and *diploid* mean, see Chapter 6.)

- **Birds:** Birds copulate, and the male bird deposits his sperm inside the female bird. The egg becomes fertilized and is deposited outside the female bird's body to continue developing until it's time to hatch. The *yolk* of an egg (the yellow part) is where the developing embryo resides, and the *albumen* (the white part) serves to nourish the embryo throughout its development.

  Just after fertilization, one spot on the yolk of an egg goes through a series of divisions called cleavage. At the end of the cleavage divisions, an embryonic disk called a blastoderm is created on one side of the yolk. The *blastoderm* is the initial cell tissue that begins to develop into a baby bird; it separates into an *epiblast,* which is the top layer, and a *hypoblast,* which is the bottom layer. The epiblast cells migrate down into the hypoblast along a line in the yolk, called a *primitive streak,* to create the *mesoderm,* which goes on to develop the rest of the bird.

- **Earthworms:** Earthworms are *hermaphrodites,* meaning they have both female and male reproductive parts (specifically ovaries, testes, seminal vesicles, vas deferens, and seminal receptacles). When earthworms copulate, they face in opposite directions and put their citella together. *Citella* are the external, smooth, nonsegmented parts of an earthworm; their job is to secrete mucus and help the sperm get from the vas deferens of one worm to the seminal receptacle of the other worm. Cocoons form and are protected by a mucus sheath created by the citella. The sperm and eggs are fertilized inside the cocoon, and the zygotes stay enclosed in it until they hatch in the soil.

- **Planarian worms:** These freshwater flatworms are fairly unique because they can reproduce both asexually (by constricting their bodies and literally splitting in two) and sexually. All planarian worms are hermaphrodites. When two of these worms reproduce sexually, they use their male organs to secrete and exchange sperm with each other. Then they use their female organs to create zygotes. The zygotes develop into small planarian worms that later hatch and mature into adults.
- **Sea urchins:** Male and female sea urchins look exactly the same on the outside — they both have a ring of genital pores at the center of their bodies. Males discharge sperm through their pores into the water, and females discharge eggs through their pores into the water. Fertilization is left to chance, but it's helped by the fact that sea urchins live in close contact, their eggs have a sticky coat to which sperm adhere, and ejaculation by any one sea urchin signals the other males to ejaculate too.

# Developing New Humans

Sexual reproduction involves the production of gametes and the act of mating to join the gametes so fertilization can occur. After fertilization takes place, the term *development* describes how the fertilized egg becomes another new organism that possesses a mix of its parents' DNA.

In the sections that follow, we walk you through the process of development, from zygote to newborn. ***Note:*** We focus on the development of human offspring because that's most relevant to your life. However, many similarities exist between the development of humans and other animals.

## From single cells to blastocyst

After the nucleus of a sperm and the nucleus of an egg fuse, fertilization is complete, and the new, diploid cell is referred to as a zygote.

The zygote begins traveling down the fallopian tube, heading for the uterus so it can implant in the uterine lining. As it travels, the zygote undergoes *cleavage,* a rapid series of mitotic divisions that result in a multicellular embryo (see Step 3 in Figure 19-4). The zygote divides into 2 cells, which divide into 4 cells, which divide into 8 cells, which divide into 16 cells. At this point, the zygote is a solid ball of cells called a *morula.*

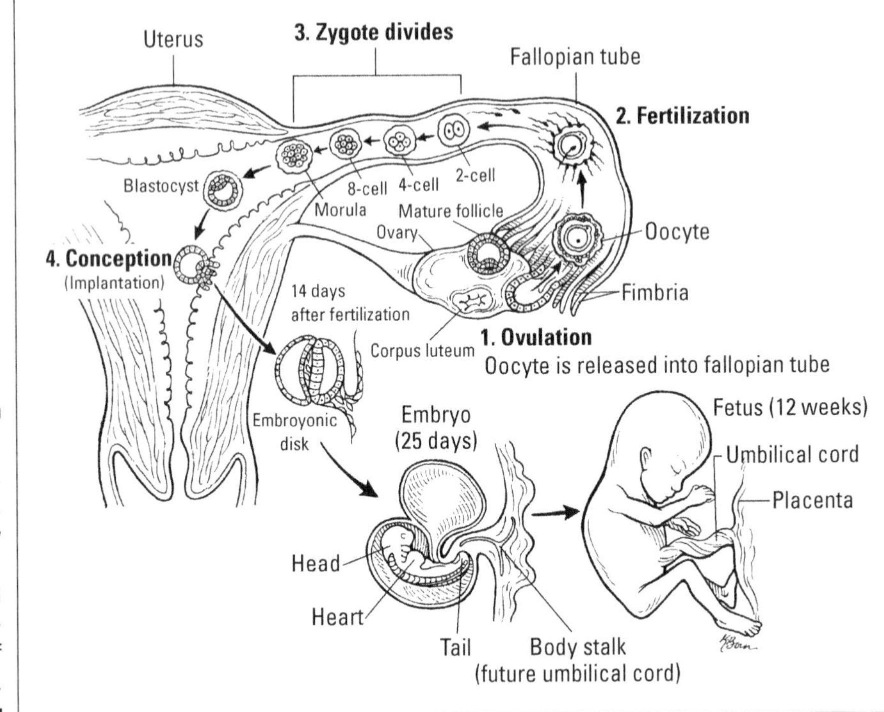


**Figure 19-4:** Fertilization, conception, and early embryonic and fetal development of humans.

Cell division continues, but the morula becomes filled with liquid, which pushes the increasing number of cells out toward the periphery of the embryo's membrane, forming a hollow ball of cells called a *blastula.* In humans, a group of cells inside the blastula become specialized to form the embryo, and the blastula becomes a *blastocyst.* Different layers of cells become specialized within the blastocyst, taking the first step toward forming specialized tissues:

- The flattened cells along the edge of the blastocyst form the *trophoblast.*
- The fluid-filled cavity is called the *blastocoel.*
- The sphere of larger, rounded cells that's destined to become the embryo is called the *embryoblast.*

The trophoblast cells secrete an enzyme that helps degrade the lining of the uterus. After the blastocyst "eats" its way into the wall of the uterus, it sinks into the wall and implants itself. *Conception* occurs when the blastocyst successfully implants itself in the wall of the uterus (see Step 4 in Figure 19-4); if the blastocyst doesn't implant in the wall of the uterus, there's no pregnancy. If conception occurs, the trophoblast cells of the blastocyst go on to form the *chorion,* which becomes part of the placenta.

After implanting itself, the developing mass of cells moves inward, forming a ball of cells called a *gastrula* that has three layers; this process is referred to as *gastrulation.* Each layer of cells in the gastrula eventually becomes a different type of tissue:

- The outer layer of the gastrula is the *ectoderm,* which develops into the skin and nervous systems.
- The middle layer of the gastrula is the *mesoderm,* which develops into the muscular, skeletal, and circulatory systems.
- The innermost layer is the *endoderm,* which gives rise to the linings of your digestive and respiratory tracts, as well as organs such as the liver and pancreas.

Conception isn't the same as fertilization. An egg can be fertilized, but a woman isn't pregnant until the blastocyst is rooted in her uterine wall where it can develop further.

## Go, go, embryo

You get more done in the 12 weeks or so that you're an embryo than at any other time in your life. That's because every organ in your body forms during the embryonic period — essentially the first trimester of pregnancy.

After gastrulation (see the preceding section), the specialized cells of the ectoderm, mesoderm, and endoderm begin migrating toward other cells with the same specialty. This cellular migration is referred to as *morphogenesis* because it gives the embryo a shape (*morph-* is Greek for form or structure).

Outside the embryo, specialized membranes develop. The chorion combines with tissues created by the mother to become the *placenta,* an organ that's filled with blood vessels and provides a large surface area for the exchange of gases, nutrients, and wastes.

A tubular structure called the *allantois* forms off of the center cavity of the blastocyst. In humans, the allantois eventually becomes the body stalk and then the umbilical cord, which connects the fetus to the placenta (both of which are shown in Figure 19-4).

The *amnion* surrounds the amniotic cavity, which fills with *amniotic fluid* that protects the developing embryo. Amniotic fluid cushions movements created by the mother and protects the developing organism from bumps. It's also thought to contain *surfactant,* a substance that coats the internal surfaces of the fetus's lungs so its lung tissues don't stick together after birth, which would prevent the infant from breathing.

The genetic material in the cells of amniotic fluid matches that of the developing embryo, so doctors can use *amniocentesis* (a procedure that uses a needle to remove a small amount of amniotic fluid for genetic testing) to see whether the embryo will have a genetic defect.

Throughout embryonic development, all other systems and structures of the body form as well. Cells that split off the neural tube and form the neural crest become the teeth, bones, skin pigments, and muscles of the skull, for example. At the end of the embryonic period, a human embryo is about an inch and a half long, and it starts to look less like a lizard and more like a human.

## Fetal development and birth

In humans, the fetal period encompasses the last six months of pregnancy, or the second and third trimesters. Fetuses are completely differentiated, meaning their cells have migrated and formed organ systems. All fetuses do in the uterus is continue to grow and develop features such as hair and nails. As the fetus gets stronger, longer, and heavier, it looks more and more like a newborn baby.

Prostaglandins and the hormone oxytocin cause the uterus to contract, but the initial production of these hormones is thought to be triggered by an as-yet-unknown chemical produced by the fetus. So, the fetus prompts its mother to start producing the hormones that initiate labor. If labor doesn't begin naturally, medical personnel give the mother prostaglandin suppositories and/or synthetic oxytocin — called *pitocin* — to induce labor.

When the fetus is finally born, the organism is called a *neonate,* meaning newborn. A life begins, and development continues. For more information about what happens during pregnancy, check out *Pregnancy For Dummies,* 3rd Edition (Wiley).

# Differentiation, Development, and Determination

We bet you never realized that a large part of who you are is controlled by three *D*'s: differentiation, development, and determination. *Differentiation* is the specialization of cells that occurs during development; it determines what the structural and functional aspects of the cell will be. *Development* is the overall process of an organism going through stages of differentiation; over time, the changes occurring at the cellular level during development become visible. *Determination* occurs when a cell commits to developing in a certain manner — that is, when it's determined to become a particular type of cell, such as a brain cell or a bone cell.

Think about this: You started out as one tiny cell. When the nucleus of your father's sperm fused with the nucleus of your mother's egg, a single cell was created containing all the genetic information you'd ever have for the rest of your life. As that first cell divided, some of its descendants became determined to be heart cells, skin cells, brain cells, and liver cells. Each of these cell types looks different and behaves differently in the body, but they all have the exact same set of genetic information. What makes each cell type different from the other isn't what kind of genetic information it has but how it uses that genetic information.

Cells become differentiated through the process of *gene regulation,* a process that controls which set of genes a particular cell uses at any given time.

If you think of the different types of cells in your body as workers that each have a different job to do, then it makes sense to think about each worker needing a different set of tools. To a cell, that means a different set of proteins. And to get proteins, cells access the genes in the DNA that contain the blueprints for those proteins. So, your heart cells use some of the genes in your DNA to build the proteins they need for contraction, whereas your skin cells use other genes in your DNA to build the proteins they need to protect you from infection (see Chapter 8 for more on gene regulation).

In the next few sections, we look at the signals that direct cells to become specialized for certain tasks in the body. We also look at some of the experiments that scientists are doing to try and figure out how to reset the programming of a cell.

## The ability to become any type of cell

Initially, each and every cell in your body (and the bodies of many other animals) has *totipotency* — the ability to develop into any kind of cell or even a whole organism. As the cells differentiate during development, they lose the ability to express all the genes they contain, which means that under normal conditions, a cell that becomes a muscle cell won't later change and become a skin cell.

Differentiation, then, occurs as a result of signals that cause cells to use only some of the genetic information they contain. Scientists are very interested in understanding these signals and discovering how to reprogram cells so they can be used to heal traumatic injury and disease. Following are just a couple of the experiments scientists have conducted to try and figure out how to reset the programming of a cell:

- Two researchers, Robert Briggs and T.J. King, tested tadpole cells to see whether they lost the ability to be reprogrammed and, if so, when they did. They found that up until the blastula stage, at which point the organism contains 8,000 to 16,000 cells, a single tadpole cell retained the capacity to develop into an entire organism. In other words, one cell from a blastula could develop into a whole new tadpole. However, when they used cells from a later period in development, they weren't successful.
- Another researcher, J.B. Gurdon, had success in "growing" normal frog embryos from adult frog skin cells. Gurdon used *nuclear transplantation* to move the nucleus from an adult frog skin cell into an egg cell from which he'd removed the nucleus. When the nucleus from the differentiated skin cell was placed into the environment of the egg cell cytoplasm, the nucleus directed the growth and development of a frog tadpole that was genetically identical to the frog that had donated the nucleus from its skin cell. As long as the egg into which the cell was transplanted wasn't past the critical point in development, a tadpole developed.

  Gurdon's experiment was the first successful attempt to *clone* an organism — produce an organism that's genetically identical to the organism that donates the transplanted nucleus. Since Gurdon's experiment, other types of animals have been cloned, the most famous of which is Dolly the sheep.

Animal cloning experiments demonstrate that the nuclei of differentiated cells retain all the genetic information necessary to become other cell types.

Although animals seem to have a point at which cells become determined to differentiate into certain types, the totipotency in many plant cells remains intact. You know how you can grow a whole plant from a cutting of another plant? The cutting doesn't contain roots, but it's able to grow them because it can still access the genes necessary for the function of root cells. It's also possible to take a few plant cells and grow an entire new plant from them.

## The factors that affect differentiation and development

When embryonic development begins, all the cells in the embryo are totipotent until signals around the cells make the cells turn into nervous system cells, muscle tissue cells, heart cells, lung cells . . . you get the idea. Most developmental changes depend upon signals in the environment of embryonic cells that tell them exactly what to do and when to do it. Here's a rundown of those signals:

- **Embryonic induction:** The influence of one group of cells on another group of cells is called *embryonic induction,* and it causes the recipients of the influence to change their course in development. Cells that wield this power are called *organizers,* and they exert their influence by secreting certain chemicals or by interacting directly with target cells through cell-to-cell contact. Embryonic induction occurs when the lens of the eye is developing. The eyes start out as bulging outgrowths (or *optic vesicles*) on the sides of the early brain. When the optic vesicles touch the ectoderm, the ectoderm thickens into the *lens placode,* which then develops into the curved lens of the eye.
- **Cell migration:** When cells move to new locations during development, that's considered *cell migration.* Cells may follow chemical trails to find their new destination. Once there, they attach to similar cells and differentiate to become a particular kind of tissue. In the developing brain, for example, primitive neural cells migrate out of the neural tube to establish the parts of the brain and then begin growing and forming connections with their target cells.
- **Homeotic genes:** Special genes that turn other genes on or off are *homeotic genes.* They work kind of like a master switch. The proteins produced from homeotic genes interact with DNA, affecting the expression of other genes. When certain genes are turned on, certain proteins are produced that contribute to development. When certain genes are turned off, the protein normally created is withheld so it can't affect development. These actions control what substances are present or absent in a developing embryo, thereby regulating the development of that embryo. The powerful effect of homeotic genes can be seen in genetic studies of fruit flies. If the homeotic genes in fruits flies are

mutated, body parts end up in the wrong places — for example, legs can appear where antennae should be.

Several years ago, a stretch of DNA about 180 nucleotides long (that's not very big in the DNA world) was found in most of the homeotic genes in many species. This short segment of genes is called a homeobox, and yes, even humans have homeoboxes. A *homeobox* is the sequence in the homeotic gene that remains unchanged generation after generation. In fact, even after so many generations, similarities exist in the homeoboxes of all living things on Earth.

- **Programmed cell death:** Also called *apoptosis, programmed cell death* causes cells to commit suicide at particular times during development in order to remove cells from the developing organism. In humans, for example, fingers are created when cells between each finger commit suicide, creating the spaces between the fingers.
- **Cytoplasmic factors:** These can create different local environments in the cytoplasm of a cell, leading to different developmental fates for the descendents of the cell. During cell division early in development, the cytoplasm is often divided unequally among secondary oocytes. So, some cells may get more of certain cytoplasmic factors than other cells. These cytoplasmic factors may cause the cell to turn on certain genes that determine its fate. For instance, differences in the cytoplasm of cells early in development establish the anterior-posterior axis in the embryo, which ultimately leads to the differentiation of the head from the rest of the body.
- **Hormones:** Hormones play a role in development after most of the vital organs (heart, lungs, kidneys, and liver) are formed and limbs and other appendages are where they belong. Then they take control of the actual appearance of the body. The production of testosterone in a male fetus, for example, leads to formation of male primary sex characteristics.

## Gender differentiation in humans

Humans are no strangers to the effects of hormones during development. In fact, human males and females are identical organisms until the time sexual differentiation occurs.

In the very early stages of development, human fetuses have two sets of ducts: one for the female reproductive system, and one for the male reproductive system. When both sets of ducts are present, the stage of development is called the *indifferent stage* (because there's no difference yet between male and female). Humans remain in this stage until about seven weeks after fertilization (about the end of the second month of pregnancy), which is why an ultrasound done any earlier than this time can't tell the sex of the developing embryo. (***Note:*** The first ultrasound of a normal pregnancy is usually performed at 16 weeks of gestation, allowing visualization of the sex of a developing fetus.)

The two sets of ducts are the *Wolffian ducts,* which eventually become the male vas deferens, epididymis (on the testes), and seminal vesicles, and the *Müllerian ducts,* which eventually become the oviducts, uterus, and vagina.

Inside the cells, the chromosomes determine whether the embryo will develop into a male or a female. Of the 46 human chromosomes, the last pair — the two Chromosomes 23 — are either two X chromosomes or an X and a Y chromosome. Two X chromosomes indicate female; an X and a Y indicate male. If two X chromosomes are inside the cells of the developing reproductive system, the female ducts develop, and the male ducts disintegrate. The reverse happens if an X chromosome and a Y chromosome are in the cells.

The following sections provide the specifics on how primary sex characteristics develop and list some of the ways that can happen incorrectly.

### How boys become boys

A gene called SRY (for *sex-determining region Y chromosome*) is the specific gene that determines maleness. SRY contains the blueprint for a protein called *testes determining factor* (TDF), which is a transcription factor that interacts with DNA to turn on the transcription of the genes necessary for testes development. (*Transcription factors* are proteins that turn genes on and off; turn to Chapter 8 to find out more.)

After the testes are formed, they begin to secrete the hormone testosterone (in the form *dihydrotestosterone,* or DHT). Testosterone supports the development of the male reproductive system and directs development of external genitalia. The tubules necessary for ejaculation of semen are complete at about 14 weeks of gestation (which is the beginning of the second trimester of pregnancy). At about that time, the penis, testes, and scrotum develop from the urogenital tubercle, urogenital swellings, and urogenital folds. The urogenital tubercle becomes the glans penis in the male, the urogenital folds become the shaft of the penis, and the urogenital swellings become the scrotum.

If SRY is absent, the primary gonad develops into ovaries, but two X chromosomes are required for the ovaries to be maintained.

### How girls become girls

The absence of DHT in a fetus is what spurs the development of female external genitalia. Without DHT, the urogenital tubercle becomes the clitoris (which is equivalent to the glans penis), the urogenital swellings become the labia majora, and the urogenital folds become the labia minora.

A female's external genitalia develop even if the internal genitalia fail to develop. External female structures are completed between 14 and 16 weeks of gestation.

### Problems with sexual development

The complex process of sexual differentiation involving genes and hormones isn't without error. Following are the problems that can occur in the hormonal stimulation of genitalia:

- **Androgen insensitivity:** A male who can't develop external male genitalia has *androgen insensitivity.* Embryos with an abnormal androgen receptor can't bind the DHT necessary to produce male genitalia. Therefore, they may be male genetically (XY), but they have female external genitalia.
- **Hermaphrodites:** People with some male and some female characteristics are *hermaphrodites.* This condition can result from hormonal imbalances. In embryos that oversecrete *adrenal androgens* (hormones that are involved in the normal synthesis of DHT and testosterone), a genetic female may have masculinized external genitalia complete with a penis but have normal ovaries and other female internal reproductive structures. Or, a genetic male may be undermasculinized.
- **Klinefelter's syndrome:** Males with *Klinefelter's syndrome* have two X chromosomes and one Y chromosome (XXY). They usually have small testes that don't produce enough testosterone. As a result, male secondary sex characteristics, such as facial hair, may not develop completely, and the male is usually infertile. Males with Klinefelter's syndrome tend to be taller; they may also have feminizing characteristics such as enlarged breasts. Treatment with hormone therapy can greatly reduce these effects and allow males to develop more normally and have normal sex lives.
- **Turner syndrome:** This genetic disorder, which leaves females infertile, can occur in two ways. First, a genetically female (XX) individual may be missing part or all of one of the X chromosomes, resulting in an XO individual (a person with only one sex chromosome, an X) that's neither completely female nor male. Second, an embryo may have an X chromosome and a Y chromosome, which normally indicates male, but a deletion occurs in the region of the sex-determining gene on the Y chromosome. This deletion prevents development of testes, so no DHT is produced. Female internal and external genitalia develop, but the ovaries fail prematurely. Women with Turner syndrome are often shorter than other women, may have extra folds of skin around the neck, and may fail to enter puberty. However, recent advances in hormone therapy have significantly reduced these effects.

# Chapter 20

# Living the Life of a Plant

### *In This Chapter*

- Looking at the structure of plants
- Acquiring what a plant needs to keep growing
- Exploring the differences between asexual and sexual reproduction

A plant's structure suits its lifestyle. After all, it has flat leaves for gathering sunlight, roots for drawing water up from the soil, and flowers and fruits for reproduction. Plants begin their lives from seeds or spores, grow to maturity, and then reproduce asexually or sexually to create new generations.

In this chapter, we present the fundamental structures of plants, how they get the energy they need to grow, and their reproductive strategies.

## Presenting Plant Structure

Like animals, plants are made of cells and tissues, and those tissues form organs, such as leaves and flowers, that are specialized for different functions. Plants have two basic organ systems: a root system (which exists underground) and a shoot system (found aboveground). The root system is responsible for anchoring the plant and also absorbing minerals and water from the soil. The shoot system ensures the plant gets enough sunlight to conduct photosynthesis; it also transports water upward from the roots and moves sugars throughout the plant.

Within their organ systems, plants have up to three types of tissues. Biologists look at the types of tissues a plant has to help them classify plants into four different groups. They also look at the different structures of plant stems.

## Plant tissues

All plants have tissues, but not all plants possess all three of the following types of tissues:

- **Dermal tissue:** Consisting primarily of epidermal cells, dermal tissue covers the entire surface of a plant. Guard cells in a plant's epidermis control the opening and closing of little holes called *stomates* that allow the plant to exchange gases with its environment (you can see a stomate in the leaf cross section in Figure 20-1).
- **Ground tissue:** This tissue type makes up most of a plant's body and contains three types of cells:
  - **Parenchyma cells** are the most common ground tissue cells. They perform many basic plant cell functions, including storage, photosynthesis, and secretion.
  - **Collenchyma cells** have thick cell walls in order to help support the plant.
  - **Sclerenchyma cells** are similar to collenchyma cells, but their walls are even thicker — so thick, in fact, that mature schlerenchyma cells die because they can't get food or water across their walls via osmosis (more about that in Chapter 4).
- **Vascular tissue:** The system of tubules inside a plant that carries nutrients around is made up of vascular tissue. Vascular tissue consists of a water-transport system called *xylem* and a sugar-transport system called *phloem.* Vascular tissue also contains the *vascular cambium,* a tissue of cells that can divide to produce new cells for the xylem and phloem. (Vascular plants make up the majority of plants on Earth. You can see the basic cells and structures of a vascular plant in Figure 20-1.)

## The types of plants

Based on the types of tissues they have and reproductive structures they make, plants can be organized into four major groups:

- **Bryophytes** are plants such as mosses that don't have a vascular system and don't produce flowers or seeds.
- **Ferns** have vascular tissue, but they don't produce seeds.
- **Gymnosperms** (also known as *conifers*) have vascular tissue and produce cones and seeds, but they don't produce flowers.

- **Angiosperms** (or *flowering plants*) have vascular tissue and produce both flowers and seeds. Two distinct groups exist among flowering plants:
  - **Monocots,** like corn and lilies, have seeds that contain one *cotyledon* (tissues within the seed that supply nutrition to the embryo and then emerge as the first leaves after the seed begins to grow; they're also referred to as *seed leaves*).
  - **Dicots,** like beans, oak trees, and daisies, have seeds that contain two cotyledons.

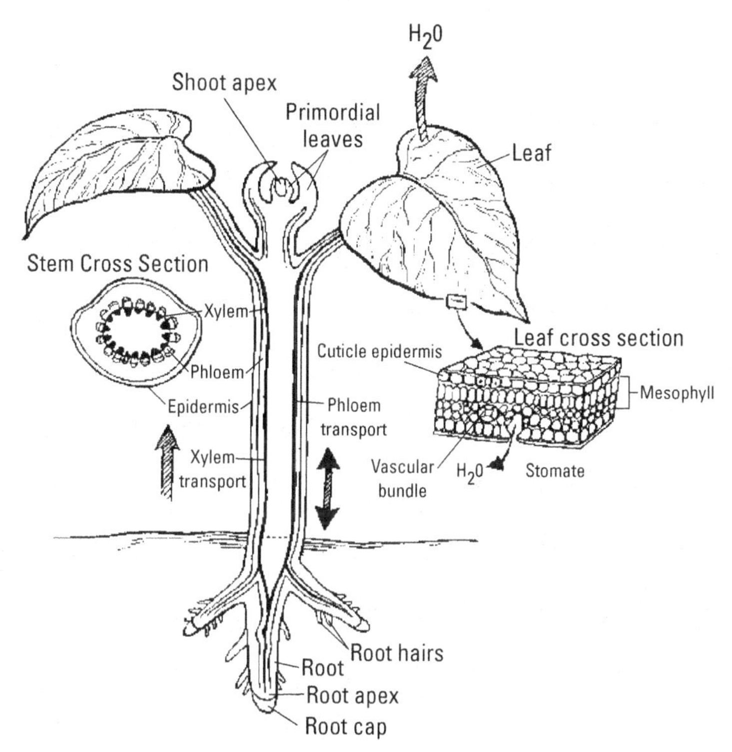


**Figure 20-1:** The basic structures of a vascular plant.

Table 20-1 presents several of the key structural differences between monocots and dicots.

**Table 20-1 Structural Differences between Monocots & Dicots**

| *Feature* | *Monocots* | *Dicots* |
|---|---|---|
| Cotyledons in seeds | One | Two |
| Bundles of vascular tissue | Scattered throughout | Form definite ring pattern |
| Xylem and phloem | Found in stem | Found in stem |
| Leaf veins | Run parallel | Form a net pattern |
| Flower parts | Are in threes and multiples of threes | Are in fours and fives and multiples of fours and fives |

## Herbaceous versus woody stems

Biologists use the appearance and feel of a plant's stem to place it into one of two categories: *herbaceous* (where the stem remains somewhat soft and flexible) and *woody* (where the stem has developed wood). All plant cells have primary cell walls made of cellulose, but the cells of woody plants have extra reinforcement from a secondary cell wall that contains lots of a tough compound called *lignin.*

Plants that survive just one or two growing seasons — that is, *annuals* or *biennials* — are typically herbaceous plants. Plants that live year after year, called *perennials,* may become woody.

The stems of herbaceous and woody dicots (see the preceding section) are organized differently. You can see these differences most clearly if you look at a *cross section* (a section cut at right angles to the long axis) of a stem. Imagine taking a hot dog and slicing it into little circles, and you've got a pretty good picture of how biologists make stem cross sections.

When you look at a cross section of the stem of an herbaceous dicot, you see that

- The center of the stem consists of *pith* (a soft, spongy tissue), which has many thin-walled cells called *parenchymal cells.* The thin walls allow the diffusion of nutrients and water between the cells.
- The vascular tissue is organized in *vascular bundles* that contain both xylem and phloem, as well as some vascular cambium (all of which are described in the earlier "Plant tissues" section). The vascular bundles are arranged in a ring around the pith.

- Outside the vascular bundle ring is the stem's cortex. It contains a layer of endodermis, additional parenchymal cells, and mechanical tissue, which supports the weight of the plant and holds the stem upright.
- On the surface of the stem are the epidermis and the cuticle.

Woody dicots start life with green herbaceous stems that have vascular bundles. As they grow, however, the bundles merge with each other to form rings of vascular tissue that circle the stem. If you were to examine a cross section of the stem of a woody dicot that was a couple years old (like the one in Figure 20-2), you'd see that

- The center of the stem consists of a circle of pith.
- The xylem tissue forms a ring around the pith. As woody plants grow, new layers of xylem are added every year, forming rings inside the woody stem. (You can count these rings in the stem of a tree to tell how old it was when it was cut.) As these rings of xylem accumulate year after year, the diameter of the woody stem increases.

  The inner part of a woody stem is called *heartwood.* It consists of older xylem tissue that's filled with material such as gums and resins and no longer conducts water. Outside the heartwood is the *sapwood,* newer layers of xylem tissue that transport water and minerals up through the stem.
- Just outside the xylem rings is a ring of vascular cambium. As the stem grows, the vascular cambium divides to produce new xylem cells toward the inside of the stem and new phloem cells toward the outside of the stem.
- Outside the vascular cambium ring is a ring of phloem. The phloem of woody plants gets pushed farther and farther outward as the xylem tissue increases in size year after year. Phloem cells are fairly delicate, and the old phloem cells get crushed against the bark as the stem grows. The only phloem that serves to transport materials through the woody plant is the phloem that's newly formed during the most recent growing season.
- Outside the phloem is the *bark,* a ring of boxy, waterproof cells that help protect the stem. Bark includes the outermost cells of the stem and a layer of cork cells just beneath that outermost layer.

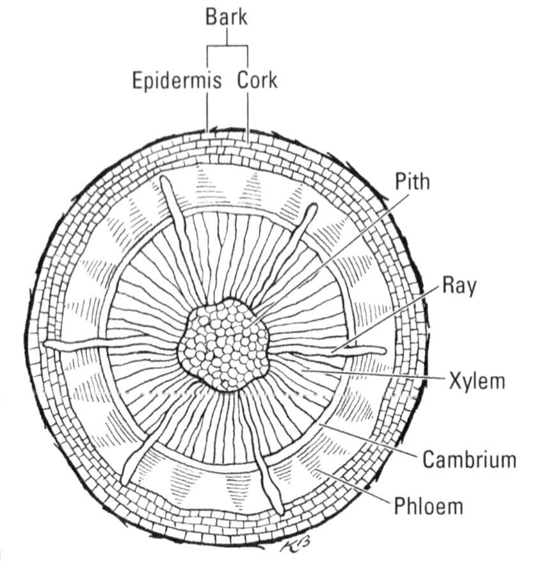


**Figure 20-2:** A cross section of a woody stem.

# Obtaining Matter and Energy for Growth

The biggest difference between plants and animals is how they get the matter and energy they need for growth. Animals have to eat other living things to get their food, but plants can produce their own food. Plants absorb sunlight and use that energy to make glucose from carbon dioxide and water during the process of photosynthesis (we describe photosynthesis in detail in Chapter 5); glucose is the food plants can use as a source of energy or matter for growth.

As you can see from the following list, plant structures are specialized to help plants get what they need for photosynthesis:

- **The shoot system helps plants capture energy from the Sun.** Shoots grow upward, bringing leaves toward the Sun. Branches spread leaves out so they can absorb light over a wider area, and many leaves are flat so they have the most surface area possible for light absorption.
- **The root system absorbs water and minerals from the soil.** Water is needed for photosynthesis and basic plant functioning. Minerals perform the same function for plants as they do for you — they improve general metabolism by helping enzymes function properly. Also, plants absorb nitrogen-containing compounds from the soil and use them, along with the carbohydrates made during photosynthesis, to construct plant proteins (flip to Chapter 11 for the scoop on how plants get nitrogen).

- **Stomates in the leaves allow plants to take carbon dioxide from the atmosphere and return oxygen to it.** The carbon dioxide provides the carbon and oxygen atoms plants need to build carbohydrates. Also, photosynthesis produces oxygen when the hydrogen and oxygen atoms in water are separated. Oxygen gas leaves plants through their stomates.

Plants extract energy from food molecules the same way animals do — by cellular respiration (see Chapter 5 for more on cellular respiration). When plants do cellular respiration, they produce carbon dioxide and use oxygen just like animals do. During the day, however, photosynthesis absorbs so much carbon dioxide and releases so much oxygen that plant respiration isn't detectable. If you were to measure gas exchange around a plant in the dark, the plant would be exchanging gases just like you.

# Going It Alone: Asexual Reproduction

Plants that reproduce asexually make copies of themselves in order to produce offspring that are genetically identical to them. The advantage to asexual reproduction is that it allows successful organisms to reproduce quickly. The disadvantage is that all the offspring are genetically identical, which decreases the ability of the population to survive changes in the environment.

The most common method of asexual reproduction is mitosis, which we describe in Chapter 6. Other ways a plant can reproduce asexually include *fragmentation,* a form of asexual reproduction that involves pieces of an individual growing into new individuals. If you break off a piece of many houseplants (the technical term is "take a cutting") and stick it in water, new roots and shoots may grow, creating a whole new plant from a piece of the parent plant. Likewise, if you cut a potato into pieces, each piece that has an "eye" can produce a new potato plant. In essence, the new plant is a clone of the parent plant: It has all the same genetic information because the cells are identical.

Other plants, such as strawberry plants, produce special structures that help them spread asexually. In addition to producing stems, the strawberry plant produces a *stolon,* also referred to as a *runner,* that spreads across the ground. Wherever that stolon starts to put roots down is where a new strawberry plant grows. Similarly, many ferns reproduce asexually by underground stems called *rhizomes.*

# Mixing Sperm and Eggs: Sexual Reproduction

Plants do have sex, believe it or not. First, they produce eggs and sperm through meiosis just like animals do (see Chapter 6 for the full scoop on meiosis). Then a sperm and an egg meet, creating offspring that have different combinations of genetic material than the parent plants.

The following sections get you familiar with the life cycle of plants that reproduce sexually along with all the little details and processes that involves, from flower structure and pollination to fertilization and the development (and protection) of plant embryos.

## The life of a plant

The life cycles of plants are a bit more complicated than those of animals. In animals, *gametes* (sperm and eggs) are usually small and inconspicuous. In plants, however, gametes can almost have a life of their own.

Plant life cycles involve the alternation of generations between two stages called sporophytes and gametophytes (see Figure 20-3). Here's a breakdown of the cycle:

1. **Meiosis in a sporophyte (a parent plant) results in the production of spores that are haploid (meaning they have half the genetic information of the parent plant).**
2. **The spores develop by mitosis into multicellular haploid organisms called gametophytes.**

   The gametophyte step of the plant life cycle is a fundamental difference between plants and animals. In animals, no development occurs until a sperm and an egg combine to produce a new organism. In plants, there's a little break between meiosis and the production of sperm and eggs. During that break, a separate little haploid plant grows.
3. **Gametophytes produce gametes by mitosis.**

   In animals, sperm and egg are produced by meiosis, but in plants, meiosis occurs to produce the gametophyte.
4. **The gametes merge, producing cells called zygotes that contain the same number of chromosomes as the parent plant (so the zygotes are diploid).**
5. **Zygotes divide by mitosis and develop into sporophytes so the life cycle can begin again.**

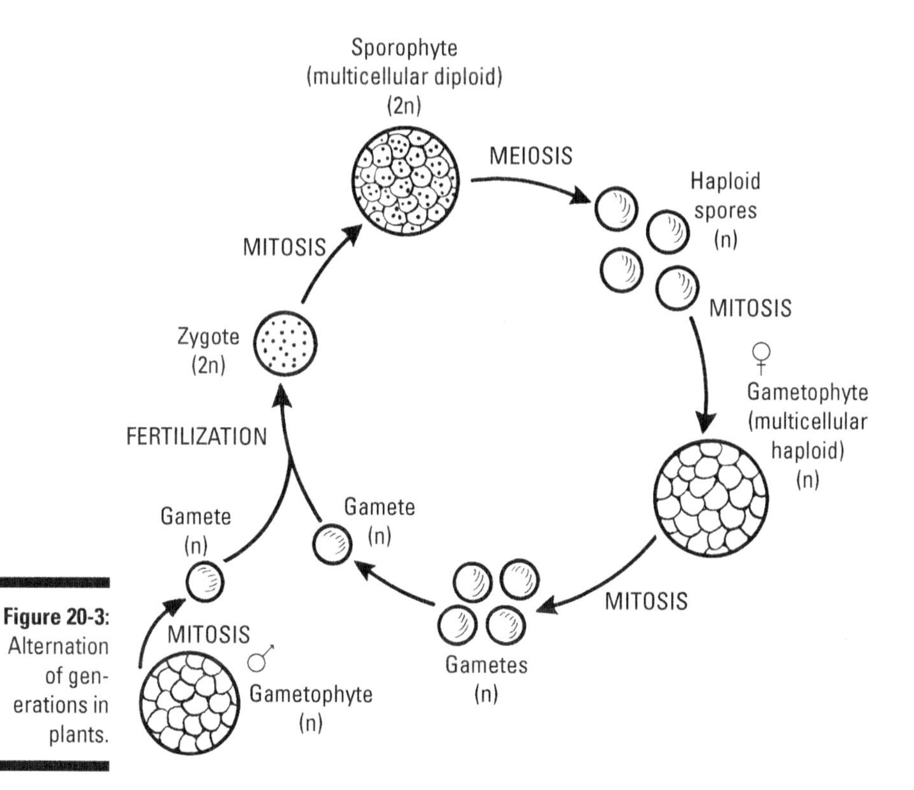


**Figure 20-3:** Alternation of generations in plants.

The plants you see when you go for a walk in the woods may be sporophytes or gametophytes; it all depends on the type of plant you're looking at.

- The mosses you see growing on trees and on the forest floor are gametophytes. If you see little structures like flagpoles sticking off the moss, then you're looking at a sporophyte. The little sporophytes grow like flags off the tops of the gametophytes. Inside the little flags, called *capsules,* meiosis is occurring to produce spores.
- Ferns you can see are sporophytes. If you look on the back of a fern's leaves, you can find little brown structures that seem dusty to the touch. These structures are where spores are being made, and the dust that comes off are the spores. Fern gametophytes are teeny — about as big as the fingernail on your pinky — making them very tough to find in the wild.
- The conifers you see in a forest are sporophytes. The gametophyte generation in conifers is very small and contained within their cones.
- Flowering plants that are visible to the eye are also sporophytes. In flowering plants, the gametophyte generation is very small and contained within the flowers.

## The parts of a flower

Whether a flowering plant's flowers are large and showy or small and dainty, sexual reproduction occurs inside. Flowers form on specialized shoots of the plant, and they have specific parts. Following are descriptions of some of these parts (you can see several of them in Figure 20-4):

- The **receptacle** is the base of the flower.
- **Sepals** are the lowest layer of leaves on the flower. They're usually green.
- **Petals** are modified leaves that are often brightly colored to attract pollinators (see the next section for more on pollinators).
- **Stamens** are the male parts of a flower. Each stamen consists of a threadlike *filament* and a little sac called the *anther.* Inside the anther, meiosis and mitosis occur to produce the male gametophyte, *pollen.*
- **Pistils** are the female parts of a flower. The ovary is located at the swollen base of the pistil. Inside the ovary, meiosis and mitosis produce the female gametophyte and the egg, which are housed inside an ovule.
- The **style** and **stigma** grow up out of the ovary. Pollen lands on the stigma and then travels down through the style to deliver sperm to the eggs inside the ovary.

## How pollination and fertilization occur

*Pollination,* the delivery of pollen to a flower's stigma, and *fertilization,* when a sperm joins with an egg, are two separate yet important events that occur within flowering plants.

Some flowering plants, like grasses, are *wind pollinated,* which means they make lots and lots of pollen and trust the wind to blow it to the right place. Other flowering plants rely on animals such as bees, wasps, birds, and even flies and bats to transfer pollen from one flower to another.

Here's how different plants attract their animal pollinators:

- Bird- and bee-pollinated plants are usually brightly colored to attract the animals; they may also provide nectar to encourage visits. Bee-pollinated flowers are often marked with lines just like little runways to direct the bees to the right place in the flower. These runways are invisible to human eyes but visible to the bees, who can see ultraviolet light.
- Some plants, like certain wasp-pollinated orchids, trick the animals into thinking that the flowers are members of the opposite sex. In this case, the wasps attempt reproduction with the flowers and get covered with pollen before moving on to the next flower to try again.

- Fly-pollinated plants smell like toilets or rotten meat in order to attract mama flies in search of a good place to lay their eggs.
- Bat- and moth-pollinated flowers are usually white so they'll be more visible at night, which is when they open.

When pollen arrives at the stigma of a flower, each pollen grain grows a long tube called a *pollen tube*. The pollen tube grows down through the flower's style so the sperm cells inside the pollen can be delivered right to the ovules inside the flower's ovary.

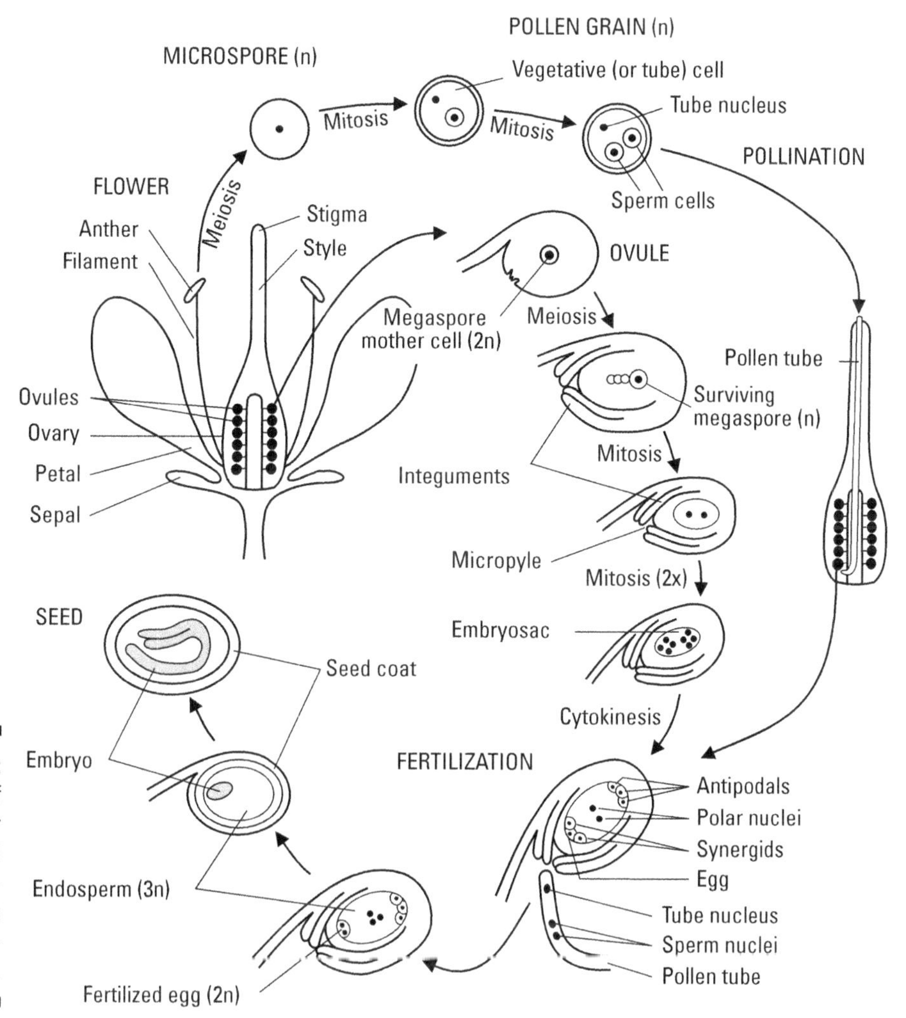


**Figure 20-4:** Parts of a flower and sexual reproduction in angiosperms.

After pollination has occurred, two sperm nuclei enter the *ovule* (the part of the ovary that contains the egg and develops into the seed after fertilization), and one fuses with an egg to form the zygote. The other sperm joins with two haploid cells inside the ovule, called the *polar nuclei,* to form a tissue called *endosperm* that helps support the developing embryo. In flowering plants, the fusion of two sperm cells, one with an egg and one with the polar nuclei, is called *double fertilization.*

## From zygote to embryo

After fertilization, the zygote divides by mitosis to produce an embryo. The first cell division produces two cells: one large, one small. Several more cell divisions occur after that, producing a line of cells called a *suspensor.* Additional cell divisions form the embryo in such a way that the cells at the bottom grow downward to become roots, and the cells at the top grow upward to become shoots.

The embryo's *hypocotyl,* which is attached to the bottom end of the suspensor, becomes the lower part of the stem and the roots. The *cotyledons,* or seed leaves, develop at the top end of the embryo; they're temporary structures that serve as nutrient storage sites for the developing plant. After the plant is growing aboveground and can start producing nutrients on its own through photosynthesis, the cotyledons shrink away.

## A little protection for the embryo: Seeds

*Seeds* are protective structures that contain plant embryos and nutritive tissue to support the embryo until it can survive on its own. The endosperm produced during double fertilization provides nutrient material to the developing embryo, tissues from the ovule harden to become the seed coat that protects the embryo, and the ovary of the flower forms a fruit around the seed. After a seed develops, it usually dries out, and its water content drops very low. This low water level keeps the embryo's metabolism at a minimal level so it can survive on stored food for a long time. So the seeds that you buy in a little packet at the local nursery are very much alive, but they're in a state similar to that of a hibernating bear.

When seeds are planted in an environment with water, the dry seeds take up water and swell. As water becomes available, the embryo's metabolism speeds up, and it begins to grow by using the stored food inside the seed. *Germination* occurs when the seedlings emerge from the seed and begin to grow into a diploid plant.

# Part V

# It's Not Easy Being Green: Plant Structure and Function

## In this part . . .

Plants have many similarities to animals: They have tissues, they circulate materials, and they reproduce sexually. Yet plants are also remarkable in their own right. Consider, for example, how well you'd do if someone buried you in the ground up to your knees and then left you there. You wouldn't fare so well, but a plant would be just fine because it can make its own food. Not only that but it may also have strategies to attract pollinators to help it reproduce and defenses to help protect it from predators. Not too shabby for an organism that can't make a sound, huh?

In this part, you get to know all about the structure and function of the green things that call planet Earth home.

# Chapter 21

# Probing into Plant Physiology

### In This Chapter

- Moving water and other nutrients through plants
- Getting sugars where they need to go
- Triggering plant responses with hormones

Photosynthesis (described in Chapter 5) isn't the only life-sustaining process carried out by plants. Below the surface, plants are busy collecting and transporting water and other nutrients upward, as well as moving sugars upward and downward to the cells that need them. Their hormones, meanwhile, are sending messages to promote growth toward sunlight and blossoming, among other things.

Consider this chapter your introduction to the physiology of plants (*physiology* is the study of the function of organisms and their parts). Here, you discover the processes plants use to transport nutrients, fluids, and sugars throughout their bodies. You also get to know the plant hormones that regulate growth and development.

## How Nutrients, Fluids, and Sugars Move through Plants

Just like you have a circulatory system that moves food and oxygen throughout your body, plants have a system to move nutrients, fluids, and sugars throughout their bodies. (Even though plants make food in their leaves by photosynthesis, the entire plant needs some of that food and the nutrients it provides.) The following sections fill you in on the different nutrients plants must absorb to stay healthy as well as how they move sugars from their leaves and water from their roots (without losing too much of it).

## Taking an inventory of the nutrients plants need to survive

All plants require carbohydrates, proteins, fats, and nucleic acids to function — the same as you do. They also need mineral elements to build their molecules and make sure their enzymes are working properly. Fortunately, plants can obtain all the nutrients they need to survive from their environment.

Plants get carbon, hydrogen, and oxygen by taking carbon dioxide from the atmosphere and water from the soil. With energy from the Sun, plants combine these molecules to form carbohydrates during the process of photosynthesis.

Plants obtain their necessary mineral elements from the soil as well. The mineral nutrients found in soil dissolve in water, so when plants absorb water through their roots, they obtain both macronutrients and micronutrients. *Macronutrients* help with molecule construction, and *micronutrients* act as partners for enzymes and other proteins to help them function. Plants generally require large amounts of macronutrients and smaller amounts of micronutrients. Table 21-1 lists the specific macro- and micronutrients plants absorb from soil.

**Table 21-1 The Essential Nutrients Plants Pull from Soil**

| *Macronutrients* | *Micronutrients* |
|---|---|
| Calcium (Ca) | Boron (B) |
| Magnesium (Mg) | Chloride (Cl) |
| Nitrogen (N) | Copper (Cu) |
| Phosphorous (P) | Iron (Fe) |
| Potassium (K) | Manganese (Mn) |
| Sulfur (S) | Molybdenum (Mb) |
| | Zinc (Zn) |

You can remember the most important elements for plants with the phrase "C. Hopkins Café, Mighty Good." This phrase stands for CHOPKNS CaFe Mg — in other words, carbon, hydrogen, oxygen, phosphorous, potassium, nitrogen, sulfur, calcium, iron, and magnesium. All of these elements are macronutrients for plants, with the exception of iron, which is considered a micronutrient.

If plants don't get enough of one of these important elements, they can't function correctly. Without carbon, hydrogen, and oxygen (from carbon dioxide and water), plants can't grow at all. And even though plants need smaller amounts of minerals, even one missing mineral can cause a specific problem. We list these problems and the mineral deficiencies they're associated with in Table 21-2.

**Table 21-2 The Effects of Mineral Deficiencies on Plants**

| *Mineral That's Missing* | *Effect of the Deficiency* |
|---|---|
| Boron | Leaves on the ends of the plant die and fall off early; plant growth is stunted; flowers and seeds usually aren't produced |
| Calcium | Leaves roll and curl; roots are poorly developed and may look gelatinous |
| Copper | Terminal shoots wilt and die; leaves appear faded in color |
| Iron | White marks show in the veins; leaves look bleached; tips of leaves look scorched |
| Magnesium | Veins look green, but leaf tissue looks white or yellow and brittle; leaves may wilt, fall off, or die |
| Manganese | Same as magnesium but stems are yellowish-green in color and often hard to the touch |
| Molybdenum | Leaves are light yellow and may not grow |
| Nitrogen | Stunted growth; leaves turn light green, then yellow, and then dry out and fall off |
| Phosphorus | Stunted growth; leaves sometimes look purplish; stems are thin |
| Potassium | Leaves have pale green or streaked yellow color and look wrinkled between the veins |
| Sulfur | Leaves appear light green to yellow in color; stems are thin |
| Zinc | Leaves die; white streaks show between the veins in older leaves |

## Transporting water and other nutrients from the ground up

Several processes work together to transport water (as well as other nutrients) from where a plant absorbs it (the roots) upward through the rest of its body. To understand how these processes work, you first need to know one key feature of water: Water molecules tend to stick together, literally. Water molecules are attracted to each other by weak electrical attractions called *hydrogen bonds.* The stickiness of water helps keep the water molecules together when you drink water through a straw — a process that's very similar to one of the methods plants use to move water through their bodies.

Water moves from the soil, into a plant's roots, and then throughout the plant thanks to a combination of three processes:

- **Osmosis:** The method plants use to draw water from the soil into the xylem cells in their roots is called *osmosis.* Root cells have a higher concentration of minerals than the soil they're in, so during osmosis, water flows toward the higher concentration of dissolved substances found in the root cells. This intake of water increases pressure in the root cells and pushes water into the plant's xylem (see Chapter 20 for the full scoop on plant structure).
- **Capillary action:** This causes liquids to rise up through the tubes in the xylem of plants. This action results from *adhesion* (when two things stick together), which is caused by the attraction between water molecules and the walls of the narrow tube. The adhesion forces water to be pulled up the column of vessel elements in the xylem and in the tubules in the cell wall.
- **Transpiration and cohesion:** *Transpiration* is the technical term for the evaporation of water from plants. As water evaporates through the stomates in the leaves (or any part of the plant exposed to air), it creates a negative pressure (also called *tension* or *suction*) in the leaves and tissues of the xylem. The negative pressure in the leaves and xylem exerts a pulling force on the water in the plant's xylem and draws the water upward. When water molecules stick to each other through *cohesion* (where like — as opposed to different — substances stick together), they fill the column in the xylem and act as a huge single molecule of water. As water evaporates from the plant through transpiration, the rest of the water gets pulled up, causing the need for more water to be pulled into the plant.

The back and forth of transpiration and cohesion is known as the *cohesion-tension theory.* It's similar to what happens when you suck on a straw. The suction you apply to the straw is like the evaporation from the leaves of the plant. Just like you can pull up a column of liquid through your straw, a plant can pull up a column of liquid through its xylem.

## Water plus sap equals . . . a dewdrop?

Those droplets of water that you see on a plant's leaves in the morning, what you think of as a bunch of dewdrops, aren't just water. They're a mixture of water and *sap,* a sugar solution from the phloem (Chapter 20 describes this and other elements of a plant's structure). These sap droplets are proof that water and minerals get pulled up from the soil and transported throughout the entire plant. (We describe how this occurs in the nearby "Transporting water and other nutrients from the ground up" section.)

## Translocating sugars upward and downward through the phloem

Phloem moves *sap,* a sticky solution containing sugars, water, minerals, amino acids, and plant hormones throughout the plant via *translocation,* the transport of dissolved materials in a plant. Unlike xylem, which can only carry water upward, phloem carries sap upward and downward from sugar sources to sugar sinks.

- *Sugar sources* are plant organs such as leaves that produce sugars.
- *Sugar sinks* are plant organs, such as roots, tubers, or bulbs, that consume or store sugars.

The specific way translocation works in a plant's phloem is explained by the pressure-flow theory, which we outline step by step in the following list:

1. **First, sugars are loaded into phloem cells called sieve tube elements within sugar sources, creating a high concentration of sugar at the source.**

   The concentration of sugars in sink organs is much lower.

2. **Water enters the sieve tube elements by osmosis.**

   During osmosis, water moves into the areas with the highest concentration of solutes (in this case, sugars).

3. **The inflow of water increases pressure at the source, causing the movement of water and carbohydrates toward the sieve tube elements at a sugar sink.**

   You can think of this like turning on a water faucet that's connected to a garden hose. As water flows from the tank into the hose, it pushes the water in front of it down the hose.

4. **Sugars are removed from cells at the sugar sink, keeping the concentration of sugars low.**

   As a sugar sink receives water and carbohydrates, pressure builds. But before the sugar sink can turn into a sugar source, carbohydrates in a sink are actively transported out of the sink and into needy plant cells. As the carbohydrates are removed, the water then follows the solutes and diffuses out of the cell, relieving the pressure.

Sugar sinks that store carbohydrates can become sugar sources for plants when sugars are needed. *Starch,* a complex carbohydrate, is insoluble in water, so it acts as a carbohydrate storage molecule. Whenever a plant needs sugar, like at night or in the winter when photosynthesis doesn't occur as well, the plant can break down its starches into simple sugars, allowing a tissue that would normally be a sugar sink to become a sugar source.

Because plant cells can act as both sinks and sources, and because phloem transport goes both upward and downward, plants are pretty good at spreading the wealth of carbohydrates and fluid to where they're needed. As long as a plant has a continuous incoming source of minerals, water, carbon dioxide, and light, it can fend for itself.

## Controlling water loss

Because water is essential to a plant's functioning, it has built-in mechanisms that help prevent it from losing too much water: a cuticle and guard cells.

The *cuticle* is a layer of cells found on the top surfaces of a plant's leaves (see Figure 21-1). It lets light pass into the leaf but protects the leaf from losing water. Many plants have cuticles that contain waxes that resist the movement of water into and out of a leaf, much like wax on your car keeps water off the paint.

Guard cells are found on the bottom of a plant's leaves, near a *stomate,* a tiny opening that you can't see with your naked eye. (An individual opening is called a stomate, or *stoma;* several openings are called stomates, or *stomata.*) Plants need to keep their stomates, shown in Figure 21-1, open in order to obtain carbon dioxide for photosynthesis and release oxygen. However, if the stomates are open too long or on a really hot day, the plant can lose too much water. To prevent such water loss from happening, each stoma has two guard cells surrounding it.

Guard cells can swell and contract in order to open and close the stomates. When the Sun is shining and photosynthesis is occurring, guard cells swell up with water like full balloons, which stretches them outward and opens the stomates. At night, when photosynthesis isn't occurring, the guard cells release some water and collapse together, closing the stomates.

## Aphids suck

Aphids, those tiny little insects that can destroy your houseplants when you're not looking, live on the sap flowing through the phloem of a plant. They have long, pointy structures called *stylets* that let them suck sap right out of the plant phloem. Insertion of the stylet doesn't harm the plant — in fact, aphids can go right into a sieve tube without the plant "feeling" a thing. An aphid can stay attached to a plant for hours, sucking out the sap all the while. It's the loss of sap — and the cumulative effect of many, many aphids on a plant — that causes damage to a plant. The aphids fill up, leaving the plant starved of its sugar mixture.

In an interesting twist, scientists have found a way to use aphids to study transport in the phloem. They allow aphids to attach to plants and insert their stylets, and then they cut the aphids away but leave the stylets embedded in the plants. Materials flowing through the phloem ooze out through the stylets and can be collected by scientists for further study.

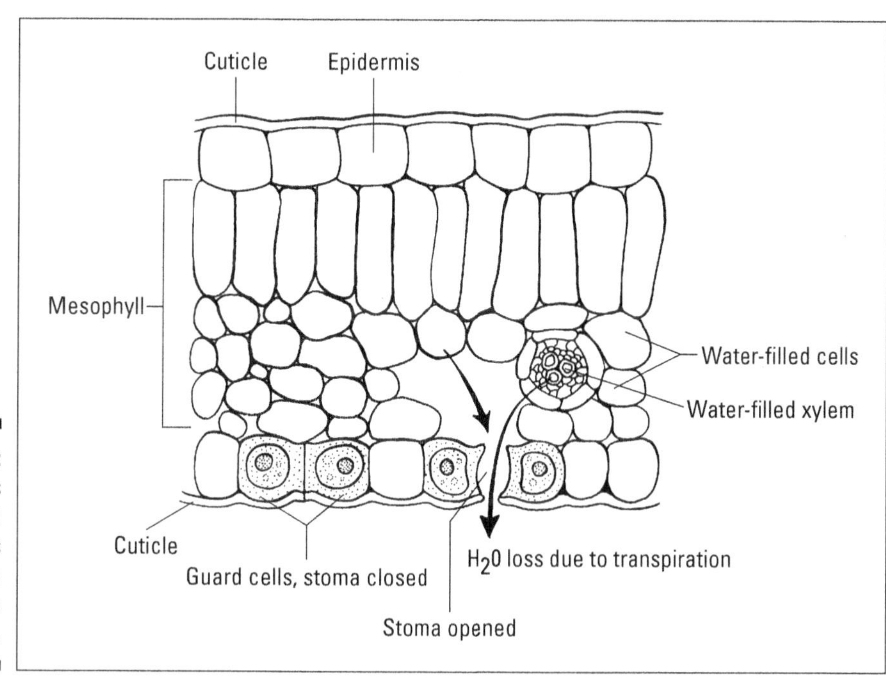


**Figure 21-1:** A plant's cuticle and guard cells keep it from losing too much water.

Some plants that live in very hot, dry environments save water by opening their stomates at night and storing carbon dioxide in their leaves. Then, during the day when it's hot and dry, they keep their stomates closed to conserve water, performing photosynthesis with the carbon dioxide they stored during the night.

# Sending Signals with Plant Hormones

Plant cells communicate with each other via messengers called *hormones,* chemical signals produced by cells that act on target cells to control their growth or development. Plant hormones control many of the plant behaviors you're used to seeing, such as the ripening of fruit, the growth of shoots upward and roots downward, the growth of plants toward the light, the dropping of leaves in the fall, and the growth and flowering of plants at particular times of the year.

Five categories of hormones control plant growth and development:

- **Auxins** stimulate the elongation of cells in the plant stem and *phototropism* (the growth of plants toward light). If a plant receives equal light on all sides, its stem grows straight. If light is uneven, then auxin moves toward the darker side of the plant. This may seem backward, but when the shady side of the stem grows, the stem, in its crookedness, actually bends toward the light. This action keeps the leaves toward the light so photosynthesis can continue.
- **Gibberellins** promote both cell division and cell elongation, causing shoots to elongate so plants can grow taller and leaves can grow bigger. They also signal buds and seeds to begin growing in the spring.
- **Cytokinins** stimulate cell division, promote leaf expansion, and slow down the aging of leaves. Florists actually use them to help make cut flowers last longer.
- **Abscisic acid** inhibits cell growth and can help prevent water loss by triggering stomates to close. Plant nurseries use abscisic acid to keep plants dormant during shipping.
- **Ethylene** stimulates the ripening of fruit and signals deciduous trees to drop their leaves in the fall. Fruit growers use ethylene to partially ripen fruit for sale.

Some of the flavor-making processes that occur in fruits happen while the fruits are still on the plant. So, even though ethylene can trigger some parts of ripening, like softening after a fruit has been picked, fruit that's picked unripe doesn't taste as good as fruit that has ripened on the plant. That's why you can buy a big, beautiful tomato at the grocery store and take it home only to discover that it doesn't have much flavor — it was probably picked unripe and then treated with ethylene.

# Part VI

# The Part of Tens

## *In this part . . .*

A *For Dummies* Part of Tens chapter is meant to contain some fun facts or useful information. In this part, we give you both. In Chapter 22, you can read about ten fascinating biology discoveries. (Of course, there are more than ten important discoveries, but we chose to pick those that contributed the most to humans' understanding of life.) In Chapter 23, we provide a list of ten pretty interesting ways that biology affects your life.

To sum things up: If you want some light-yet-informative reading, you're in the right part!

# Chapter 22

# Ten Great Biology Discoveries

### In This Chapter

- Figuring out the secrets of DNA structure, cell processes, and more
- Experimenting to create vaccinations, antibiotics, and treatments for genetic defects

Get ready to dive into ten of the most important biology discoveries to date. We list them in no particular order because they've all made a significant impact on the advancement of biology as a science and increased what people know and understand about the living world.

## Seeing the Unseen

Before 1675, people believed the only living things that existed were the ones they could see. That year, a Dutch cloth merchant named Antony van Leeuwenhoek discovered the microbial world by peering through a homemade microscope. Van Leeuwenhoek was the first person to see bacteria, which he described as little animals that moved about here, there, and everywhere. His discovery of a previously unseen universe not only turned people's worldviews inside out but also laid the foundation for the understanding that microbes cause disease.

## Creating Penicillin, the First Antibiotic

People had very few tools to combat bacterial infections until Alexander Fleming discovered the antibacterial properties of penicillin in 1928. Fleming was studying a strain of staphylococcus bacteria when some of his petri dishes became contaminated with *Penicillium* mold. To Fleming's surprise, wherever the *Penicillium* grew on the petri dish, the mold inhibited the growth of the staphylococcus bacteria.

The compound penicillin was purified from the mold and first used to treat infections in soldiers during World War II. Soon after the war, the "miracle drug" was used to treat infections in the general public, and the race to discover additional antibiotics was on.

# Protecting People from Smallpox

Would you believe that the idea of vaccinating people against diseases such as smallpox, measles, and mumps originated in ancient China? Healers there ground up scabs taken from a smallpox survivor into a powder and blew this dust into the nostrils of their patients. Gross as this may sound, these ancient healers were actually inoculating their patients to help prevent the spread of the disease.

# Defining DNA Structure

James Watson and Francis Crick figured out how a code could be captured in the structure of DNA molecules, opening the door to an understanding of how DNA carries the blueprints for proteins. They proposed that DNA is made of two nucleotide chains running in opposite directions and held together by hydrogen bonds between the nitrogenous bases. Using metal plates to represent the bases, they built a giant model of DNA that was accepted as correct almost immediately.

# Finding and Fighting Defective Genes

On August 24, 1989, scientists announced their discovery of the first known cause of a genetic disease: They found a tiny deletion from a gene on Chromosome 7 that resulted in the deadly genetic disease cystic fibrosis. This identification of a genetic defect, and the realization that this defect causes a disease, opened the floodgates of genetic research. Since that fateful day, the genes for other diseases, such as Huntington's disease, inherited forms of breast cancer, sickle cell anemia, Down syndrome, Tay-Sachs disease, hemophilia, and muscular dystrophy, have been found. Genetic tests for these diseases are available to detect whether an unborn baby has a defective gene or whether two potential parents would likely produce an affected baby. And knowing what causes the diseases enables researchers to focus on ways to possibly cure the diseases.

## Discovering Modern Genetic Principles

Gregor Mendel, a mid-19th century Austrian monk, used pea plants to perform the fundamental studies of heredity that serve as the basis for genetic concepts to this day. Because pea plants have a number of readily observable traits — smooth peas versus wrinkled peas, tall plants versus short plants, and so on — Mendel was able to observe the results of cross-pollinating and growing various varieties of pea plants.

Through his experiments, Mendel was able to establish that genetic factors are passed from parents to offspring and remain unchanged in the offspring so that they can be passed on again to the next generation. Although his work was done before the discovery of DNA and chromosomes, the genetic principles of dominance, segregation, and independent assortment that Mendel originally defined are still used to this day (and explained in detail in Chapter 7).

## Evolving the Theory of Natural Selection

Charles Darwin's study of giant tortoises and finches on the Galapagos Islands led to his famous theory of natural selection (also known as "survival of the fittest"), which he published in his 1859 book titled *On the Origin of Species.* The main point of Darwin's theory is that organisms with traits that are better suited to the conditions in which they live are more likely to survive and reproduce, passing on their traits to future generations. These better-suited variations tend to thrive in the given area, whereas less-suited variations of the same species either don't do as well or just die off. Thus, over time, the traits seen in a population of organisms in a given area can change. The significance of Darwin's theory of natural selection can be seen today in the evolution of antibiotic-resistant strains of bacteria.

## Formulating Cell Theory

In 1839, zoologist Theodor Schwann and botanist Matthias Schleiden were talking at a dinner party about their research. As Schleiden described the plant cells he'd been studying, Schwann was struck by their similarity to animal cells. The similarity between the two types of cells led to the formation of *cell theory,* which consists of three main ideas:

- All living things are made of cells.
- The cell is the smallest unit of living things.
- All cells come from preexisting cells.

## Moving Energy through the Krebs Cycle

The *Krebs cycle,* named for German-born British biochemist Sir Hans Adolf Krebs, is the major metabolic process that occurs in all living organisms. This process results in the transfer of energy to ATP, which all living things use to fuel their cellular functions. Defining how organisms use energy at the cellular level opened the door for further research on metabolic disorders and diseases.

## Amplifying DNA with PCR

In 1983, Kary Mullis discovered the *polymerase chain reaction* (PCR), a process that allows scientists to make numerous copies of DNA molecules that they can then study. Today, PCR is used for

- Making lots of DNA for sequencing
- Finding and analyzing DNA from very small samples for use in forensics
- Detecting the presence of disease-causing microbes in human samples
- Producing numerous copies of genes for genetic engineering

## Chapter 23

# Ten Ways Biology Affects Your Life

### In This Chapter

- Seeing how biology provides you with the essentials (food, clean water, and life)
- Discovering how humans manipulate organisms to create designer genes, medicines, and more

Sometimes science seems like something that happens in a lab somewhere far removed from everyday life. That may be, but the effects of scientific research have a huge impact on your day-to-day existence, from the food you eat to the energy that powers your home. Following is a rundown of ten important ways that biology affects your life. Most are good; others aren't so good. Either way, you just may be surprised by a couple of them.

## Keeping You Fed

First off, if plants didn't produce their own food, you wouldn't have anything to eat — period. So you can thank the process of photosynthesis (covered in Chapter 5) the next time you sit down to a luscious-looking salad or steak dish.

What you may or may not be aware of is that plants aren't the only organisms that make food. People do too. They make foods such as yogurt, cheese, bread, sausages, pickles, tempeh, and more in part by fermenting bacteria and yeast.

## Putting Microbial Enzymes to Work

Microbes aren't just for making foods; they have a wide variety of industrial applications too. Manufacturers put bacterial enzymes in laundry detergent to help break down greasy stains and meat tenderizers in meats to help break down proteins. If you take vitamin C, chances are that vitamin was produced by a fungus. If you drink a protein shake regularly, the amino acids in that shake probably also came from bacteria. So you see, not all microbes are to be feared. Some of them actually improve your life by simplifying tasks and keeping you healthy.

## Designing Genes

The food you eat could very likely contain *genetically modified organisms* (GMOs) — living things whose genes have been altered by scientists in order to give them useful traits. For example, crop plants may be engineered to better resist pests, and animals may be treated with hormones to increase their growth or milk production.

## Obtaining Fossil Fuels for Energy

The fossil fuels that power modern society are the remnants of photosynthesis from long ago. Way back in the Carboniferous period, about 350 million years ago, green algae, plants, and bacteria harvested energy from the Sun and transformed that light energy into the chemical energy stored in their cells. When these living things died, they were deposited in such a way that their remains converted into coal, natural gas, and oil.

Of course, people are now facing a problem as the reserves of these fossil fuels start to run low. But maybe one solution to the problem lies in mimicking the green organisms that stockpiled this energy in the first place — people could act like plants and go solar!

## Causing and Treating Infectious Disease

Whenever you get sick from an infectious disease, such as a cold or strep throat, you're dealing with the reproduction of an alien invader. Your immune system springs into action, activating the cells necessary to fight the invasion and keep the infectious virus or bacteria from replicating itself any further. Also, whenever you take an antibiotic, you're taking a medicine made by an organism such as a fungus or a bacterium.

## Staying Alive

Every minute of every day, your cells are quietly working away, digesting your food, sending signals that control your responses, transporting oxygen around your body, contracting so you can move, and making all of your other bodily processes happen. If your cells weren't functioning, your tissues, organs, and organ systems wouldn't be either.

## Providing You with Clean Water

You have wetlands to thank for the clean water you enjoy. *Wetlands* are areas that are saturated by water most of the time. They act like natural sponges, holding onto water and slowly filtering it around the plants that live there. As water slowly filters through wetlands, plants and microorganisms have time to absorb human wastes such as fertilizers and sewage, cleaning the water and making it safer for humans and other animals to consume. All life on Earth needs water — clean, fresh water — in order to be healthy, so wetlands are pretty important to your quality of life. Unfortunately, wetlands are under incredible pressure from development and oil exploration, and they're disappearing at a rapid rate.

Another way living things help keep water clean is through sewage treatment. Bacteria break down the organic matter in sewage, helping to clean the water before it's released back into the environment.

## Changing Physically and Mentally

Chances are that at some point in your life you either were or will be "ruled" by your hormones. Case in point: You meet someone you're attracted to, signals cause hormones to be released, and suddenly your conscious mind isn't making all the decisions. If that example doesn't convince you of the power of hormones, just think back to puberty. During that time, your body went through an incredible transformation based solely on the signals from these potent chemical messengers.

## Creating Antibiotic-Resistant Bacteria

When people use antibiotics, the susceptible bacteria die first, leaving behind the most resistant cells. These super-resistant cells multiply and take over the available space. As this scenario repeats over time, populations of bacteria eventually become resistant to antibiotics. This fact explains why sometimes doctors don't have the drugs to help people who are infected with an antibiotic-resistant bacteria, such as *MRSA* (which stands for *methicillin-resistant Staphylococcus aureus*).

## Facing Extinction

Perhaps you don't think about extinction much, but it's something worth being aware of. If you need an example, consider the case of polar bears. As global temperatures rise, the polar ice is melting, leaving polar bears with less and less habitat. Not quite so noticeable, but also endangered, are 1,900 other species of plants and animals.

As humans convert more land and resources to their own uses, less and less habitat is available for the other organisms on Earth. Each species needs certain conditions and resources to thrive, and the sheer number of humans on Earth is threatening to overwhelm many ecosystems. That spells bad news for humans because we depend upon the health of ecosystems for our own survival. For example, as humans develop coastal regions, we're drastically reducing the area of our estuaries, which are important breeding grounds for many species of fish. This decrease in breeding grounds results in fewer fish in the oceans, which is bad for marine life and people. (Up to 80 percent of fish species that are harvested commercially spend some part of their lives in estuaries.)

# Index

## • A •

## • B •

## • D •

## • E •

## • G •

## • H •

## • I •



## • J •



## • K •



## • L •

## • M •

## • N •

## • O •



## • P •

• R •

## • S •

## • T •



## • U •



## • V •

## • W •

CPSIA information can be obtained
at www.ICGtesting.com
Printed in the USA
LVOW04*1723270317
528623LV00014BA/405/P

9 781119 173786